Climate Change Management

Series editor

Walter Leal Filho, Hamburg, Germany

More information about this series at http://www.springer.com/series/8740

Walter Leal Filho
Editor

Climate Change Adaptation in Pacific Countries

Fostering Resilience and Improving the Quality of Life

Springer

Editor
Walter Leal Filho
Faculty of Life Sciences
Hamburg University of Applied Sciences
Hamburg
Germany

ISSN 1610-2010 ISSN 1610-2002 (electronic)
Climate Change Management
ISBN 978-3-319-50093-5 ISBN 978-3-319-50094-2 (eBook)
DOI 10.1007/978-3-319-50094-2

Library of Congress Control Number: 2016960267

Printed on acid-free paper

This Springer imprint is published by Springer Nature
The registered company is Springer International Publishing AG
The registered company address is: Gewerbestrasse 11, 6330 Cham, Switzerland

Preface

The Pacific region is affected by climate change at different levels. According to the 5th Assessment Report by the Intergovernmental Panel on Climate Change (IPCC), current and future climate-related drivers of risk for small islands during the twenty-first century, include sea-level rise, tropical and extratropical cyclones, increasing air and sea surface temperatures, and changing rainfall patterns, among others.

The future risks associated with these drivers, according to the IPCC, include loss of adaptive capacity and damages to ecosystem services critical to lives and livelihoods in small islands. In addition, sea-level rise is mentioned as posing one of the most widely recognized climate change threats to low-lying coastal areas on islands and atolls. Furthermore, given the dependence of island communities on coral reef ecosystems for a range of services including coastal protection, subsistence fisheries, and tourism, there is high confidence that coral reef ecosystem degradation will negatively impact island communities and livelihoods. Given the inherent physical characteristics of small islands, AR5 reconfirms the high level of vulnerability of small islands to multiple stressors, both climate and non-climate. These elements illustrate the fact that in addition to the necessary measures in the field of environmental mitigation, adaptation approaches are urgently needed.

The above state of affairs illustrates the need for a better understanding of how climate change affects the Pacific region and for the identification of processes, methods, and tools which may help the countries in the region to adapt. There is also a perceived need to showcase successful examples of how to cope with the social, economic, and political problems posed by climate change in Pacific countries.

This book, which contains a set of papers presented at the Symposium on Climate Change Adaptation in the Pacific Region, held in Fiji in July 2016, serves the purpose of showcasing experiences from research, field projects, and best practice in climate change adaptation in Pacific countries, which may be useful or implemented in other countries and regions. A further aim of this book is to document and disseminate the wealth of experiences available today.

This book is divided into four parts:

Part I, titled "Implementing Climate Change Adaptation in Rural Areas and Communities," entails a set of papers showing how climate change is being handled in urban and rural areas.

Part II, titled "Climate Change Adaptation, Resilience and Hazards," describes a variety of initiatives showing how Pacific countries are handling many problems associated with climate change and many hazards associated with it.

Part III, titled "Information, Communication, Education and Training on Climate Change," describes some of the education and information initiatives taking place across the region.

Part IV, titled "Trends on Climate Change Adaptation," contains some papers outlining some of the operational aspects and their implications for policy-making.

We thank the authors for their willingness to share their knowledge, know-how and experiences, as well as many peer reviewers, who have helped us to ensure the quality of the manuscripts.

Enjoy your reading!

Hamburg, Germany
Winter/Spring 2017

Walter Leal Filho

Contents

Part I
Implementing Climate Change Adaptation in Rural Areas and Communities

Chapter 1
Implementing Climate Change Adaptation Interventions in Remote Outer Islands of the Pacific Island Region

Gillian Cambers, Pasha Carruthers, Titilia Rabuatoka, Sanivalati Tubuna and Juliana Ungaro

Background

Climate change is one of the most serious threats to sustainable development in the Pacific Island countries, and indeed threatens the survival of atoll countries and communities. For the purposes of this paper climate change is defined as per the Intergovernmental Panel on Climate Change (IPCC) and includes changes in the climate due to anthropogenic emissions of greenhouse gases as well as climate variability (IPCC 2014).

The Pacific Island countries span a large area of the Pacific Ocean from around 15°N to 23°S of the equator and are immensely diverse in terms of their history, geography, climate, natural resource base and culture. Geologically they range from high volcanic islands to low lying atolls just a few metres above sea level. Many of the countries consist of archipelagos spread across several degrees of latitude and longitude, e.g. the Federated States of Micronesia (FSM) has 607 islands which extend from latitude 1°S to 14°N, and longitude 135°E to 166°E. The 14 Pacific Island countries are considered Small Island Developing States (SIDS), a group of countries that share similar sustainable development challenges including small populations, limited resources, remoteness, susceptibility to natural disasters, vulnerability to external shocks, dependence on international trade, and fragile environments. SIDS was first recognized as a distinct group of developing countries at the United Nations Conference on Environment and Development in June 1992. Table 1.1 illustrates some of the variability between the countries.

G. Cambers (✉) · P. Carruthers · T. Rabuatoka · S. Tubuna · J. Ungaro
Geosciences Division, Pacific Community (SPC), Suva, Fiji
e-mail: gillianc@spc.int

T. Rabuatoka
e-mail: titiliar@spc.int

W. Leal Filho (ed.), *Climate Change Adaptation in Pacific Countries*,
Climate Change Management, DOI 10.1007/978-3-319-50094-2_1

Table 1.1 Characteristics of the Pacific Island countries (Adapted from: Australian Bureau of Meteorology and Commonwealth Scientific and Industrial Research Organisation 2011)

Country	Land area (km^2)	Number of islands larger than 10 km^2	Exclusive economic zone area (million km^2)	Maximum height above sea level (m)	Population (2010 estimate except Vanuatu)	Languages	Government
Cook Islands	240	6	1.800	652	11,400	English, Maori	Self-governing country in free association with New Zealand
Federated States of Micronesia	702	7	2.978	791	111,364	English, Chuukese, Pohnpeian, Yapese, Kosraean, Ulithian, Carolinian	Independent with free association arrangements with USA until 2023
Fiji	18,333	37	1.300	1,323	844,420	English, Fijian, Hindi	Independent state
Kiribati	811	18	3,600	87	100,835	I-Kiribati, English	Democratic republic
Marshall Islands	181	6	2.131	3	54,439	Marshallese, English	Republic in free association with the USA until 2023
Nauru	21	1	0.320	70	9,976	Nauruan, English	Republic with parliamentary system
Niue	259	1	0.390	68	1,470	Niuean, English	Free association with New Zealand
Palau	488	4	0.629	214	20,518	Palauan, English	Republic in free association with USA
Papua New Guinea	462,243	20	3.120	4,697	6,744,955	Pidgin, English + more than 700 other languages	Independent state
Samoa	2,935	2	0.120	1,860	183,123	Samoa, English	Independent state
Solomon Islands	28,785	26	1.340	2,447	549,574	English, Pidgin + 87 other languages	Independent state
Tonga	649	10	0.700	1,030	103,365	Tongan, English	Independent kingdom
Tuvalu	26	0	1.300	5	11,149	Tuvaluan, English	Independent state
Vanuatu	12,281	30	0.664	1,700	234,023 (2009)	Bislama, French, English + 105 other languages	Republic

Yellow shaded countries are those discussed in this paper

People living in the Pacific Island countries are already experiencing changes and variability in their climate such as shifts in rainfall patterns, higher air and sea surface temperatures, changes in extreme events and rising sea levels (Australian Bureau of Meteorology and Commonwealth Scientific and Industrial Research Organisation 2011). For example in the Pacific Island countries, air temperatures rose between 0.08 and 0.20 °C per decade over the past 50 years; sea level rise varied across the Pacific and ranged between 3 and 12 mm year^{-1} over the period 1993–2009, which is higher than the global average; ocean acidification increased (with aragonite saturation values generally falling to around or below 4 since the mid-1990s).

These changes are affecting peoples' lives and livelihoods, as well as important industries, such as agriculture, fisheries and tourism. In recognition of the seriousness of the adverse effects of climate change, the Pacific Island countries and territories developed the Pacific Islands Framework for Action on Climate Change (PIFACC) 2006–2015 (SPREP 2006) to build their resilience to the risks and impacts of climate change. A new framework for the region is currently under development for the post-2016 period.

The PIFACC developed an action plan around the following principles:

- Implementing adaptation measures
- Governance and decision-making
- Improving our understanding of climate change
- Education, training and awareness
- Contributing to global greenhouse gas reduction
- Partnerships and cooperation

The PIFACC provided an overall framework for the many climate change activities, projects and programmes implemented by national governments, regional and international organisations, non-governmental organisations (NGOs) and civil society over the decade. A review of the PIFACC was presented at the Pacific Climate Change Roundtable in May 2015 (SPREP 2015).

For the purposes of this paper climate change adaptation is defined as "The process of adjustment in natural or human systems in response to actual or expected climatic stimuli or their climate and its effects, which moderates harm or exploits beneficial opportunities" (IPCC 2014). Over the period of the PIFACC there have been numerous climate change adaptation interventions ranging from community-based to multi-country initiatives and important outcomes have been achieved and lessons have been learnt.

This paper will focus particularly on some climate change adaptation activities and outcomes of the Global Climate Change Alliance: Pacific Small Island States (GCCA: PSIS) project over the period 2011–2015 (GCCA: PSIS 2015). These activities will be described and discussed through the lens of outer islands development.

The GCCA: PSIS project was funded by the European Union (EU) and implemented by the Pacific Community (SPC) in collaboration with the Secretariat of the Pacific Regional Environment Programme (SPREP). The overall objective of the GCCA: PSIS project was to support the governments of nine Pacific smaller island states, namely Cook Islands, FSM, Kiribati, Marshall Islands, Nauru, Niue, Palau, Tonga and Tuvalu, in their efforts to tackle the adverse effects of climate change.

The project approach was to assist the nine countries design and implement practical on-the-ground climate change adaptation demonstration projects in conjunction with mainstreaming climate change into line ministries and national development plans. The rationale was that this would help countries move from an *ad hoc* project-by-project approach towards a programmatic approach underpinning an entire sector. The nine countries chose their own sectors based on their national priorities and plans.

Six of the nine countries chose to focus their demonstration projects in outer islands. National partners felt that outer islands had been relatively disregarded in the past with most of the projects being centred in the main island of an archipelago. Thus outer islands, where the need was often greater, had been somewhat neglected.

This paper discusses several on-the-ground climate change adaptation interventions that were implemented in outer island communities in different countries through the GCCA: PSIS project and to analyse the challenges and how they were addressed. The purpose of this paper is to show to development partners working in the Pacific and to the research community that climate change adaptation interventions can be successfully implemented in outer island communities provided the constraints are fully recognised and accommodated in the conceptualisation and planning stages.

Methodology

The methodology for designing and implementing the demonstration projects consisted of a multi-step process. At the start of the project climate change profiles were prepared for each country; these included the physical, historical and socio-economic background for each country, the national and sector planning policies, the most up-to-date climate change projections for the country, and climate change adaptation interventions undertaken in the previous five years. These country profiles were used to analyse gaps and needs. This was followed by a consultation phase during which countries used their strategic and development planning policies, plans and budgets to identify a specific sector for the project focus. Project concept notes were prepared and reviewed, followed by further consultations during which national stakeholders designed the elements of their demonstration project. Finally project design documents were prepared describing the activities, and including logframes, budgets, risk matrices and exit strategies. This entire process took on average 12–18 months.

Once the project design document was approved, implementation began. Typically this included recruitment of project staff; procurement of equipment, goods and services; implementation of the intervention; and ongoing monitoring and evaluation cumulating in the capture and exchange of lessons learnt. On average this phase took two or more years.

One of the major limitations of this paper is that there has been insufficient time to monitor the impact of the interventions, as implementation in most countries continued right up to the project end date. A further 3–5 years are required before the impact of the interventions can be fully assessed. This is a major challenge experienced by many development projects in the Pacific and elsewhere, in that implementation usually continues right up to the end of the project, and in many cases beyond, thereby requiring the need for project extensions. Despite this lack of objective and thorough impact analysis, it is still useful to discuss the interventions in terms of the lessons learnt and what were perceived as best practices during the implementation.

Implementing Climate Change Adaptation Activities in Pacific Outer Islands

Five case studies covering six different countries will be described to illustrate the challenges involved and how they were met in delivering project activities to outer islands.

Managing Marine Resources in the Northern Group of the Cook Islands

The Cook Islands is made up of 15 islands lying between 8°S and 23°S in the south Pacific Ocean and divided into a southern group and a northern group. The GCCA: PSIS climate change adaptation project focused particularly on the Manihiki atoll in the northern group. Up until the beginning of the 21st century there used to be a vibrant pearl farming industry in the Manihiki Lagoon (Ponia et al. 2000), however, following a severe cyclone in 1997 and outbreaks of pearl oyster disease relating to intense El Niño and La Niña events, export production of pearls fell from a peak of US$ 9 million in 2000 to US$ 3 million in 2003 (Ponia 2010; Diggles and Hine 2001). The decline was also related to poor farming practices.

The project sought to strengthen real time environmental monitoring, including water quality so as to provide information that will assist pearl farmers improve their farming practices, and avoid disease outbreaks and stress to the oysters due to present environmental conditions and future projected conditions under climate change.

One of the planned activities was to establish an automatic water quality monitoring buoy in the lagoon to provide real time information to the pearl farmers. While this was achieved to a limited extent it was not as successful as hoped. One of the reasons was the remoteness of Manihiki, which lies 1,200 km from the main island, Rarotonga, and is served by an infrequent and somewhat irregular shipping schedule and a small 8-seater plane which only flies the 4 h trip once every two weeks. Added to which Cook Islands, with a population of around 15,000 people, does not have the skills in-country for maintaining and servicing such equipment. Routine maintenance, such as calibration of sensors, involves the equipment being sent off-island with resultant long delays and breaks in data records.

An alternative approach was adopted which proved to be more successful and this involved stationing a marine biologist from the Cook Islands Ministry of Marine Resources full-time on Manihiki to work with the communities and pearl farmers to regularly monitor water quality with mobile probes and to advise on practical ways to improve farming practices on a day-to-day basis. Monitoring results are regularly posted on community noticeboards and water quality alerts are sent out to the farmers via emails and text messages on mobile phones e.g. when water temperatures are high and the oyster shells should not be disturbed.

This simpler type of technology combined with a more personal approach is providing some positive results e.g. one pearl farmer noted that receiving the information not to handle the shells between January and April 2016 because of the high sea surface temperatures helped her plan to spend that period in Rarotonga and focus on pearl marketing.

Providing Clean Drinking Water to Outer Island Communities in Kiribati

In Kiribati, which consists of 33 scattered low-lying islands, mainly atolls, dispersed over 3.5 million km^2 in the central Pacific Ocean, national stakeholders selected climate change adaptation and health as the focus for its climate change adaptation project. Changing rainfall patterns as a result of climate change and population pressures are leading to the contamination of the freshwater lens in the islands and this is the main source of drinking water. As a consequence Kiribati has one of the highest rates of childhood mortality from diarrhoea in the Pacific (WHO and UNICEF 2015).

In the main centre of population, South Tarawa, a low-cost method for water disinfection, named SODIS (Solar Disinfection), was trialled through the GCCA: PSIS project and is currently being rolled out in the outer islands where it is particularly appropriate. Solar water disinfection involves placing contaminated water in plastic PET (polyethylene terephthalate) 1.5 l bottles on a reflective surface in direct sunlight for 6 h, after which the water is drinkable. It is particularly effective in reducing rates of diarrhoea (Ungaro 2016), and results in considerable cost savings—up to AU$ 300 per household per year—as residents no longer have to boil water. This approach was initially trialled in one of the poorest communities in Bairiki, the capital of South Tarawa. After it was shown to be successful, from both a social and scientific perspective, SODIS was launched nationally in March 2015.

With the help of trained community water champions, SODIS was introduced in Bairiki over a six month period starting in October 2014 and continuing through to March 2015. Figure 1.1 shows that there was a substantial drop in the number of reported diarrhoeal cases after the introduction of SODIS.

Since then SODIS has been rolled out in three outer islands, by the Ministry of Health and Medical Services and the Kiribati Adaptation Programme—Phase III (funded by the World Bank) which has integrated SODIS into their planned project interventions. This also illustrates the importance of partnerships for sustainability of project activities beyond donor timeframes.

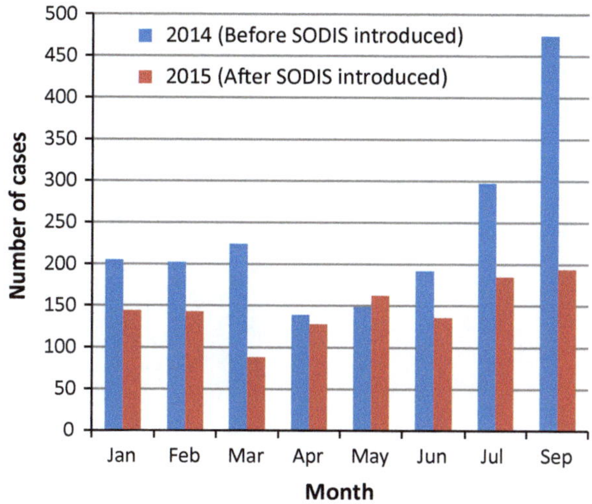

Fig. 1.1 Number of cases of diarrhoea *before* and *after* the introduction of SODIS (*Data source* Bairiki Health Clinic, Kiribati, 2015)

Enhancing Water Security in the Outer Islands of FSM and Palau

Palau is located in the northwest tropical Pacific, 800 km east of the Philippines and has over 500 islands, most of which are the small, uninhabited Rock Islands. Only nine islands are currently inhabited and these are divided into 16 states. FSM lies to the northeast of Palau, and as already described, has over 600 islands divided into four states, and around 65 islands are populated.

In both countries the GCCA: PSIS project focused on enhancing water security in outer islands: five outer islands in Palau and one in FSM. Many of the smaller islands in particular are dependent on rainwater harvesting; and some have small freshwater lenses from which water can be pumped, although often the groundwater is contaminated and cannot be used for drinking.

One of the major challenges with implementing projects in such outer islands is the absence of protected landing sites. The islands are usually surrounded by coral reefs with very small gaps through the reef, which in many cases can only be accessed using small open dinghies. This was the case in Fais Island in Yap State of FSM, where 46 large (5,000 l) plastic water tanks had to be unloaded from a freighter one by one onto a small dinghy which was then navigated through a gap in the reef and eventually unloaded by hand onto the beach, see Fig. 1.2. Obviously such hazardous unloading requires calm sea conditions and very skilled operators.

Similar situations were encountered in Palau where water tanks were delivered to the outlying island states of Sonsorol and Hatohobei, more than 500 km south of the main port in Koror, and involving up to three days at sea each way. For these islands similar unloading challenges exist, compounded by irregular shipping schedules and few windows of opportunity each year when sea conditions are favourable.

Fig. 1.2 Unloading water tanks in Fais Island, Yap State, FSM, April 2015

Such factors add significantly to the cost and duration of a planned climate change adaptation intervention in an outer island. In FSM and Palau, however, some economies of scale were achieved by combining the shipment of tanks from their point of manufacture to the main port in each country.

Protecting Low-Lying Atolls in the Marshall Islands

Coastal erosion is a serious issue in many Pacific islands, e.g. there has been a landward retreat between 5 and 50 m on the eastern coast of Kosrae in FSM (Gombos et al. 2014). Pacific atolls have historically undergone coastal changes (Webb and Kensch 2010) as the small sandy islands shift their position on the reef in response to the waves and currents—often resulting in erosion on one side of the island and accretion on the other. However, there is growing observational evidence in many of the populated Pacific atolls of increasing coastal erosion, which is being attributed, in part, to rising sea levels (Ramsay 2011).

Such changes are affecting lives and livelihoods of the islanders living in these distant atolls, for example in Ailinglaplap Atoll in the Marshall Islands, Woja Island was in danger of being cut into two parts, and conditions were so serious that students from one part of the island had to wait on the tide each day to cross to the other part of the island and as a result they sometimes missed school.

As part of the GCCA: PSIS project a 70 m causeway was constructed to strengthen the narrow strip of land joining the two parts of Woja Island. This involved coastal and marine studies as well as extensive feasibility and design work. One of the contributing factors making this project feasible was that armour stone for the rock causeway existed in abundance on the reef flat having broken off the outer edge of the reef platform during large wave events and then thrown up onto the reef flats. Over time the rocks had been reduced to smaller sized boulders suitable for the armour stone. The existence of this natural rock for the armour stone meant that the cost was within the project budget; otherwise the cost of shipping in armour stone would have been prohibitive.

Information about coastal changes and climate change is often poorly understood in the Pacific Islands and especially in remote outer islands. This is largely due to the fact that information on technical aspects of climate change is written in English and technical terms are often hard for people to understand in their native language, and even more so in a foreign language. To begin to overcome this impediment, the Office of Environmental Planning and Policy Coordination (OEPPC) in the Marshall Islands, with support from the GCCA: PSIS project, prepared a Climate Change Glossary giving the Marshallese and English descriptions for technical climate change and environmental terms (OEPPC 2015). This glossary is being trialled by some Marshallese teachers in 2016 and it is hoped it will help the Marshallese people better understand the challenges posed by climate change and the adaptation options available.

Augmenting Food Security in the Islands of Tuvalu

Tuvalu, in the southern Pacific Ocean, consists of four raised atolls and five raised reef islands, all less than 5 m in elevation. Especially in the outer islands there is a subsistence economy with farming and fishing being the major sources of employment. However, over recent decades there has been a decline in subsistence agriculture and increasing dependence on imported foods such as rice and flour. This has been linked to increasing incidence of non-communicable diseases such as diabetes (WHO 2013).

There has been wide research on agroforestry in the Pacific (e.g. Wilkinson and Elvitch 2000). Agroforestry practices consist of integrating understory crops with tree crops as shown in the schematic in Fig. 1.3. This technique, although widely used in several Pacific countries in the past, has in Tuvalu declined over the past few recent decades.

Through the GCCA: PSIS project, traditional agroforestry farming practices were revived and combined with new technology, specifically the use of innovative “climate-ready” crops and trees (sourced from the SPC Pacific Centre for Crops and Trees), which had been bred for enhanced resilience to climate change. These new and old farming technologies were demonstrated in both the main and one outer island of Tuvalu on underutilized land. Coconut plantations were thinned, senile trees removed, and a variety of “climate-ready” crops and trees planted to maximize land use and promote additional production beyond subsistence needs. Farmers and landowners developed the demonstration sites and at the same time were trained in agroforestry design, compost making, plant grafting and breeding techniques, through hands-on technical training and local language communication materials.

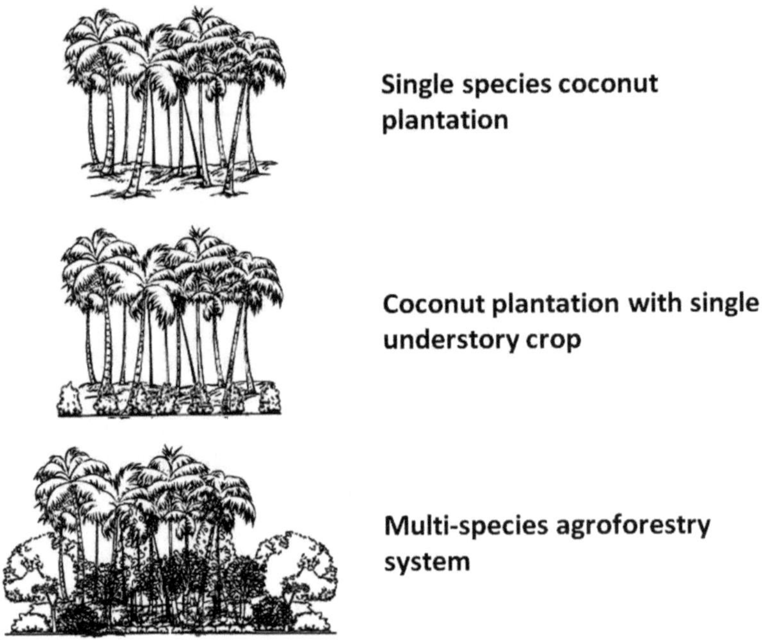

Fig. 1.3 Stages in agroforestry in Tuvalu

One year after the demonstration sites were completed, they have been productive, with yams, bananas, breadfruit, figs, sweet potato and coconuts available for market. The trees and crops planted are recorded in a database so that each variety can be assessed for its effectiveness in the Tuvaluan atoll environment. This will provide valuable information for future replication of this integrated agroforestry farming technology.

As part of the project women were targeted for home gardening activities through the Tuvalu National Council of Women. Training and equipment was provided to women's groups in each island to develop their own home gardens. A home garden competition in 2015 proved to be a strong motivating factor, and this approach is planned to be replicated annually by the National Council of Women.

Discussion

This paper has described several case studies involving the implementation of climate change adaptation interventions in several sectors (marine resources, coastal protection, food security, water security and health) in outer islands. Similarities have emerged and these are discussed below.

Adding New Technology and New Knowledge to Traditional Practices

Approaches that combine new and practical technology and traditional practices may be more appropriate for the outer island situation. For example in the Cook Islands, a relatively simple method to measure water quality combined with the full time stationing of a marine biologist at the remote outer island, is providing immediate benefit to the pearl farmers, in contrast to the more sophisticated real time technology.

In Tuvalu, combining an accepted agricultural practice, agroforestry, with new technology, the supply of "climate-ready" crops, is beginning to provide positive results for farmers by helping them move from subsistence to market production.

In the Marshall Islands the coastal areas adjacent to the new causeway were stabilised with coastal plants. Instead of using plastic bags to germinate the seedlings, coconut leaves were woven into small baskets and these served the same purpose. Another traditional technique was to use copra sacks filled with coral rock and sand and held in place along the beach berm with ironwood stakes to protect the newly planted seedlings from inundation by the sea.

The limited supply of good quality water is a serious challenge facing those living in outer islands. While there are sophisticated technologies that can augment supply such as reverse osmosis plants, enhancing traditional practices such as rainwater harvesting may be more suitable for the outer island situation where maintenance and the procurement of spare parts is often difficult. However, it is very important to understand the specific circumstances of each island, e.g. in Fais Island in FSM, the houses have very low roofs, so it was necessary to procure low rectangular tanks that fitted below the roof overhang instead of the typical cylindrical tanks.

In many atoll countries water quality is another challenge and the usual way to purify water is to boil the water. As was seen in Kiribati, research and practice has shown that a simple "new" technology, SODIS, is appropriate to the outer island situation and can reduce diseases such as diarrhoea and save money through reduced fuel costs.

Meeting the High Cost of Climate Change Interventions in Outer Islands

Implementation of projects in outer islands face very high costs and delays due to their remoteness, infrequent shipping schedules, and absence of sheltered and secure unloading facilities. These constraints need to be built into project planning and design, e.g. when planning a project in an outer island it is recommended to undertake all the necessary feasibility, design and costing studies taking into account local knowledge and best practice and then to apply the "times 2 rule of thumb." So that if the investigative and planning studies show the project will take one year to implement and cost $100,000, then budget and plan for 2 years and $200,000. Furthermore in some very special circumstances it may be necessary to increase the "times 2" factor.

Particularly with regional projects, there may be opportunities for economies of scales as was seen in the case studies for FSM and Palau where the water storage tanks were combined in one shipment the cost of which was then split between the two countries. Such measures need advance planning.

Sourcing local materials and applying local methods may also be a way to create cost saving. The use of boulders washed up onto the reef flat for armouring the causeway in Woja Island, in Marshall Islands, is one such example.

Effective Communication and Addressing the Required Behavioural Change

All of the cases studies have illustrated the need for communicating information about project activities in the local language. This can be exceedingly complex for many Pacific Island countries as there are several local languages. For example with the water security project in FSM, it was necessary to use the language commonly used in Yap State, Yapese, and the language of a target outer island, Ulithian, requiring more effort and investment in translation.

Climate change is a relatively new concept for most people and trying to understand English terms such as "greenhouse gases" can be difficult. Local language glossaries such as the one developed by the Marshall Islands are a good first step to addressing this challenge, although the complexity of selecting and agreeing on words and spelling for the new terms should not be underestimated.

Exchanging experiences through south-south cooperation is a very effective way to learn about others' applied knowledge. An exchange visit by Nauru's Water Strategy Manager to the Environmental Health Unit of the Kiribati Ministry of Health and Medical Services in December 2015 and February 2016 is one such example. "The visit to Kiribati was particularly useful because water resources in Kiribati are extremely scarce very like the situation in Nauru, so it was good to work with another country that experiences similar challenges" (Agir in GCCA: PSIS 2016).

Experience from the GCCA: PSIS and other projects have shown the benefits of adopting a behavioural change approach. This approach, which is based upon change theory, involves working with and understanding the intended audience to identify and address clear, measurable behavioural objectives (UNICEF 2005).

In Kiribati three workshops were held over a period of 6 months to research the situation and the audience. Reducing child mortality by preventing outbreaks of diarrhoeal diseases was identified as the objective, and the means to achieve this included: (i) encouraging parents of children less than 5 years old to boil and treat water, and (ii) promoting hand washing. Further research identified SODIS as a suitable technique to treat water. This was trialled in one community. Six community water champions, who took part in the workshops, were hired to help interested households in the community properly use SODIS, supported by communication tools ranging from videos to card games. Results showed the number of cases of diarrhoea decreased by more than 30% after 1 year (Ungaro 2016) and 4 months after the trial campaign finished, the end line survey indicated 85% of households were still practicing SODIS on a regular basis.

Customising Development Indicators for Outer Islands

Outer islands in the Pacific with their small populations, limited communication with the outside world, remoteness and subsistence economies face special challenges when it comes to measuring how aid is distributed. For example, one of the standard indicators, dollars per capita, which consists of dividing the total project cost by the number of people benefitting, provides an incomplete picture of the real situation, as it does not reflect the special needs and challenges of outer islands.

In Fais Island in FSM, for example, less than 10% of the population earn an income and the majority of the working population depends on a subsistence economy. Yet the Fais Island community contributed significantly to the project by providing many weeks of labour to install the water tanks.

Furthermore outer island residents have important knowledge to share. In the southern group of the Cook Islands, senior residents were surveyed about the environmental changes they had witnessed in their lifetimes. The results were analysed to advance the understanding of climate variability in the Cook Islands (Rongo and Dyer 2015).

There is a need for further research on indicators for outer islands that show how external finance is used and distributed and whether it is effective.

Conclusion

There are many challenges to the effective delivery of climate change response measures to outer islands including small populations with skewed demographics, high costs, infrequent and unreliable transportation, and technology and communication limitations. These limitations may not easily fit in with funded project priorities and timelines, and therefore outer islands are sometimes overlooked.

Yet outer islands have special needs that should be considered in the selection and delivery of climate change response measures. These include an absence or very few interventions in outer islands, subsistence economies with little cash available, small to micro-size markets, and a heavy reliance on the natural environment.

As described here, countries and their development partners need to consider the limitations faced in implementing climate change adaptation interventions in outer islands and develop appropriate timeframes, budgets, indicators, technologies, approaches and reporting mechanisms to enable effective delivery.

Future prospects for climate change adaptation interventions in outer island communities include:

(i) Developing new indicators that take the special needs and circumstances of outer island communities into account. For example, outer islands are often able to effectively contribute to development or adaptation initiatives, but often this contribution is in-kind.

(ii) Redesigning the scope of project timeframes beyond the typical 3–5 years duration by adding a second phase of 3–5 years with a much smaller level of funding (e.g. 10% of the phase 1 funding). This could provide for full impact evaluation, monitoring and maintenance, and overall sustainability of the intervention.

(iii) The GCCA: PSIS and the Pacific Adaptation to Climate Change (PACC 2012) projects, among others, have shown the usefulness and benefits of practical climate change adaptation demonstration projects. They have also paved the way for future interventions to roll out different approaches and technologies which might include working directly with private sector enterprises. For example, instead of using international procurement procedures to purchase water security items like rainwater tanks, first flush devices and solar pumps, to work with existing private sector contractors and businesses at the national level to see if there are ways in which they can cost effectively stock and promote use of these best practice technologies. Including in-country private enterprises in ongoing behavioural change is a promising approach for new interventions and sustaining existing ones into the future.

Acknowledgements The authors would like to recognise the stakeholders in the GCCA: PSIS project countries of Cook Islands, FSM, Kiribati, Marshall Islands, Nauru, Niue, Palau, Tonga and Tuvalu for their dedication and hard work in designing and implementing the project as well as all members of the GCCA: PSIS team and colleagues in SPC.

In addition, the advice, assistance and support provided by the European Union are gratefully acknowledged.

Disclaimer

The contents of this paper in no way reflect the views of the European Union.

References

Australian Bureau of Meteorology and Commonwealth Scientific and Industrial Research Organisation (CSIRO). (2011). *Climate change in the Pacific: Scientific assessment and new research. Volume 1: Regional overview, Volume 2: Country reports*. Melbourne, Australia: Australian Bureau of Meteorology and Commonwealth Scientific and Industrial Research Organisation. Retrieved April 4, 2016, from http://www.pacificclimatechange.net/index.php/eresources/documents?task=view&id=564&catid=87 and http://www.pacificclimatechangescience.org/wp-content/uploads/2013/09/Volume-2-country-reports.pdf.

Bairiki Health Clinic. (2015). *Diarrheal data—Bairiki Health Clinic*. Tarawa, Kiribati: Ministry of Health and Medical Services.

Diggles, B. K., & Hine, P. M. (2001). *Mortality of black-lip pearl oysters (Pinctada margaratifera) in Manihiki Lagoon*. National Institute of Water and Atmospheric Research client report prepared for the Ministry of Marine Resources, Government of the Cook Islands, Wellington, New Zealand.

GCCA: PSIS. (2015). *Climate change adaptation in the Pacific—Project overview*. Suva, Fiji: Secretariat of the Pacific Community. Retrieved April 4, 2016, from http://www.pacificclimatechange.net/index.php/eresources/documents?task=view&id=2318&catid=229.

GCCA: PSIS. (2016). *South-south exchange to inform water quality monitoring in Nauru*. Suva, Fiji: Secretariat of the Pacific Community. Retrieved April 6, 2016, from http://capacity4dev.ec.europa.eu/gcca-community/blog/south-south-exchange-inform-water-quality-monitoring-nauru.

Gombos, M., Ramsay, D., Webb, A., Marra, J., Atkinson, S., & Gorong, B. (Eds.). (2014). *Coastal change in the Pacific Islands. A guide to support community understanding of coastal erosion and flooding issues* (vol. 1). Pohnpei, Federated States of Micronesia: Micronesia Conservation Trust.

IPCC. (2014). Annex II: Glossary. In K. J. Mach, S. Planton, & C. Von Stechow (Eds.). *Climate change 2014: Synthesis report. Contribution of Working Groups I, II, and III to the Fifth Assessment Report of the Intergovernmental Panel on Climate Change* (Core writing Team, Pachauri, R. K., and Meyer, L.A. (Eds.)). Geneva, Switzerland: IPCC.

Office of Environmental Planning and Policy Coordination. (2015). *Glossary of climate change terms: English—Kajjin Majel*. Majuro, Republic of the Marshall Islands: Office of Environmental Planning and Policy Coordination. Retrieved April 20, 2016, from http://www.pacificclimatechange.net/index.php/eresources/documents?task=view&id=2410&catid=229.

PACC. (2012). *Pacific Adaptation to Climate Change overall introduction*. Apia, Samoa: SPREP. Retrieved April 14, 2016, from http://www.adaptation-undp.org/resources/project-brief-fact-sheet/pacc-overall-introduction-july-2012.

Ponia, B. (2010). *A review of aquaculture in the Pacific Islands 1998–2007: Tracking a decade of progress through official and provisional statistics*. Noumea, New Caledonia: Secretariat of the Pacific Community.

Ponia, B., Napari, T., Ellis, M., & Tuteru, R. (2000). *Manihiki Atoll black pearl farm census and mapping survey*. SPC Pearl Oyster Information Bulletin (December 14, 2000). New Caledonia, Noumea: Secretariat of the Pacific Community.

Ramsay, D. (2011). *Coastal erosion and inundation due to climate change in the Pacific and East Timor—Synthesis report*. Hamilton, New Zealand. Retrieved April 11, 2016, from http://www.environment.gov.au/system/files/resources/d5ab25b9-5e53-425b-a974-015bc1182302/files/coastal-erosion.pdf.

Rongo, T., & Dyer, C. (2015). *Using local knowledge to understand climate variability in the Cook Islands*. Rarotonga, Cook Islands: Climate Change Cook Islands, Office of the Prime Minister. Retrieved April 8, 2016, from http://www.adaptation-undp.org/resources/document/using-local-knowledge-understand-climate-variability-cook-islands-january-2015.

SPREP. (2006). *Pacific Islands framework for action on climate change 2006–2015*. Apia, Samoa: Secretariat of the Pacific Regional Environment Programme. Retrieved April 3, 2016, from http://www.sprep.org/climate_change/pycc/documents/PIFACC.pdf.

SPREP. (2015, May). *Lessons learnt from the ten-year implementation of the Pacific Islands framework for action on climate change (2006–2015)*. Presentation at the 2015 Pacific Climate Change Roundtable. Apia, Samoa: Secretariat of the Pacific Regional Environment Programme.

Ungaro, J. (2016). *Improving water quality in Kiribati with solar disinfection (SODIS)*. Poster presentation at UNISDR Science and Technology Conference on the Implementation of the Sendai Framework for Disaster Risk Reduction 2015–2030, January 2016. Geneva, Switzerland.

UNICEF. (2005). *Strategic communication for behaviour and social change in southeast Asia*. Kathmandu, Nepal: UNICEF Regional Office for South Asia. Retrieved April 14, 2016, from http://www.who.int/immunization/hpv/communicate/strategic_communication_for_behaviour_and_social_change_unicef_rosa_2005.pdf.

Webb, A. P., & Kench, P. S. (2010). The dynamic response of reef islands to sea level rise: Evidence from multi-decadal analysis of island change in the Central Pacific. *Global and Planetary Change*, *73*(3), 234–246. Elsevier. Retrieved April 6, 2016, from https://researchspace.auckland.ac.nz/handle/2292/17029.

WHO. (2013). *World Health Organization—Health profiles Tuvalu*. World Health Organization, Western Pacific Region. Retrieved April 3, 2016, from http://hiip.wpro.who.int/portal/Countryprofiles/Tuvalu/HealthProfiles/TabId/203/ArtMID/1060/ArticleID/117/Default.

WHO & UNICEF. (2015). *Sanitation, drinking-water and health in Pacific Island countries: Progress and ways forward*. Manila, Philippines: WHO Western Pacific Regional Publications.

Wilkinson, K. M., & Elvitch, C. R. (2000). *Integrating understory crops with tree crops: An introductory guide for Pacific Islands*. Hawaii, USA: Agroforestry Guides for Pacific Islanders #4. Permanent Agriculture Resources.

Chapter 2
Customary Land and Climate Change Induced Relocation—A Case Study of Vunidogoloa Village, Vanua Levu, Fiji

Dhrishna Charan, Manpreet Kaur and Priyatma Singh

Introduction

The South Pacific being the hub of climate change associated environmental and social developments is irrefutably one of the world's most predisposed regions when it comes to the climate and weather induced disasters (Boege 2011). Particularly susceptible are the several of the low-lying coral islands (Nunn 2012). The livelihoods of majority of the Pacific Islanders which revolve around the Pacific Ocean is being acutely affected due to rising sea levels, increased coastal erosion, inundation, flooding and salinization of coastal aquifers (Ferris et al. 2011). For several of the communities in the South Pacific, adaptation has become an immediate necessity for survival. The pressing need to acclimatize to climate change adversities has escalated over the last couple of years and the issue of climate change taking its toll in many island nations has surfaced in recent discourses (Barnett and Campbell 2010).

On the onset, Fiji's marine and coastal ecosystems endow considerable physical, financial, societal, ecological and cultural benefits to approximately half of the country's estimated 902 964 population (Govan 2009). Yet, the repercussions of climate change on the coastal ecosystems are threatening the way of life of the coastal inhabitants and for the residents of Vunidogoloa in the province of Cakaudrove in Vanua Levu, relocation has emerged as a reality for more than three decades.

D. Charan · P. Singh (✉)
Department of Science, The University of Fiji, Saweni, Lautoka, Fiji
e-mail: priyatmas@unifiji.ac.fj

D. Charan
e-mail: dhrishnac@unifiji.ac.fj

M. Kaur
Department of Language, Literature and Communication, The University of Fiji, Saweni, Lautoka, Fiji
e-mail: manpreetk@unifiji.ac.fj

W. Leal Filho (ed.), *Climate Change Adaptation in Pacific Countries*,
Climate Change Management, DOI 10.1007/978-3-319-50094-2_2

In February 2014, the village was the first in Fiji to reposition; moving 2 km inland after years of inundation, storm surges, coastal abrasion and unwarranted flooding had made their village susceptible to the impacts of climate change (United Nations Office for the Coordination of Humanitarian Affairs 2014). The traditional responses of disaster relief were no longer protecting the village community despite thousands of dollars spent on the construction of sea walls. Community relocation was the only cogent solution to safeguard the inhabitants of Vunidogoloa (Edwards 2012, p. 3). Conversely, this was an enormously emotional and harrowing headway for the villagers especially since they had to retreat from their customary land which has been part of their culture and identity for their entire life.

Relocation may be the last resort but also one of the best adaptation responses for several of the coastal Fijian villages currently facing similar tribulations as Vunidogoloa (Rubelli 2015). This also indicates that quite a few of these vulnerable villagers will be experiencing similar limitations as faced by the people of Vunidogoloa. Some of the drawbacks are the availability of land for settlement, governance and funding and perhaps the most intricate of all is the traditional and emotional place attachment. Disputes over land rights as well as loss of social and communal cohesion will highly likely create some of unconstructive effects of population relocations (Ferris et al. 2011). According to Wewerinkle (2013), the cultural identity of the people is likely to be impeded by the loss of customary land that is anticipated to occur as a result of climate change. A report by Nurse et al. (2014) explains that barriers to taking action have also been attributed to endogenous factors such as traditional values and awareness.

In many indigenous communities access to land depends on membership in a specific clan. For the iTaukei (indigenous Fijians) the ownership of land is vested in the *mataqali* (Fijian clan or landowning unit) (Fonmanu et al. 2003). Land offers not only livelihood but it is also the source of the traditional and spiritual wellbeing for many of the island communities. This is why despite the distressed situation on the islands there are still people who do not want to relocate (Boege 2011). Generation gap also influences the decision to relocate. In the Vunidogoloa resettlement case, it was particularly the elderly who did not want to move, while members of the younger generation were keen to move.

Developing countries also have a major limitation in capacity making adaptation difficult. Limitations include both human capacity and financial resources. The lack of funding available in various forms and difficulties in accessing the funds which are available represents a major barrier for adaptation, particularly for local community action (United Nations Framework Convention on Climate Change 2006). Climate-induced population displacement entails a governance and policy framework that can holistically respond to communities challenged with harsh impacts of climate change. Lack of proper awareness and institutional capacity also limits adaptation process (Amundsen et al. 2010). Relocation of Vunidogoloa village provides an opportunity to address the multiple societal issues to foster long term sustainability in the process of relocating communities.

The document on Peninsula Principles on Climate Displacement within States (Displacement Solutions 2013) forms a preliminary guiding framework and premise for policy and lawmakers, based on current international law. Myriad doctrines such as community engagement and consent, provision of affordable housing, land solutions, basic services and economic opportunities to those affected, have been experiential in Vunidogoloa.

The purpose of this paper is to consider how social, cultural, financial and environmental factors can form barriers to the process of climate change induced relocation. The paper also aims to provide recommendations for assimilation of socioeconomic and customary elements in the much anticipated institutional relocation strategies. In addressing this purpose, the study as exemplified by testimonies and series of in-depth semi structured interviews from some people of Vunidogoloa village and the government administrators provides a synopsis of some of the fundamental challenges encountered by the people of Vunidogoloa village from the inception till the completion of the entire resettlement. In particular, it accentuates the intricacies surrounding the socio-cultural aspects of the relocation process. The loss of Fijians customary land that is projected to occur as a result of climate change is plausible to impede with their cultural identity and associated climate induced repositioning. The challenges to relocating the community manifestly exhibits the exigency for new policies and procedures that specifically respond to climate induced relocation. This paper concludes by proposing some strategies that can be applied to accomplish an improved transition that suits closely knit Fijian communities as a whole and also cares for the various socio-cultural facets that embrace the community.

Method

Many forecasts have been made to predict the number of climate change induced migration with the International Organization for Migration (2009) estimating the number of environmental migrants to range from 25 million to a billion by the year 2050. In the past, more than 15 million people have been estimated to be displaced due to natural disasters annually around the globe with these numbers projected to increase significantly due to the increasing risk brought upon by climate change (Yonetani 2014). The government of Fiji, in recognition of the eminent dangers and urgency of climate induced relocation, is currently engaged in a design and consultation process of establishing a relocation guideline to assist communities forced to migrate due to climate change impacts. The vulnerability assessments carried out by the Fijian government has identified as many as 830 communities that are at risk from climate change with 45 of these being recommended for relocation (Turagaiviu 2015).

This article examines Vunidogoloa village relocation with respect to the challenges faced by the villagers, essentially in terms of their attachment to the land, cultural and community cohesion and governance and funding. The challenges faced by the villagers and their coping mechanisms will be highlighted in this article with the expectation it will provide a repository of experiences that contains lessons and guidelines for other villages in Fiji. Therefore this research is oriented towards unearthing and documenting the experiences of the villagers to form recommendations that can be used by villagers who will experience similar relocation due to sea-level-rise. This research forms a key element to tackle the adverse impacts of climate change by proactively learning about climate induced relocation and adaptive responses in an effort to increase the resilience of those affected.

The sentiments surrounding relocation is complex and as such a mixed methodology was employed to gain an insight into the experiences of the Vunidogoloa villagers during their relocation. Key informant interviews and village survey were conducted to gather primary data on the relocation from those who had first-hand knowledge about the entire relocation process. The key informant interviews and survey were conducted during the month of January, 2016. Literature review was also conducted using electronic databases, relevant websites and online reports related to the research topic. The review of existing research, related to this topic, was used to explore and define how this study fit in the work being carried out on climate induced displacements regionally and globally.

Key informant interviews using a structured questionnaire was conducted with government officials and other professions who were directly involved in the Vunidogoloa relocation process. The ten key informants interviewed provided an insight which served as a basis for future information gathering as the study progressed. The village survey was carried out using questionnaires via face-to-face interviews with Vunidogoloa villagers. The researchers interviewed 20 villagers. Non proportional quota sampling with categories of gender, age, social status and membership in community sub-groups (such as women's group) was used to identify the survey participants within the village. Semi-structured interviews were carried out, exploring the following themes: demographic characteristic; level of awareness on climate change and its related adaptation activities; coping mechanisms employed to overcome flooding and sea level rise at their previous village site; challenges faced during/after the relocation. Each interview usually lasted between 30 and 45 mins.

Communication in iTaukei language was carried out and interpreted by a translator who was engaged to assist the interviewers. The language barrier was a major limitation that was faced during this research. The villagers, especially the elderly had difficulty speaking in English and a translator was involved in this research project, to overcome this problem. Since translation is an interpretive act which depends on the interpretation and understanding by the translator, there is always a risk that meaning may get lost in the translation process. For instance, it was difficult at times to capture the strong sentiments expressed by the villagers especially in relation to the relocation day, by translating the iTaukei language they used to English. The villagers also had difficulty understanding some of the scientific terms used in the questionnaire as it was difficult to translate some scientific jargons in the iTaukei language. However, constant discussion and consultation between the translator and the researcher during each step of the interview process ensured that translation related problems were reduced whenever possible.

Focus of Vunidogola Relocation

The Vunidogoloa village is located in Cakaudrove Province in Fiji's second largest, northern island of Vanua Levu. The original site of the village was located on a tract of land overlooking the Natawa Bay in the rural town of Savusavu with the houses sited only a few meters inland from the coast. A report by Nurse et al. (2014) predicts with certainty, acceleration in sea level rise for the smaller island countries in the Pacific and forecasts severe sea floods and erosion for low-lying coastal areas. This holds true for the previous site of the Vunidogoloa village, where heavy rain and high tides continuously coalesced causing inundation and salt water intrusion, rendering sustainable local community gardens extremely difficult to manage and causing widespread damage with floodwater incessantly threatening the safety and health of the villagers (McNamara and Des Combes 2015). Following abortive attempts to adapt by building seawalls which easily succumbed to raging waves, in 2014 the villagers moved 2 km inland from their original village to a new site which fell within the customary land boundaries of the community. The villagers named this new site, as *Kenani*—a Fijian word for the biblical word *Canaan f*or "promised land".

Undoubtedly, the issue of relocation is a sad predicament facing many communities around the world. More so, many villagers like those in Vunidogoloa in their "promised land" are increasingly and innocently being subjected to complex problems such as forced land abandonment and likely socio-environmental changes while having a very minute contribution to anthropogenic climate change and associated sea level rise. The paper by Mitchell et al. (2015) proposes a joint response by the government and community to ensure the success of such relocations. The government of Fiji largely organized the Vunidogoloa relocation as an integrated action across different government ministries.

Dynamics of Land, Identity and Adaptation

Irrefutably, Fijians have always shared a special relationship with land, regarding it as a foundation of their identity, a place that defines them as the rightful owners of their land, closely associated to the natives' wealth, status and placement in their respective *mataqali's*. Hence, land is revered nowhere more fervently than in Fiji. For the natives of Fiji, land is viewed as being more than a resource, with islanders possessing an instinctive and spiritual attachment to their land (Edwards 2014). An iTaukei community refers to their land as *vanua*, a term which unites the concepts of personhood and land ownership as an inseparable entity where the ownership of an area of *vanua* is translated to mean the land area one is identified with.

Majority of land in Fiji is owned in trust by natives with the remainder being state land and freehold (Boydell 2001). Currently, 87% of the land is owned by traditional Fijian *mataqali's* and according to the laws of the country, this land is prohibited from being sold, exchanged or sub-let by the *mataqali*. The iTaukei Land Trust Board (TLTB) oversees all the indigenous land and is also responsible for liaising with various *mataqali's*. Depending on the size, a village can have several land-owning *mataqali's* within its boundaries. As a result of such strong land laws, the Fijians remain a proud race, forming strong spiritual and family connections with the land (Boydell and Reddy 2000) which is an integral component of a Fijians' communal life style.

Boydell and Shah (2003) compacts this relationship fittingly by suggesting that a customary land belongs to a communal stewardship rather than an individual with the *mataqali* entrusted with the responsibility of taking care of the land for the spirits of one's ancestors, use of one's life and protection and to ensure sustainability of the land for one's descendants. Due to these reasons, relocating becomes psychologically stressful for the entire village. They collectively view the process of relocation as agonizing, poignant and one that is robbing them of their communal identity. This shared identity of belonging to a *mataqali* and people of a respective *vanua* is a pride that they uphold as landowners and landowning units in their village. Hence, acquiring alternative land for relocating is a primary obstacle and this is even true for communities where such land is available as villagers habitually associate leaving their ancestral homes as potentially losing their identity and becoming a displaced, landless people (Edwards 2014).

Such is the overwhelming case of the Vunidogoloa village, which moved to a land owned by the same *mataqali*. Since the community moved to a land within its boundaries, the transition was smoother than what the villagers would have experienced if a suitable resettlement site was not available. Noteworthy is the fact that even though the relocation was within the vicinity of the existing place, the prospect of leaving their place of birth, the very land that defines them and solidifies their identity, has a pessimistic bearing on the villagers.

The unwillingness to relocate was echoed by a villager who blatantly responded, "We were trying to adopt by our own so that we don't have to leave our land and each time the sea came to our doorsteps, we moved a little away from it until it became so worse that we knew we had to relocate" (Vunidogoloa village interviewee, personal communication, 28 January 2016, Savusavu, Fiji). Sovaraki (2014) quoted a village spokesman of Vunidogoloa who revealed that plans to relocate had begun in 1956 but due to the reluctance of the elders to leave their ancestral grounds/boundaries, they had to wait till 2006 to start the process of relocating. The hold of the land was so great that it bound the villagers for almost half a century, delaying their process of adaptation till the sea burst its bounds. Moving away from their land inevitably affects an individual's physiological and mental health, particularly in the context of disaster recovery and extreme weather events (Frumkin et al. 2008). Social networks and community connections are likely to be fractured as a result of climate induce displacement.

The prime contributing factor of the Vunidogoloa relocation stemmed from the constant salt-water inundation, which coupled with the consequent relocation to another land, cogenerated significant disruptions to the social and mental health of the villagers. The emotional distress and anxiety of leaving their ancestral land was expressed by a villager who vividly recalled, “Initially relocating was not an option to us at all but climate change came like an enemy that chased us away by taking our land, taking our food, taking everything” (Vunidogoloa village interviewee, personal communication, 28 January 2016, Savusavu, Fiji). The connection to the land and the environment, traditions and customs associated with it are very profound for a Fijian community, forming part of their identity. Relocation due to rising sea level is likely to cause communities to have some physiological impacts while they try to re-build their lives in a new location after their upheaval from the ancestral lands.

The Cultural-Spiritual Dimension of Resettlement

The findings also recount the emotional ordeal which obscures the process of relocation. As recollected by one of the villagers in an interview, “Movement to new site was very painful and upsetting” (Vunidogoloa village interviewee, personal communication, 28 January 2016, Savusavu, Fiji). This is a precise indication of the level of attachment and sentimental value upheld by people who are in the process of relocation. As shared by the villagers, it was a difficult situation to leave the old site where they had lived all their lives, and to go through the harrowing decision to disentomb and shift the remains of their ancestors to a new burial site. In an interview with another village representative, it was noted, “We didn’t want to leave the cemetery where it was, to be washed away, so the church arranged for the burial site to be moved. Sadly the first burial at the new site was that of a still-born child which is interpreted as a bad omen from our ancestors” (Vunidogoloa village interviewee, personal communication, 28 January 2016, Savusavu, Fiji).

The Fiji Police Force facilitated the relocation process and in a conversation with one of the police officers, it was recorded that, “the spiritual connection to the land was so strong that some families were adamant not to move despite being fully aware of the severity of climate change impacts” (Fiji Police Force interviewee, personal communication, 28 January 2016, Savusavu, Fiji). All the villagers interviewed during field study indicated that climate change and the need to relocate had always been a topic of discussion at community gatherings, church assemblies, and even at homes amongst family members and such deliberations had often ended with sundry responses from the people. It seemed that the youngsters were willing to relocate while the elders of the village could not come to terms with having to uproot an entire community and relocate—it meant detachment from their identities, parents and cultural ambiance. In an interview with an administrator from Cakaudrove Provincial Council, it was recorded, “People were not ready and discussions concerning relocation took several years" (Cakaudrove Provincial Council interviewee, personal communication, 25 November 2015, Savusavu, Fiji).

For the people of Vunidolgoa, relocation would not have eventuated if it was not utterly necessary. The old site was linked to some exceptional and momentous nostalgic significance. The day of actual relocation was the most difficult moment in the lives of the people of Vunidogoloa. A representative of the Vunidogoloa Women's group described the movement, "as if a funeral procession was underway" (Vunidogoloa village interviewee, personal communication, 27 January 2016, Savusavu, Fiji). Although, the relocation for Vunidogoloa village was planned and executed with the consent of the villagers the entire process from its inception to completion created a profound spiritual predicament. An increased feeling of alienation and stress associated with relocation may pose threat to the health, and the overall welfare of the people (Fresque-Baxter and Armitage 2012).

The Vunidogoloa relocation is the first successful project of its kind in the South Pacific and for Fiji. However, in an interview with an administrator of Ministry of iTaukei Affairs it was revealed that climate change adaptation through relocation may not be so easy to achieve for many of the other vulnerable Fijian villages (Ministry of iTaukei Affairs interviewee, personal communication, 26 November 2015, Suva, Fiji). Piazza and Bolalevu (2014) elucidate that the most important factor is the willpower of the people especially since the cultural and spiritual significance that indigenous people attach to their lands and territories goes far beyond any monetary or productive value or even the value of their life.

In the case of another village, Vunisavisavi, also located in the Cakaudrove Province, the relocation is impeded by traditional obligation of the villagers to the *Tui Cakau* (High Chief of Cakaudrove Province in Fiji). Silaitoga (2016) quoted a village spokesman of Vunisavisavi village who revealed that the cultural obligation towards the *Tui Cakau* was an ancestral tradition and upheld great values. The villagers believe in customary punishment succeeding relocation which will affect their livelihood at the new location. The residents of Vunisavisavi treasure their land and associated cultural values so much that despite being strongly advised to relocate, the villagers are not willing to budge (Frontline Truths by the Pacific Climate Warriors 2015).

In an interview with a Vunidogoloa villager representing the chiefly clan, it was highlighted that there is a need to develop the faith of the people. He mentioned "detachment from customary land is heartbreaking and it is important to move people together with their church and faith to make relocation a success" (Vunidogoloa interviewee, personal communication, 29 January 2016, Savusavu, Fiji). People's opinion of climate change and its impacts is greatly influenced by the church, which is a significant barrier to adaptive capacity (Kuruppu and Willie 2015). The Pacific Conference of Churches, village church leaders and the village chiefs play a pivotal role in ensuring that the traditional values are integral during and after the repositioning process. For indigenous societies, every aspect of climate change is mediated by culture. Community cohesion and place attachment are also key elements in sustaining indigenous societies actions against climate change (Adger et al. 2013).

Adaptation Limits—From a Governance and Funding Perspective

The initial consultations between the village headman and the Fijian Government officials occurred in year 2006. It was not until year 2012 that some events started to transpire. During this period the villagers waited tensely for the government to reciprocate (United Nations Office for the Coordination of Humanitarian Affairs 2014). In the 2012 national budget, the Fijian Government sanctioned an allocation of FJD $1.0 million to the Ministry of Provincial Development and National Disaster Office for Disaster Risk Reduction measures (The Fijian Government 2012). The total for Vunidogoloa relocation project cost is estimated at around FJD $980,000 of which FJD $740,000 was government contribution while an approximate sum of $240,000 was subsidized by the community in the form of timber used for construction (Cakaudrove Provincial Council interviewee, personal communication, 29 January 2016, Savusavu, Fiji). Timber was provided from their forest reserves. Relocation is undeniably costly in many ways and is often considered as a last resort for vulnerable communities (McNamara and Des Combes 2015). Community contribution is essential as it eases the financial burden endured by the government. However, several coastal Fijian communities have limited resources and may not be able to make significant contributions which will further exhaust government funding.

The legislative issues coupled with community consultations also complicates the relocation process and results in delays. This concern was noted in an interview with a Fijian government official, "It is not easy for government to relocate a community, as witnessed in the case of Vunidogoloa, the process takes a lot of time as a lot of government departments are engaged, and there is a lot of paper work" (Fiji Government interviewee, personal communication, 28 January 2016, Savusavu, Fiji). Similar concern was noted in another interview with an iTaukei provincial council official, "It takes a lot of time to consult with iTaukei administrators and ensure that indigenous protocols are observed" (Cakaudrove Provincial Council interviewee, personal communication, 26 January 2016, Savusavu, Fiji). There is an immediate need not only for Fiji but for other Small Island Developing States to reinforce synchronization between various departments responsible for climate change adaptation and disaster management (Kuruppu and Willie 2015).

Fiji's current climate change policy does not specifically address relocation as an adaptation measure. There is no clear mandate to accommodate the relocation process and therefore, a national relocation guideline is being conscripted, however, it is still a work in progress (Pareti 2015). There is no clear indication as to much longer it would take to establish the guideline which must support a factual and participatory approach to planning and execution of relocation (McNamara and Des Combes 2015). A lack of legal and policy frameworks, procedural and human resources and financing for relocation impedes progress of the guideline (Wilsons 2014). Customary land plays a fundamental role in identity, way of life, communal unity and source of revenue for Pacific islanders, and forced detachment from the land is a sensitive issue which further complicates relocation policies. Although relocation is considered to be adaptive response to climate change, many Pacific Islanders oppose the notion of resettlement and this has significant repercussions for national policy that supports relocation as an adaptive strategy (Campbell and Warrick 2014).

For the people of Vunidogoloa, the preferred new site was available about 2 km inland on higher ground and was owned by the same *mataqali* who owned the old site. This avoided any land issues which would have emerged if additional customary land were to be acquired from a different *mataqali* (Mitchell et al. 2015). In Fiji, majority land is owned by different *mataqali's* and while some are prepared to accommodate new occupants at a small cost, some may still engage in extensive negotiations and sizeable remuneration (Wilson 2014). Cultural obligations also plays a significant role in shaping people's perceptions of climate risks at the local level in the Pacific and a thorough relocation policy would have to adequately address this issue. This concern was noted in an interview with one of the Vunidogoloa villager, "government must consider relocating people together with their church". The reluctance of Vunisavisavi villagers to move to new site is also an example of a cultural issue that can hinder adaptation. Several government ministries provided their input in the relocation process. Some of the authorities provided funding while some provided significant resources for the provision of essential services and new livelihoods. The multi-sectoral approach is vital in maintaining the socio-economic status of the relocated families and must be prioritised in the policy framework.

National governments do not have the capacity to offer displaced people infinite financial support (Edwards 2013, p. 2). Developing nations also experience complications in accessing global climate change fund. The prerequisites to acquiring such funds are sometimes difficult to achieve and as a result, developing nations fail to benefit from them. The process of accreditation is a lengthy one which is prompting some developing nations to opt for alternative route. The Fijian government managed to access some grant from the Green Climate Fund through a partnership with the Asian Development Bank, which has already attained accreditation (Takehiko 2015). The Fijian government's relocation policy must emphasize on the concept of 'relocation with dignity', since relocation has countless strings attached to it for the person, family or community who has or need to be relocated.

Community Cohesion

Fijian settlement is extensively identified as a closely-knit community due to its communal way of living. This is ingrained in the minds of the people who in many Fijian villages around the country practice and even today are unable to view themselves as a separate entity from their *mataqali*. During the field survey almost all the interviewees stressed that the villagers deal with issues collectively and not as individual units. One villager shared the reason for Vunidogoloa settlement taking so much time to finally relocate, "Our people of the *Vanua* view this place as *kece* (everyones) and as a community we deal with problems together" (Vunidogoloa interviewee, personal communication, 29 January 2016, Savusavu, Fiji). In recognition of the important role that community social protection plays in Fiji, efforts to relocate the people to improve targeting, inclusiveness and governance of community-level is a huge challenge here in Fiji. Due to the nature of family units, a strong bonded Pacific community with its cultural roots defining them as one people makes it rather complicated to talk communities for relocation (Ferris et al. 2011).

In addition, according to one villager interviewed during field study, he echoed that "Elders of Vunidogoloa were really against relocation that the younger generation decided not to move unless all the elders within certain age category finally died" (Vunidogoloa interviewee, personal communication, 29 January 2016, Savusavu, Fiji). The respect and solidarity displayed by the younger generation to not hurt the feelings of the elders is noteworthy. The support system in place in this village is a testimony of community unity and to what extent these common people can go for each other during times of tribulation. Together these ideologies of shared community and one people of the *vanua* can buffer these communities against vulnerabilities such as climate change and its impacts. For many years the villagers lived at the same place that continued to threaten their livelihoods, rip them off their prospect of progress, only to live in harmony at their place of birth, a place that defines them as a unique clan. This is one reason it took them so long to relocate even after fully understanding the threat their current place of residence posed on their lives and their families.

Awareness and Monitoring

The Pacific Island countries (PICs) are projected to be one of the first to experience the impacts of climate change. This calls for a proactive approach by the PICs government to create awareness amongst its citizens to ensure that they are equipped to tackle the looming dangers. In the Pacific Islands, however, the reactive approach as compared to a more proactive approach, is more common as the awareness follows after the first hand experiences with the effects of climate change (Ferris et al. 2011). This is clearly demonstrated by the relocated villagers of Vunidogoloa who were initially caught unaware that the receding shore and the advance of the sea waves were signs of climate change. This led to a period of inactivity with little effort by the villagers to adopt using soft adaptation techniques before the seawall construction and the subsequent relocation were finally ensued after 40 years of first detection in the 1950s. Awareness and community outreach programs by civil organisations and the government is of paramount importance as such incentives spark behavioural changes in local people that enhance social resilience and enable them to better adapt to the impacts of climate change.

Capacity building has been highlighted as a priority in developing countries to ensure access to information, knowledge and participatory principles (Kumamoto and Mills 2012). In the case of Fiji Islands, this becomes even more imperative since the national government is usually the contact point and the recipient of international adaptation financing. Studying the case of the Vunidogoloa relocation has made it clear that awareness of the cause of their inundation issues was essential in advancing the relocation and in enhancing the adaptive capacity of the villagers. Government monitoring of communities that are particularly vulnerable and raising awareness at the initial detection of vulnerability is critical to ensure that community members actively search for options to adapt rather than be caught unaware until the full effects of climate change come to bear on them.

Conclusion

The findings from the survey unveil some valuable qualitative insights into people's perception of climate change and also establish the fact that relationship between land and culture is inherent in the context of climate induced relocation. For the Vunidogoloa village community the call for support from government to consider relocating the village was genuine and necessary. The findings further disclose that in relocating vulnerable communities, it becomes fundamental to contemplate not only the financial and bureaucratic conditions surrounding their relocation, but also the numerous socio-cultural features that encompass the community. Fijian communities are diverse and entrenched in cultural sensitivities of the environment which as realized through this study can be a significant barrier to climate change adaptation. A number of awareness programs are needed to stimulate behavioural changes in people that can increase their capacity to better adapt to the impacts of climate change. The outcomes of this exploratory study also irrefutably support the notion that relocation must be conducted in a holistic manner that befits communities and conserves quality of life. Indubitably, it is essential to note that many such communities embrace distinctive sets of traditional knowledge and abilities that provide opportunities for adaptation both by staying put and by relocating. The Vunidogoloa relocation being the first effective relocation project in Fiji, and the South Pacific sets the premise for future climate change induced relocations by recognizing a list of potential complications that Fijian government must address in their national relocation policy. Fijian government has identified a number of communities that need to be relocated in the next decade and is also developing a national relocation guideline. This paper through its findings recognizes the exigent need for an official relocation policy document which encompasses all the legislative issues and planning framework and is meticulously prepared in consultation with all the relevant stakeholders including research and academic institutes. This is to ensure that climate change induced relocation is aligned to national strategic plan and government primacies. Relocation as a climate change adaptation strategy must be considered as a last resort particularly because it involves an increased cost that encumbrances the national budget. Communities differ in the level of vulnerability and a systematic preliminary assessment must be carried out to determine the most appropriate adaptation strategies for each community to reduce their vulnerability. Procuring land for relocation may prove to be problematic since majority of the land in Fiji is customary and protected by the laws of the country. A legitimate approach and advance planning and consultation is thus obligatory for securing land for successful relocation. The bonding to land is also strongly embedded in the lifestyle of the Fijian community and a national relocation plan must be able to subtly tackle this issue. Culture, communal unity, and traditional obligations to ancestors play a colossal role in deliberations relating to relocation and in extreme cases even prevent communities to move. It is thus essential to incorporate sociocultural parameters in a relocation policy. In-depth consultations and community outreach programs with vulnerable villagers' necessitates for them to fathom that customary livelihoods may be enhanced through relocation than by staying back and guarding an existing settlement. Socio-economic quandary embraces even greater sway in decisions related to community resettlement and for this reason it is imperative to adopt a cross-sectoral and a participatory approach that is inclusive of environmental consultants, pertinent government and non-government officials, church leaders, community residents, social researchers, academics and local community specialists.

References

Adger, W. N., Barnett, J., Brown, K., Marshall, N., & O'Brien, K. (2013). Cultural dimensions of climate change impacts and adaptation. *Nature Climate Change, 3*(2), 112–117.

Amundsen, H., Berglund, F., & Westskog, H. (2010). Overcoming barriers to climate change adaptation—a question of multilevel governance? *Environment and Planning C: Government and Policy, 28*(2), 279–289.

Barnett, J., & Campbell, J. (2010). *Climate change and small island states: Power, knowledge, and the South Pacific*. London: Earthscan.

Boege, V. (2011). *Challenges and pitfalls of resettlement measures: Experiences in the Pacific region*. Centre on Migration Citizenship and Development Workers Paper Series, Bielefeld.

Boydell, S. (2001). *Philosophical perceptions of Pacific property—land as a communal asset in Fiji*. Paper presented at the 7th Pacific rim real estate society annual conference, 21–24 January 2001, Adelaide. Retrieved February 20, 2016, from http://prres.net/Papers/Boydell_Philosophical_Perceptions_of_Pacific_Property.pdf.

Boydell, S., & Reddy, W. (2000). Contemporary land tenure issues in the Republic of Fiji. In *Conference papers, 6th Pacific Rim Real Estate Society (PRESS) Conference*, 21–24 January 2001, Sydney (pp. 23–27).

Boydell, S., & Shah, K. (2003). *An inquiry into the nature of land ownership in Fiji*. Paper presented at the International Association for the study of common property (IASCP), September 7–9, 2003, Brisbane. Retrieved February 19, 2016, from https://dlc.dlib.indiana.edu/dlc/bitstream/handle/10535/2266/Boydell_%26_Shah_inquiry_into_land.pdf?sequence=1.

Campbell, J., &Warrick, O. (2014). *Climate change and migration issues in the Pacific*. Report, United Nations Economic and Social Commission for Asia and the Pacific. Retrieved December 10, 2015, from http://www.unescap.org/sites/default/files/Climate-Change-and-Migration-Issues-in-the-Pacific.pdf.

Displacement Solutions. (2013). *Peninsula principles on climate displacement within states.* Retrieved February 13, 2016, from http://displacementsolutions.org/wp-content/uploads/2014/12/Peninsula-Principles.pdf.
Edwards, J. (2012). Location, location, relocation, the bula bulletin, April–May, p. 3. Retrieved December 5, 2016, from http://www.methodist.org.uk/downloads/wcr-julia-edwards-fiji-newsletter-aprilmay2012.pdf.
Edwards, J. (2013). *Financing the future wave of relocations.* Retrieved December 5, 2015, from http://www.methodist.org.uk/media/757401/wcr-pcc-relocation-policy-fiji-jan13.pdf.
Edwards, J. (2014). Climate justice and the Pacific conference of churches: Moving the relocation agenda in the Pacific. In L. Andrianos, J. W. Sneep, G. Kerber, & R. Attfield (Eds.), *Sustainable alternatives for poverty reduction and eco-justice*, (Vol. 1, p. 209). Newcastle, England: Cambridge Scholars Publishing.
Ferris, E., Cernea, M. M., & Petz, D. (2011). On the front line of climate change and displacement: Learning from and with Pacific island countries. In *Proceedings of the International Disaster and Risk Conference* (1st edn.). New York: Brookings Institution.
Fonmanu, K. R., Ting, L., & Williamson, I. P. (2003). Dispute resolution for customary lands: some lessons from Fiji. *Survey Review, 37*(289), 177–189. Retrieved December 5, 2015, from http://hdl.handle.net/11343/33965.
Fresque-Baxter, J. A., & Armitage, D. (2012). Place identity and climate change adaptation: A synthesis and framework for understanding. *Wiley Interdisciplinary Reviews: Climate Change, 3*(3), 251–266.
Frontline Truths by the Pacific Climate Warriors. (2015). *Frontline truths—Fiji.* Retrieved January 29, 2016, from http://350pacificclimatewarriors.tumblr.com/post/122267652905/george-nacewa-fiji-islands.
Frumkin, H., Hess, J., Luber, G., Malilay, J., & McGeehin, M. (2008). Climate change: The public health response. *American Journal of Public Health, 35*(5), 401–402–445. Retrieved December 18, 2015, from http://www.sciencedirect.com/science/article/pii/S0749379708007071.
Govan, H. (2009). Achieving the potential of locally managed marine areas in the South Pacific. *SPC Traditional Marine Resource Management and Knowledge Information Bulletin, 25*, 16–25. Retrieved January 7.
International Organization for Migration. (2009). *Migration, environment and climate change-assessing the evidence.* Retrieved January 5, 2016, from http://publications.iom.int/system/files/pdf/migration_and_environment.pdf.
Kumamoto, M., & Mills, A. (2012). What African countries perceive to be adaptation priorities: Results from 20 countries in the Africa adaptation programme. *Climate and Development 4*(4), 265–274. Regional Studies Association.
Kuruppu, N., & Willie, R. (2015). Barriers to reducing climate enhanced disaster risks in least developed country-small islands through anticipatory adaptation. *Weather and Climate Extremes, 7*, 72–83.
McNamara, K. E., & Des Combes, H. J. (2015). Planning for community relocations due to climate change in Fiji. *International Journal of Disaster Risk Science, 6*(3), 315–319.
Mitchell, D., des Combes, H. J., Myers, M., & McEvoy, D. (2015). Addressing land issues in disaster risk management in the Pacific island countries. *Land Tenure Journal, Climate, 1*, 36–69.
Nurse, L. A., McLean, R. F., Agard, J., Briguglio, L. P., Duvat-Magnan, V., Pelesikoti, N., et al. (2014). Small islands. In *Climate change 2014: Impacts, adaptation, and vulnerability. Part B: Regional aspects. Contribution of working group II to the fifth assessment report of the intergovernmental panel on climate change* (pp. 1613–1640).
Nunn, P. (2012). Climate change and Pacific island countries, Asia-Pacific human development report background papers series 2012/07. Bangkok: United Nations Development Programme.
Pareti, S. (2015). *Fiji shares its 'do's' and 'don'ts' on relocation.* Retrieved February 15, 2016, from https://www.sprep.org/climate-change/fiji-shares-its-dos-and-donts-on-relocation.

Piazza, B., & Bolalevu, A. (2014). Rising sea levels have prompted the Fiji government to relocate one of the country's most vulnerable seaside communities. *World News Australia Radio*, radio transcript, 1 February. Retrieved March 5, 2016, from http://www.sbs.com.au/news/article/2014/01/31/rising-sea-levels-prompt-relocations-fiji.

Rubelli, E. (2015). Escaping the waves: A Fijian village relocates. *The Sydney Morning Herald*, 3 October. Retrieved December, 12, 2015, from http://www.smh.com.au/good-weekend/escaping-the-waves-a-fijian-villages-forced-relocation.

Silaitoga, S. (2016). Traditional Obligations to Fiji Chief Prevents Climate Relocation. *Fiji Times*, 5 February. Retrieved March 10, 2016, from http://pidp.eastwestcenter.org/pireport/2016/February/02-05-15.htm.

Sovaraki, A. (2014). Vunidogoloa relocation mooted in the 50s. *Fiji Sun*, 20 August. Retrieved December 15, 2015, from http://fijisun.com.fj/2014/08/20/vunidogoloa-relocation-mooted-in-the-50s/.

Takehiko, N. (2015). Scaling-up for sustainable development in Asia and Pacific: Stronger, better and faster ADB. *Asian Development Bank*, speech transcript. Retrieved February 15, 2016, from http://www.adb.org/news/speeches/scaling-sustainable-development-asia-and-pacific-stronger-better-and-faster-adb.

The Fijian Government. (2012). *Fiji Budget Estimates*. Retrieved January 14, 2016, from http://www.fiji.gov.fj/Budget/2012.aspx.

Turagaiviu, E. (2015). 830 Fijian communities at risk from climate change. *Fiji Broadcasting Corporation*, 29 October. Retrieved February 12, 2016, from http://www.fbc.com.fj/fiji/34377/-830-fijian-communities-at-risk-from-climate-change.

United Nations Framework Convention on Climate Change. (2006). *Climate change: Impacts, vulnerabilities and adaptation in developing countries report*. Retrieved March 1, 2016, from https://unfccc.int/resource/docs/publications/impacts.pdf.

United Nations Office for the Coordination of Humanitarian Affairs. (2014). *Fiji: Building resilience face climate change*. Retrieved February 18, 2016, from http://www.unocha.org/top-stories/all-stories/fiji-building-resilience-face-climate-change.

Wewerinke, M. (2013). A right to enjoy culture in face of climate change: Implications for "climate migrants". CGHR Working Paper 6/4CMR Working Paper 7, Centre of Governance and Human Rights, University of Cambridge. Retrieved January 11, 2016, from https://www.repository.cam.ac.uk/bitstream/handle/1810/245116/CGHR_WP_6_2013_Wewerinke(rev).pdf?sequence=4.

Wilson, C. (2014). *Fiji leads Pacific region on climate adaptation efforts: Vunidogoloa: An exercise in inclusive adaptation*. Retrieved February 15, 2016, from http://www.ipsnews.net/2014/05/fiji-leads-pacific-region-climate-adaptation-efforts/.

Yonetani, M. (2014). *Global estimates. 2014: People displaced by Disasters*, International Displacement Monitoring Centre, Geneva, Switzerland.

Chapter 3
Climate Change and Migration in the Maldives: Some Lessons for Policy Makers

Johannes Luetz

The Maldives: Geographic, Demographic and Climate Change Issues

Drawing on field research conducted in the Maldives in December 2011 and January 2012, this case study examines the linkages between climate change and human movement with a view to raising policy options for more equitable human migration. The significant level of government coordinated migration makes the Maldives a useful microcosm for the study of migration relevant success factors. Although at present the majority of migration across the Maldives is internal and not climate change related, useful lessons can be learned from how the government has planned and implemented macro-managed migration.

The Maldives, officially the Republic of Maldives, is an archipelagic nation made up of two long chains of a total of 26 atolls located southwest of India and Sri Lanka. With a population density of approximately 1,053 people per sq km of land the Maldives is grouped among the most densely settled nations in the world, even by small island state standards (World Bank 2011). Comprising an estimated 1,190 coral islands which are scattered over a distance of more than 850 km of ocean (Godfrey 2007, p. 9), and with 99.9% of the nation's territorial area (90,000 km^2)

Preamble: This paper is based on Ph.D. research conducted at the University of New South Wales, with the unabridged Maldives case study available as Chap. 6 in the Ph.D. thesis entitled: "Climate migration: Preparedness informed policy opportunities identified during field research in Bolivia, Bangladesh and Maldives" http://handle.unsw.edu.au/1959.4/52944.

A short background video to this research appeared in *The Guardian* on 6 August 2015: "Climate refugees: the communities displaced by global warming" http://gu.com/p/4ba7t/sbl.

J. Luetz (✉)
CHC Higher Education, Brisbane, Australia
e-mail: jluetz@chc.edu.au

W. Leal Filho (ed.), *Climate Change Adaptation in Pacific Countries*,
Climate Change Management, DOI 10.1007/978-3-319-50094-2_3

consisting of water (Masters 2006, p. 16), the Maldives is also numbered among the most geographically dispersed countries in the world.

Given that the archipelago's geospatial properties, general shortage of land, human migration, rapid urbanisation and population consolidation trends are working in concert to put a growing premium on what limited land is available for settlement and development, it is not hard to appreciate why policy makers are increasingly taking steps to more proactively guide the nation's demographic evolution. Already housing appears to be unaffordable for parts of society, "with prices in Malé exceeding 12 times the average annual income" (GOV MAL 2009, p. 64). It is in this context that the government is seeking to address congestion, urban sprawl, housing shortages, rental increases and "[w]orsening standards of housing within the existing units, sometimes reaching slum conditions" (MHUD 2008, p. 15).

Additionally, "[t]he rapid growth in the number of expatriate workers—estimated to be around 80,839 (2008) representing more than 25% of the Maldivian population and just less than 80% of total Maldivian employment is becoming an area of concern. There are economic and social implications of such a large contingent of expatriate workers" (Baum 2012, p. 25).

It appears that the Indian Ocean Tsunami that struck multiple Asian countries on 26 December 2004 provided the nation an additional impetus to see the government strategy of population consolidation implemented more rapidly (MEEW NAPA 2007; MHUD 2008). The influence that the tsunami, and more generally, the Maldives' vulnerability to natural disasters and climate change have had on internal migration across the archipelago is significant.

> Global warming and the associated sea level rise threaten the fragile ecosystems of the Maldives where 80% of islands are less than 1 m above mean sea level. The tsunami of 26 December 2004 truly exposed the vulnerability of the Maldives. Most of the islands that suffered damage had little or no coastal protection. The islands are fully exposed to the dangers of wave action, erosion and flooding. While a tsunami of the magnitude experienced in December 2004 is an extremely rare event, with the predicted sea level rise, flooding may become a more frequent phenomenon. (MPND 2007, p. 9)

Godfrey (2007) notes that the tsunami "affected every island in the Maldives … and highlighted the economic and environmental vulnerability of the country" (p. 13). According to a needs assessment paper jointly prepared by the World Bank, Asian Development Bank and United Nations, the tsunami waves ranged "from 4 to 14 feet" in height and were noted "in all parts of the country", leaving 83 people "confirmed dead,"[1] "significantly" damaging 39 islands, "completely" destroying 14, and "severely" affecting one third of the country's population (ADB UN WB 2005, p. 4). In total the tsunami caused the displacement of more than 20,000 islanders (8,500 intra-island and nearly 12,000 inter-island). Thus, with 7% of the population displaced, and with estimated total damages equal to 62% of the nation's

[1]An additional 25 people are missing, assumed dead (ADB UN WB 2005, p. 4).

Gross Domestic Product, plus with the livelihoods of thousands compromised, "the Maldives experienced a disaster of national proportion" (ibid, pp. 3–4).

The tsunami also appears to have served as a powerful reminder just how vulnerable the country is to climate change induced sea level rises (ADB UN WB 2005; MEEW NAPA 2007; Vince 2009a). According to the National Adaptation Program of Action (NAPA) prepared by the Ministry of Environment, Energy and Water, "[m]ost islands in Maldives are barely 1 m above sea level. Under the predicted worst case sea level rise scenario, large areas of Maldives could be inundated" (MEEW NAPA 2007, p. 19). Vulnerabilities are exacerbated by the small size of many islands: "96% of the islands are less than 1km2 in area [which] forces people to live next to the sea. […] 44% of the settlement footprints of all islands are within 100 m of coastline" (ibid, pp. 18–19).

Recent scientific sea level rise projections have raised the submersion and implied forced human emigration from islands in the Maldives as a possibility (Hansen 2007, 2008; Hansen et al. 2016), with a number of studies (Ananthaswamy 2009; Rahmstorf 2012; Rahmstorf et al. 2012; Schaeffer et al. 2012; Solomon et al. 2009; World Bank 2012) corroborating the view that preparedness seems prudent in the context of contemporary emissions trajectories now on course to produce global average temperature increases on the order of 4 °C by 2100 (Hamilton 2010, pp. 190–208; New et al. 2011; Nicholls et al. 2011; Schellnhuber 2008; Stewart and Elliott 2013; Vince 2009b; WBGU 2009a; World Bank 2012, 2014).

The implications of such high temperature increases could have far-reaching consequences for many small island states. Research by Schaeffer et al. (2012) suggests that limiting global warming to below 1.5 or 2 °C would produce 75–80 cm sea level rise (above the year 2000) by 2100, and 1.5 m (1.5 °C warming) and 2.7 m (2 °C warming) respectively, by the year 2300 (Schaeffer et al. 2012, p. 867).

Notwithstanding, sea level rise is not the only climate change related pressure potentially thwarting the perpetuity of the Maldives, and perhaps not even the most severe peril.

The warming of the ocean itself, along with implicit detrimental consequences for coral formation, would be a fate far worse—and suffered much earlier—than the longer-term and slower-onset threat posed by sea level rises, a concern that is well-documented in the literature (Barnett and Campbell 2010, pp. 11–15; De'ath et al. 2009; Frieler et al. 2013; Hoegh-Guldberg 2011; Nunn 2009). According to the UNEP's global Millennium Ecosystem Assessment, "coral reefs buffer land from waves and storms and prevent beach erosion" (UNEP 2006, p. 14), but worldwide 20% of coral reefs are "severely damaged and unlikely to recover; 70% are destroyed, critical, or threatened" (ibid, p. 11).

The Maldives is no exception to this global trend. The report "State of the Environment Maldives 2011" indicates that all across the small island state coral bleaching "[o]ccurred in 1977, 1983, 1987, 1991, 1995, 1997, 1998" (SOEM 2011, p. 67), and that "bleaching events were also reported in 2010" (ibid, p. 131). It states that following the bleaching event in 1998 live coral cover was "reduced to 5% after the event" (ibid, p. 66), and that "[t]he coral bleaching events of 1999 reduced coral cover to a mean of 2.1% compared to the pre-bleaching level of 30–40%" (ibid, p. 131). Bleaching events in the Maldives have "caused devastating damage to reefs and their inhabitants" (ibid, p. 131), "caused decline of live bait" (ibid, p. 67), and "caused the disappearance of two reef fish species" (ibid, p. 67). The Maldives Ministry of Environment and Energy anticipates that bleaching incidences will "increase with the rise in temperature" (ibid, p. 70), an outlook that already appears to be realised (Slezak 2016).

Recent research suggests that coral reefs are even more acutely sensitive to sea-surface temperature increases than previously thought, and that even "limiting warming to as little as 1.5 °C may not be sufficient to protect reef systems globally" (Frieler et al. 2013, cited in World Bank 2012, p. 61). Importantly, climate system inertia implies that based on past emissions of greenhouse gases humanity is already committed to 1.3–1.5 °C global warming (Hansen 2008, 2009; UNDP 2007, p. 4; WBGU 2009a, b; Woo et al. 2011; World Bank 2012, 2014). Further, a number of studies have indicated that approximately 4 °C global warming could be reached by around 2100 in the absence of successful climate change mitigation leading to a largely and rapidly decarbonised global economy (Hamilton 2010, pp. 190–208; New et al. 2011; Nicholls et al. 2011; Schellnhuber 2008; Stewart and Elliott 2013; Vince 2009b; WBGU 2009a; World Bank 2012).

In view of such adaptation challenges now facing sensitive reef systems globally it might appear questionable whether coral cover may still be relied upon for coastline protection in the future[2] (IPCC 1992, p. 107). One expert interviewed for this case study cautioned that the effect of rising sea levels should not be expected to be countervailed by concurrent coral growth:

> You can't expect an atoll island, unless there are exceptional circumstances, to survive under a sea level that's 1.2 m higher than today. A lot of people in the early days thought that coral reefs would suddenly start to grow up and continue to protect islands, but there's no evidence that coral reefs have ever done that because it requires a huge ecological change … in addition to everything else—so many reefs are degraded. (Q33/Exp/UNSW-Sydney/20120216)

It would therefore seem that as the growing conditions for coral reefs are gradually eroding, so are realistic hopes of artificially "planting" coral cover for future coastline protection (IPCC 1992, p. 107; Nunn 2009; Vince 2009a, pp. 37–39; Slezak 2016).

Although erosion cannot be simplistically attributed to climate change as a monocausal effect of the same (Mimura et al. 2007, p. 698; SOEM 2011, p. 27), there is anecdotal evidence by islanders and experts alike that erosion in the Maldives could be getting worse. According to the Ministry of Environment and Energy, "[o]ne of the most serious environmental issues the Maldives islands face is

[2]"If the rate of sea-level rise exceeds the maximum rate of vertical coral growth (8 mm/yr), then inundation and erosive processes begin to dominate, leading to the demise of the coral atoll. However, if the rate of sea-level rise is small, then coral growth may be able to keep pace." (IPCC 1992, p. 107).

beach erosion. The sand at the beach and the shore line are being washed off at a greater rate than it is accreted on many islands" (SOEM 2011, p. 27). Moreover, "[t]he tsunami appeared to have worsened chronic shoreline erosion problems in many islands" (ibid, p. 177), and climate change may further amplify erosion related difficulties in coastal regions: "[s]ea-level rise is expected to exacerbate … erosion and other coastal hazards, thus threatening vital infrastructure, settlements and facilities that support the livelihood of island communities (very high confidence)" (Mimura et al. 2007, p. 689; also IPCC 2007, p. 52).

Low-lying islands have a very narrow freshwater lens, also referred to as the Ghyben-Hertzberg lens, which is easily contaminated by rising sea levels and storm surges (Barnett and Campbell 2010, pp. 13, 172; Lubna 2012; SOEM 2011, pp. 67, 74, 84, 176). Hence as climate change impacts are progressively borne out the shrinking availability of drinking water is another possible cause for preoccupation (Figs. 3.1 and 3.2). This concern also appears to be conform with synthesis projections made by the Intergovernmental Panel on Climate Change (IPCC), which suggest that "[t]here is strong evidence that under most climate change scenarios, water resources in small islands are likely to be seriously compromised (very high confidence)" (Mimura et al. 2007, p. 689).

Climate change induced hydrological changes and extreme weather events spawning stronger storms and higher surge heights are other climate change related issues of relevance. There is agreement among researchers that climate change will

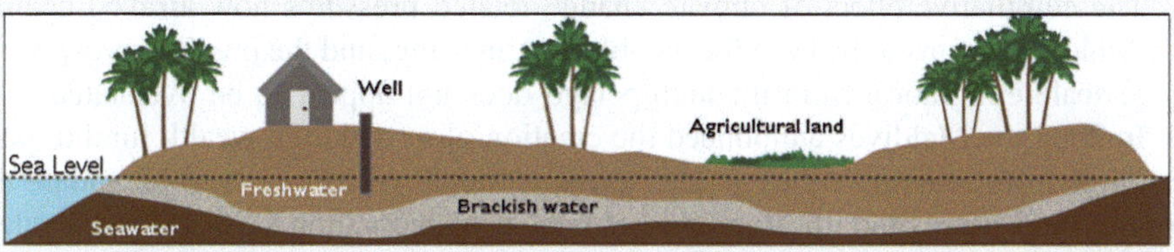

Fig. 3.1 Schematic representation of island subsistence (normal sea level) *Illustration* © World Vision, quoted from Luetz (2008), p. 23; adapted from Aung et al. (1998, p. 97)

Fig. 3.2 Schematic representation of progressive island submergence (rising sea level) *Illustration* © World Vision, quoted from Luetz (2008), p. 23; adapted from Aung et al. (1998, p. 97)

exacerbate storm surge risk: "by far the most certain aspect of climate change that will influence surge characteristics is global-mean sea-level rise. [...] The overall conclusion is that the surge hazard will evolve significantly throughout the 21st century" (Nicholls 2006, pp. 89–90, linked to Church et al. 2001).

All things considered it seems that natural pressures common in small island environments are being compounded by climate change (Barnett and Campbell 2010). The Maldives National Adaptation Program of Action advances the following analysis and recommendation: "[t]he scarcity of land in the Maldives, the smallness of the islands and extreme low elevation makes retreat inland or to higher grounds impossible. [...] Unless expensive coastal protection measures are undertaken the human settlements face the threat of inundation" (MEEW NAPA 2007, p.22).

According to Ali Rilwan of Environmental NGO Bluepeace, artificially designed islands patterned after the Hulhumalé blueprint and raised even higher to 3 or 5 m above sea level could one day accommodate the country's entire population: "[o]ver the longer term, in order to accommodate the climate IDPs (internally displaced persons), at least seven adapted contingency islands need to be developed in seven different regions across the archipelago of the Maldives" (Bluepeace 2008). In an expert interview Rilwan conceded that implementing the idea of Contingency Adapted Raised Islands (Figs. 3.7 and 3.8) would certainly "cost millions" but reasserted that the concept itself was technologically feasible, and importantly, that it would guarantee that massive-scale emigration from the archipelago would not remain the only policy option in the future (Q2/Exp/Malé/20111222).

The cumulative effect of climate change related pressures now arrayed against the Maldives seems to be both formidable and growing, and the possible prospect of large-scale emigration from the archipelago does not appear to be overstated.

In 2008 the Maldives announced the creation of a sovereign wealth fund to pool earnings from tourism to buy land elsewhere should rising sea levels inundate the country and necessitate the forced resettlement of the nation's 300,000+ islanders (BBC 2008; Henley 2008; Ramesh 2008; Schmidle 2009; UNDP 2009, p. 45, attributed to Revkin 2008; Vince 2009a, pp. 37–39).

During fieldwork this author was unable to obtain an independent confirmation of the existence or intended creation of such an insurance mechanism, neither from the Minister of Housing and Environment (Q29/Exp/Malé/20120102), nor from the Department of National Planning (Q30/Exp/Malé/20120102). Some respondents even speculated that the announcement may have been more of a "publicity stunt" by former President Mohamed Nasheed (cf, Lang 2009), rather than a formal declaration of official government migration policy (Q3/Exp/Hanimaadhoo/20111224; Q31/Exp/Hulhumalé/20120102).

Notwithstanding, the President's announcement seems to underscore the perceived priority that the archipelago ascribes to a forward-thinking posture of policy "preparedness" in relation to human migration, and already the country has the advantage of an impressive track record of government sponsored intra- and inter-atoll migration working in its favour (MHUD 2008; GOV MAL 2009; MPND 2006, 2007; Mohamed 2002).

The benefits of early preparedness and planning in research are quite well established, both in relevant literature (e.g. Blanco et al. 2009; Foresight 2011), and by pertinent UN agencies: "[h]oping—and working—for the best while preparing for the worst, serves as a useful first principle for adaptation planning" (UNDP 2007, p. 198); "early preparedness could also help avert a humanitarian catastrophe by promoting orderly movements of affected populations and increasing the viability of the move." (UNHCR 2009, p. 3; see also Leighton 2012, pp. 703, 718).

This research represents an inquiry into what lessons may be learned from present-day intra-atoll and inter-atoll human migrations in the Maldives. While the majority of this movement is not presently associated with climate change, valuable lessons can be learned from the government's management of it. These lessons are highly relevant for future climate migration scenarios, both in the Maldives itself, as well as in other Small Island Developing States (SIDS). In point of fact, the Maldives is a particularly useful case study because much of the internal migration can be characterised as macro-managed and pre-planned.

This study extends previous research by expressly inviting the participation of migrants and host communities. In soliciting these unique grassroots perspectives this research aims to identify and encourage more anticipative and congenial migration processes and outcomes.

Research Design, Methodology, and Data Collection and Analysis

Data were sourced in the Maldives during December 2011 and January 2012, and are supported by observations made and conversations conducted during that time, as well as during months of preparation and follow-up. The Maldives field research was also aided by a preceding research visit in May 2009 during which tsunami related human resettlement was studied. This earlier research visit also enhanced familiarity with a number of pertinent issues discussed in this case study (e.g. cultural, environmental, socioeconomic vulnerabilities) and laid the foundation for strategic partnerships in areas of translation/interpretation, logistics, and overall research support. Given this earlier research visit it also appeared sensible to interview some of the same experts again to gauge if/how matters in the country had changed.

Participants were sourced from three groups of respondents: "migrants" (those who had already moved), "non-migrants" (those living in communities affected/surrounded by human movement, including host communities), and "experts" (those professionally involved with migration, climate change, environmental issues and/or relevant preparedness policy and practice.). Interview subjects were identified and recruited through a "non-probability sampling" technique, and "snowball sampling".

A face-to-face semi-structured on-site interview process was used. On rare occasions follow-up questions were asked by e-mail, phone or via skype. Participants remained anonymous unless attribution by name was expressly agreed or requested by the respondent/s.

An interview questionnaire was used to guide conversations into seven key areas of interest, of which this paper will discuss the following six: (1) interviewee demographics; (2) the migration situation generally; (3) migration push factors; (4) migration pull factors; (5) preferred migration destinations; (6) preferred migration management and/or solutions.

The first section on interviewee demographics included data such as interview date and location, respondent name, sex, age, nationality, ethnicity, religion, marital status, number of children, household size and highest level of education.

The second section asked respondents to describe the general migration situation in their communities, what they had personally seen, heard and/or experienced, and why people were moving and where to, and what role environmental factors played in shaping people's decisions to stay or leave, etc.

The third section included two tables listing potentially relevant environment and non-environment related pressures ("push factors") that may or may not have been contributing to migration in areas of investigation. Respondents were asked to rate these potential push factors as either "unimportant" OR "important" OR "very important" in terms of each factor's role in contributing to migration. Respondents were also prompted to add and rate additional factors as relevant.

The fourth section included a table listing possible migrant aspirations ("pull factors") and sought to gauge what relative priority respondents ascribed to each potential attraction in terms of its contributing role played in inducing human movement. Respondents were asked to reflect on what migrants want/seek most and then rate each listed potential aspiration as having a "very high priority" OR "high priority" OR "medium priority" OR "low priority" OR "very low priority". Respondents were also prompted to add and rate additional migrant aspirations as relevant.

The fifth section sought to establish where people go, went, and/or wish to go when compelled to move. Questions about favoured migration destinations included both de facto destinations (where people actually go/went), and hypothetical destinations (where migrants would elect to go in the absence of restrictions, e.g. financial, political or otherwise).

The final section solicited qualitative data on preferred migration management options and solutions and asked respondents to think about what would make the experience of moving more positive, how policy makers, NGOs, migrant and/or host communities could help, whether migrants valued maintaining the integrity of their community (vis-à-vis the prospect of experiencing assimilation/dissolution in the new place of destination), and what respondents (if their migration experiences had the benefit of hindsight) had personally learned from the process, before, during and after the move. In short, the final section asked respondents to reflect on personal lessons learned and to suggest better migration management approaches and solutions as identified from their unique individual vantage points.

The same questionnaire was used during interviews with "migrants" (those who had already moved), "non-migrants" (those living in communities affected/surrounded by human movement, including host communities), and "experts" (those professionally involved with migration, climate change, environmental issues and/or relevant preparedness policy and practice). Given that a standard questionnaire tool guided conversations with respondents from widely divergent educational backgrounds it was important that questions be "worded as simply as possible" (Berdie et al. 1986, p. 31) whilst retaining the capability to extract responses characterised by a level of complexity.

To minimise the potential for interpretative bias, and to enable genuine responses that were not guided into predetermined cognitive frameworks, the issue and language of "climate change" was mostly avoided or de-emphasised, and "environmental factors" highlighted instead. Moreover, questions were posed in a "neutral" way and with "no hint as to what the 'right answer' should be." (Sedlack and Stanley 1992, p. 226).

It was important that latent qualitative variables (e.g. perceptions, attitudes, priorities, etc.) were accurately translated into quantitative measures which could be plotted on scales. The questionnaire development process broadly followed the five steps outlined by Radhakrishna (2007), and validity was ensured by accommodating feedback and input from the research, ethics, and pilot communities.

Testing of the interview questionnaire occurred during a 17-day pilot study in 2010 on two atolls in Bougainville/Papua New Guinea and led to the incorporation of simplifying features into the final questionnaire design.

Key informant interviews sought to establish not only the extent to which climate change (including related disasters) may be implicated in the migration and displacement of human populations in the Maldives, but also how inter- or intra-island migrations may be managed to minimise negative impacts on communities and individuals. Additionally, the research sought to appreciate the migration experience over time and space across the whole spectrum of movement, from initial displacement or departure to temporary or permanent relocation. Hence fieldwork in the Maldives aimed to engage respondents in both island communities of "origin" and "destination" so as to gain a comprehensive understanding of migration across the archipelago. Holistic approaches and perspectives are taken to be crucial if preparedness policy, practice and advocacy relating to future human migration management are to be appropriately informed.

Island communities in the Maldives were chosen based on recommendations of local key informants, and interviews were primarily conducted on islands where government coordinated in-migration was the dominant process of demographic change. Communities of destination included human settlements in the following atolls and islands: Malé Atoll: Malé, Hulhumalé; Haa Dhaaluu Atoll: Hanimaadhoo, Nolhi-varanfaru, Kulhudhuffushi; Ihavandhippolhu Atoll: Huvarafushi. Human settlements were also re-visited by migrants on the abandoned Island of Hathifushi Island (Ihavandhippolhu Atoll), which may be classified as a community of origin. In addition, three experts were interviewed in Sri Lanka, Singapore, and Australia.

With a total of 33 semi-structured on-site interviews conducted in a face-to-face manner, the aim of conducting at least 30 semi-structured interviews was slightly exceeded. Given that individual interviewees were frequently encountered in groups, seven planned and impromptu "focus group" discussions were conducted. With between three to five respondents participating in each conversation the total number of respondents involved in this field research is estimated to be 90. The availability of locally sensitive interpreters significantly assisted in moderating these group discussions.

Key commentaries follow a simple in-text referencing system that ensures respondent anonymity[3] while retaining pertinent interview details: upon specifying the questionnaire number (e.g. Q14) the in-text referencing system distinguishes between migrants (Migr), non-migrants (NMigr), host communities (Hst), and experts (Exp); communities of origin (Orig) and destination (Dest); and interview locations (e.g. Hanimaadhoo) and interview dates in the format YYYYMMDD. This system makes it possible for both researcher and reader to perceive contextually significant interview nuances with ease while concurrently reducing footnotes or cross-references.

Of all respondents queried, 88% had Maldivian nationality, of whom 100% were Muslim, and 21% were female. Eleven key informant interviews (33.3% of the sample) were conducted with "experts", including a senior sea level researcher, island chief, professional tour guide, co-founder of an environmental NGO, writer and civil society activist, council office director, senior education administrator, court office assistant, senior coastal researcher, parliamentarian, and senior public administrator.

Interviews generally lasted about 1 h. Of all interviews conducted, 27 were carried out in the country's lingua franca Dhivehi with the help of a local guide and interpreter. The remainder was conducted in English. Respondent ages ranged from 20–82 years, with 43.4 years as the average age.

Data analyses followed a mixed methods paradigm (Creswell and Plano Clark 2011), which comprised approaches informed by grounded theory (Charmaz 2006), comparing (Punch 2014; Tesch 1990), phenomenography (Marton 1981), and ethnomethodology (Silverman 1993).

Results and Key Findings

While island abandonment is neither new nor unique to the Maldives (e.g. Arenstam Gibbons and Nicholls 2006), the element of government planning provides particular insight into macro-managed migration success factors. Emergent insights can inform both present and future management of climate change related migrations and may enable policy makers to assume a more forward thinking policy

[3]Unless attribution by name was expressly agreed or requested by the respondent.

posture in order to make adequate and appropriate preparations before these are needed (UNDP 2007, p. 198; see also UNISDR-UNDP 2012).

From the research a number of key findings have been synthesised and consolidated below.

(1) *Population consolidation and internal migration are likely to continue*

There are strong indications that both the 2004 Tsunami and the threat of future climate change have been catalysts for the government to push harder for population consolidation measures to be implemented more rapidly. This research suggests that more inaccessible islands with low populations may likely be abandoned in the future, with the overall number of inhabited islands in the Maldives declining further.

The ambition to maintain fewer inhabited islands has been expressly stated by a senior policy maker who declared his intention to see the number of inhabited islands "reduced by half" (Bosley 2012) or possibly even decreased to "25 to 30 islands" (ibid, para. 15). This demographic consolidation response may be understandable in the context of the significant vulnerability of many islands to sea level rise and other climate change impacts (MPND 2007, pp. 24–29; SOEM 2011).

The economic rationale for spatially reducing human and asset exposure to ocean hazards through demographic consolidation was dramatically demonstrated by the 2004 Tsunami which in a matter of hours inflicted macro-economic damage on the order of 62% of the archipelago's GDP (ADB UN WB 2005, pp. 3, 22).

Moreover, it is the view of several key informants that population consolidation is necessary in light of the growing need to implement expensive and expansive technological "hard" reinforcements that can only be afforded at a smaller scale and cannot be rolled out across the entire archipelago.

In summary, population consolidation in the Maldives may perhaps be likened to the process of rural-urban migration seen elsewhere in the world: sparsely settled islands/regions are being abandoned for densely settled ("urbanised") ones. This predominant trend can be expected to continue as increasingly people reach for goods, services, and opportunities not available to them in remote/inaccessible island locations where it is "difficult to import essential commodities" (Q5/Migr/Dest/Hanimaadhoo/20111225). Moreover, recognising that a significant amount of migration is already underway for reasons other than climate change may help to reinterpret, reconceptualise and realign "climate migration" within the psyche of collective perception (Diamond 2005, pp. 435–436) as natural, normal and positive instead of extraordinary, foreign and frightening (Bettini 2013).

(2) *Migration is experienced as positive adaptation to change*

There is strong observational evidence that the experience of government coordinated migration is largely perceived as positive both in terms of processes and outcomes. This view persists both among migrating islanders and welcoming host communities, and was similarly and variously affirmed throughout fieldwork by a majority of migrants, hosts and experts. For example,

Aspirations of forced migrant communities	Ascribed importance
Better or higher education	5.0
Better health care	4.8
Employment/making money	4.64
Moving together *with* the family	4.57
Same country	4.4
Same culture/ethnicity	4.1
Same language	3.9
Reproductive/family planning	3.8
Similar climate	3.7
Proximity to the *origin* community	3.2

Fig. 3.3 Migrant Aspirations Aggregate Index (MAAI). *Ascribed importance* weighted average calculated from assigned values: very high priority = 5 points; high priority = 4 points; medium priority = 3 points; low priority = 2 points; very low priority = 1 point

> we find life here in Huvarafushi better compared to life in Berinmadhoo… in all aspects, economically, socially, we have access to better facilities and resources. My birthplace and my original island was Berinmadhoo. However, since moving to Huvarafushi on 1 March 2006 we have not experienced any unpleasant thing or misery in Huvarafushi. I would say life is better in all aspects. If the population is higher people will be more satisfied than living in an island with a low population. (Former Island Chief and migrant from abandoned Berinmadhoo Island; Q25/Exp/Migr/Dest/Huvarafushi/20111231)

(3) *Responding to aspirational pursuits can influence migration outcomes*

There is robust empirical evidence that the three key aspirations impacting on migration in the Maldives are education, health care and employment (Fig. 3.3).

This finding confirms similar results previously reported in the literature: "This [development in Malé] triggered migration from islands towards Malé for services, mainly education, health, and employment" (GOV MAL 2009, p. 64). By keeping these aspirational priorities in view it may be possible to mobilise migration in such a way that social services are more evenly distributed and/or more equitably enjoyed.

Respondents also recurrently recommended that more job opportunities should be created "before people are moved" (Q21/Exp/Nolhivaranfaru/20111230), and that the development of new resorts, for example, should become less reliant on foreign immigrant workers[4] but instead draw on local labourers. In short, by proactively responding to aspirational pursuits (i.e. health care, education, and employment) policy makers can create conditions which are both conducive to macro-demographic decentralisation and concurrently encourage more equitable

[4]According to Ali Rilwan of Maldives environmental NGO Bluepeace there are approximately 100,000 expatriates living and working in the Maldives (Q2/Exp/Malé/20111222).

migration outcomes. Moreover, there is evidence that many islanders desire a strong government coordination and leadership role in achieving successful migration outcomes.

(4) *The special role and promise of expansive education*

There is overwhelming evidence that "better education" may hold the strongest promise to enable more positive macro-managed migration outcomes. At present, the high level of prioritisation respondents ascribed to education seems second to no other aspiration. The very high esteem in which education is held was made recurrently apparent not only in quantitative results which highlighted "insufficient educational facilities" as the most important "push driver" (Fig. 3.4), and "better/higher education" as the most important aspirational "pull factor" (Fig. 3.5), but was also underscored in numerous key commentaries which raised education as a highly promising long-term strategy by means of which the nation's overall resilience could be significantly bolstered.

Naturally, education raises options, and inversely, lack of education limits them. Hence the strong desire expressed by large parts of the population to benefit more from better education opportunities can be seen as a core component and precondition for enhancing human resilience: it holds the dual promise of strengthening both "hard system" engineering solutions and "soft system" social adaptation programmes. Evidently, expanding the domestic knowledge base and building up in-country expertise will enhance hardware related adaptation designs; and

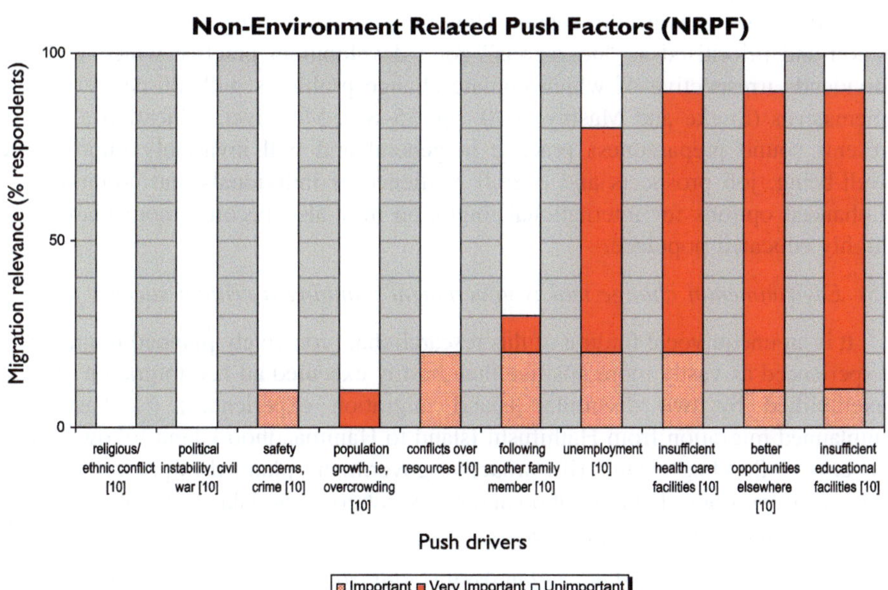

Fig. 3.4 Non-Environment Related Push Factors (NRPF). Important and very important non-environment related pressures to migrate; Q = 33 FG = 90 N = [ ]

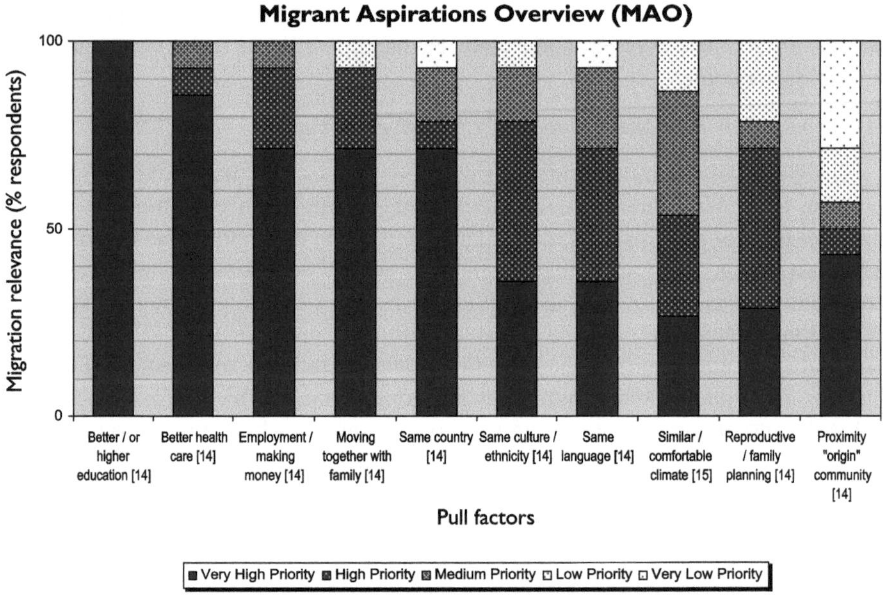

Fig. 3.5 Migrant Aspirations Overview (MAO). Comparative levels of prioritisation as ascribed by respondents; Q = 33 FG = 90 N = [ ]

educating people to understand climatic, environmental and social changes will strengthen the software dimension. Upscaling education may therefore be understood and prioritised as "no regrets" good development practice which will pay dividends irrespective of which climate change problems will ultimately present themselves (Starke and Mastny 2010, pp. 55–82). Moreover, education also represents sound preparedness practice in general and will invariably improve the well-being, job prospects and overall resilience of individuals and communities. Enhanced options for international migration may also become more viable to a highly educated populace.

(5) *Environmental change makes government planning a critical success factor*

It is an unequivocal finding of this research that proactively planned migration is experienced as vastly more positive than hastily executed ad hoc migration. This is exemplified by two dissimilar island migration experiences: (i): hasty and unplanned migration from Hathifushi Island to Hanimaadhoo Island following two successive flooding events; (ii): proactive government planned migration from the Islands of Faridhoo, Kunburudhoo and Maavaidhoo to the Island of Nolhivaranfaru following six years of preparation.

(i) *Hasty and unplanned migration from Hathifushi Island to Hanimaadhoo Island*

The example of the Island of Hathifushi which was hastily abandoned by its resident community of 295 islanders on 5 July 2007 following storm surges in June

2007 dramatically illustrates both the storm surge risk to low-lying islands, and the social and economic cost of unanticipated migration. According to one migrant from Hathifushi the island is naturally low-lying: during a regular high tide only "six inches [15 cm] of land is above the water" (Q6/Exp/Migr/Dest/Hanimaadhoo/20111226).

According to anecdotal accounts by Hathifushi islanders, even though the 2004 tsunami saw 90% of the island area covered (Q6/Exp/Migr/Dest/Hanimaadhoo/20111226) to a height of 30–50 cm with ocean water (Q24/Migr/Orig/Hathifushi/20111231), islanders regarded the storm surges in 2007 as the more frightening and economically ruinous disaster event since they occurred at night, lasted much longer (two days), and covered the entire island with water reaching up to a height of 1 m (Q24/Migr/Orig/Hathifushi/20111231).

"We suffered huge losses … we had to leave all items from all houses … We did not take anything with us … There was no sufficient time" (Q24b/Migr/Orig/Hathifushi/20111231).

Importantly, although the community of Hathifushi islanders appears to have re-quested relocation as early as in the 1960s, and as late as in 2004, it was not moved by the government until *after* the devastating storm surges triggered a swift and final evacuation.

> We were having thoughts of moving to another island since 1963 or 1968. Our forefathers wanted to move but could not achieve that but we have now been able to move. Similar to many other islands of Maldives, we were experiencing erosion of the island and when the tsunami of 2004 came, water swept the whole island and flooded 90% of the island. Several Maldivians suffered from the tsunami. We faced several difficulties and we discussed very sincerely with the government to prioritise the needs of our people. But a decision had not been made before storm surges swept the island around June 2007. The whole island was flooded on a worse scale than the tsunami. The government then decided to move us … on 5 July 2007. […] The entire population was moved to … Hanimaadhoo. We started living in houses of Hanimaadhoo people, and we are still living in their houses […] We don't know about the status of the housing project. The work is being carried out on this island to build housing units. We have no idea if those housing units [that are presently under construction] are for the people who moved from Hathifushi or not. Neither the Council Office nor do we have any information about that so we can't comment on that. (Q6/Exp/Migr/Dest/Hanimaadhoo/20111225)

At the time of this research, the majority of Hathifushi islanders continued living under the roofs of Hanimaadhoo "host families" who agreed to offer temporary accommodation to evacuees in exchange for modest government subsidised monthly rental payments of 500 Rufiyaa[5] per accommodated person per month. This impromptu "host family" arrangement was still operational at the time of this research visit to Hanimaadhoo (Q9+10/Migr+Host/Dest/Hanimaadhoo/20111227).

(ii) *Proactive government planned migration from the Islands of Faridhoo, Kunburudhoo and Maavaidhoo to the Island of Nolhivaranfaru*

---

[5]Equivalent in value to approximately US$ 32 (i.e. on the day the interview took place).

In stark contrast, the three sparsely settled Islands of Faridhoo, Kunburudhoo and Maavaidhoo were proactively abandoned in January 2011 following six years of preparation, and the three island communities collectively and quasi-simultaneously resettled on the "host" Island of Nolhivaranfaru ("community of destination") on three consecutive days in January 2011 (Haveeru Daily 2011; TPO 2011; SAMN 2011). One migrant and island administrator recounts pertinent lessons and experiences:

> We made considerable efforts in collaboration with the Island Development Committee that existed at that time to facilitate the migration of people from three islands by including that in the government's policies. [The government] was planning and implementing a housing project to move people from Faridhoo, Kunburudhoo and Maavaidhoo. If I remember correctly, 325 people from Kunburudhoo, 230 people from Faridhoo and 546 people from Maavaidhoo moved. [...] At that time the population of Nolhivaranfaru was 587. Now the total population of the four islands reached 1,757. Through this process I have realised how difficult it is to move an island population to another island and develop a community ... in the initial days after the migration, there [were] some issues that caused dissatisfaction. However, our belief is that in the future Nolhivaranfaru will be a convenient island to live ... After we migrated to Nolhivaranfaru we have received several benefits and advantages. Those include easy access to [the regional hub] Kulhudhuffushi and other islands and access to a better harbour. In all aspects we see Nolhivaranfaru as a better island to live. The problem is that the population [on many islands] is too small. There is no adequate education or health, there are no jobs available, it is so difficult to use the harbour ... However, now that we have moved here ... transportation will be easier [and] health too, it will be a better place. Hence, in all aspects islands with larger populations will develop faster. (Nolhivaranfaru Council officer and migrant from Faridhoo Island; Q15/Exp/Migr/Dest/Nolhivaranfaru/20111229)

Despite the fact that migration to Nolhivaranfaru was not experienced as altogether perfect,[6] arriving migrants had nonetheless all moved into brand new government funded houses which had been specifically built and prepared for the migrant community, and which heads of families received on a chance basis when two students from Nolhivaranfaru "drew lots to determine who would get which house" (Q18a/Migr/Dest/Nolhivaranfaru/20111229) while around "500 people stood by and watched near the island's central tree" (Q18b/Migr/Dest/Nolhivaranfaru/20111229). This allocation method was perceived as a "fair and good process" (ibid). Moreover, the three relocating island communities which arrived successively in Nolhivaranfaru over the course of three days (one island per day) were expectantly welcomed on the beach by their new hosts who offered them an "island community welcome" (Q21/Exp/Dest/Nolhivaranfaru/20111230) which

[6]Some migrant respondents from Maavaidhoo and Kunburudhoo regretted that the government had not made good on its promise to compensate them for fruit-bearing trees that they had abandoned in their islands of origin (Q17/Q18/Q19/Migr/Dest/Nolhivaranfaru/20111229). Moreover, with the in-migrating islanders all receiving brand new government funded houses there was a shared sentiment among parts of the host community that the original "Nolivaranfaru islanders were missing out" (Q15+16/Exp/Nolhivaranfaru/20111229).

entailed “welcome drinks, food parcels, waiting ambulances for the sick” (ibid), as well as subsequent visits to the arrivals’ new houses “to see if anybody had missed out” (ibid).

For the purposes of this case study the two aforementioned migration experiences appear to highlight two very important points. First, natural disasters and environmental change—of which climate change is a part and to which it contributes—can combine with other problems to swiftly overwhelm a community’s collective coping capacity. The resultant situation may be experienced as so frightful and exasperating as to trigger unplanned ad hoc migration responses, which are lacking in critical preparation, coordination, and funding (i.e. Hathifushi). Second, policy maker foresight and anticipatory preparedness can promote proactive migration in such a way that critical preparations can enable a more benign migration experience where the element of force is more muted. The important point is that critical preparations are made before these are needed (i.e. Nolhivaranfaru).

While a number of factors had long prompted the Hathifushi island community to seek government sponsored relocation (the causes), it was the storm surges (the triggers) that finally compounded the vulnerabilities and became the tipping element that pushed the whole community “over the edge”, resulting in their rapid ad hoc migration response (Q32/Exp/EOS-NTU, Singapore/20120106). Therefore, progressive climate change and creeping environmental change appear to make proactive migration planning an increasingly important priority for policy makers to focus on.

A further lesson that has emerged from this study is that long-term migration planning needs to be unencumbered by electoral cycles and/or partisan politics. Incidentally, less than 5 weeks after the end of this fieldwork the government in Maldives changed again (Aljazeera 2012; BBC 2012a, b and c), triggering the dissolution and reconfiguration of the Ministry of Housing and Environment, along with a raft of other changes. In the absence of long-term strategic migration planning, ad hoc migrants can easily fall through the cracks of day-to-day policy maker business, or even become pawns in the hands of politicians (Q24b/Migr/Orig/Hathifushi/20111231; Q9+10/Migr+Host/Dest/Hanimaadhoo/20111227).

(6) *Waste, infrastructure and environmental management*

Despite a number of significant environment related migration pressures (Fig. 3.6) it appears that in the minds of most Maldivians, for now at least, mass emigration from the archipelago does not seem to constitute an immediate threat. At the same time it is the view of some respondents that most islands are likely to undergo unprecedented topographical and geomorphological transformation this century (Q29/Exp/Malé/20120102), and/or that “most islands will disappear” by 2100 (Q2/Exp/Malé/20111222). This possibility appears to make the piloting of viable adaptation measures indispensable.

In this context Hulhumalé may perhaps already be regarded as a pilot project by means of which to assess the viability of computer designed and artificially created “Contingency Adapted Raised Islands”, which may offer both long-term protection

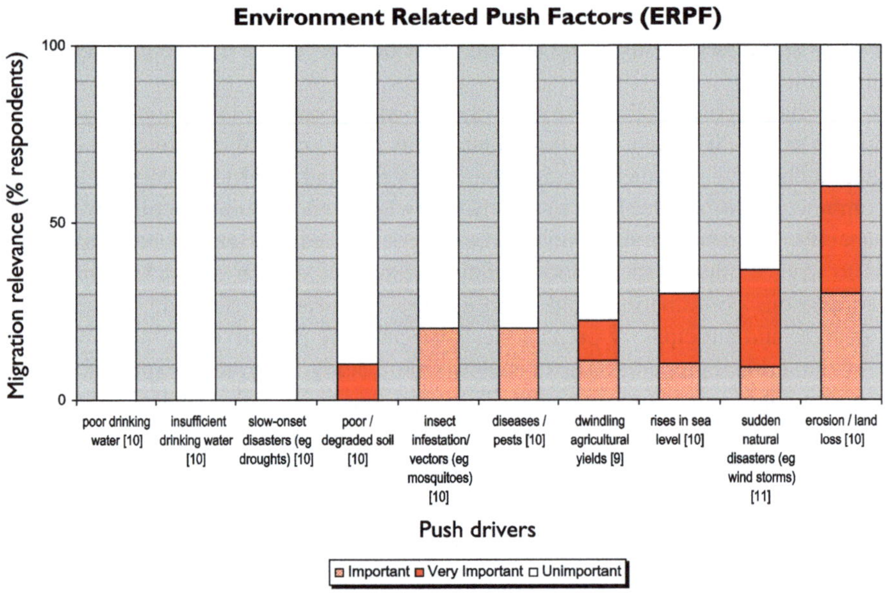

Fig. 3.6 Environment Related Push Factors (ERPF). Important and very important environment related pressures to migrate; Q = 33 FG = 90 N = [ ]

Fig. 3.7 An average Maldivian island protrudes less than 1 m above sea level. *Illustration* © Bluepeace Maldives

from climate change related sea level rises, and act as an in situ insurance policy against future forced human emigration scenarios (Figs. 3.7 and 3.8). However, at a cost of multiple millions of dollars the affordability and replicability of such artificially designed and raised island reclamation projects has been called into question (Q30+31/Exp/Hulhumalé/20120102). Moreover, a duplication or multiplication of the Hulhumalé blueprint may be financially prohibitive in the absence of international climate change adaptation or compensation financing—a funding stream which seems notoriously unreliable, if past UN climate summits are to be seen as a reliable guide (Harrabin 2012a, b). Finally, the longer-term affordability of massive-scale engineering bulwarks would seem very much dependent on enduring foreign exchange earnings from tourism and fisheries—a funding source which

Fig. 3.8 Contingency Adapted Raised Island with 3–5 m elevation. *Illustration* © Bluepeace Maldives

could dry up if climate change impacts should cause the country's coral reef or fish stocks to erode or collapse (SOEM 2011, cf, Slezak 2016).

Irrespective, it would be inappropriate to explore "hard" engineering solutions without drawing attention to environmental knock-on effects. Referring to lagoon dredging and island reclamation projects (e.g. Hulhumalé, Gulhifalhu) a number of respondents highlighted serious environmental damages which hard engineering measures inflict through encroachment, habitat destruction, biodiversity decimation and species extinction (Q3/Exp/Hanimaadhoo/20111224). Respondents also cautioned that land reclamation and artificial harbours impact on sedimentation processes which can lead to "increased erosion" (Q4/Migr/Dest/Hanimaadhoo/20111225) in adjacent areas or islands, or cause "sand [and] debris to float and settle in reef areas" with consequent "adverse effects in dive sites" (Q3/Exp/Hanimaadhoo/20121228; Q14/Exp/Hanmaadhoo/20111228). Respondents also advocated that "mangroves should be protected" (Q3/Exp/Kulhudhuffushi/20111227), that sand should be mined "from the bottom of the sea [rather] than from the lagoon" (Q31/Exp/Hulhumalé/20120102), and that "marshlands should be preserved" instead of being filled in to make room for expansive land reclamation (Q3/Exp/Kulhudhuffushi/20111227). Respondents emphasised these protective measures as important both to maintain the Maldives' natural richness in bird and fish species and to ensure that human exposure to ocean hazards does not increase further (Q3/Exp/Hanimaadhoo/20111224).

Given that the natural resilience of low-lying reef islands diminishes considerably after subjection to human modification (hard adaptation) makes the conservation of mangroves and environmental shelterbelts (soft adaptation) even more important (Kench et al. 2006, p. 177; Luetz 2008, pp. 60–65; Mimura et al. 2007, p. 698).

(7) *Sharing expertise, lessons learned and best practices*

There seem to be at least four elements that make planned migrations comparatively successful in the Maldives: (1) many islanders want to move; (2) there is significant consultation and communication between the government and the affected islanders, including "host" and "guest" communities; (3) relocation processes involve anticipatory planning and relatively long lead times[7]; and (4) there are generous government benefits and compensations for migrating islanders.[8]

Given the Maldives' significant know-how in coastal management and its financial resource base from tourism and fisheries, it is the view of this researcher that the Indian Ocean archipelago may consider playing a more proactive leadership role in assisting other less resourced Small Island Developing States (SIDS), especially in the Pacific, adapt to climate change impacts this century—of which sea level rise may be among the more benign.

Discussion

At present three migration scenarios can be appreciated across the Maldives. First, inhabited islands are completely abandoned with the migrating islander populations integrating into host communities on other inhabited islands (government sponsored migration by assimilation). Second, inhabited islands are completely abandoned, with the migrating islander populations collectively relocating to other heretofore uninhabited islands which have been specifically prepared and developed by the government for the purpose of settlement (government sponsored migration by non-assimilation). Third, individual islanders move (temporarily or permanently) to population centres in other islands for reasons of intermarriage or to avail themselves of job opportunities or services not available to them in their islands of origin (non-government sponsored assimilation migration).

For years Maldivian governments have sought to consolidate the spatially dispersed population to ensure that resources are not spread out too thinly across the vast area of the archipelago. According to the National Population Policy submitted by the Ministry of Planning and National Development (MPND) in 2005, the Maldives' "extremely dispersed population distribution and the small size of many island communities are important concerns for cost-effective provision of basic infrastructure and social services. Although significant progress has been made in this area, the costs are extremely high, often 4–5 times the level recorded in continental developing countries" (MPND 2005, pp. 10–11).

"[W]ith the sparsely dispersed population, the education sector is challenged to be creative in finding cost effective ways to cater for the education needs of the smaller islands. Double-shift system is the norm in all schools which limits providing the full curriculum for the primary and secondary cohorts" (MPND 2007, p. 126).

---

[7]For example, it took the government six years to implement the migrations of the islander communities Faridhoo, Maavaidhoo and Kunburudhoo to Nolhivaranfaru.

[8]For example, each migrating family typically receives a free government funded house in the place of destination.

The "[s]mallness, remoteness and wide dispersal of island communities" (MPND 2007, p. 8) represents a well-known challenge to prospects of sustainable development in the Maldives. Public expenditure and macro-economic investments need to be strategically allocated if economies of scale are to be realised, and if high quality services are to be consistently offered and equitably enjoyed. This calls for creative macro-managed migration approaches.

Against this background it can be reasonably expected that

> [population consolidation in the Maldives] will continue… Even though there are about 200 inhabited islands now, some of those islands have a [small] population of 200. […] It is impossible to spend from the government budget to provide services to islands with about 200 people. It is not possible … Some people might have such views [that it is pleasant to live on a small island]. For example, a tourist spending a holiday in a resort might have such thoughts, however when a person starts actually living [in a small island they see a different reality]. (Nolhivaranfaru Council officer and migrant from Faridhoo Island; Q15/Exp/Migr/Dest/Nolhivaranfaru/20111229).

According to the Ministry of Planning and National Development recent years have seen a policy departure from the former Atoll Development Programs which "aimed to develop infrastructure and services on all inhabited islands on an equitable basis" (MPND 2007, p. 26), a policy that essentially failed because it was "not sustainable due to budget limitations and diseconomies of scale in the provision of resources" (ibid). The Ministry also sees the failure of that policy evidenced by unabated migration to the country's capital Malé: "one third of the country's population now lives in Malé and of this nearly 58% are migrants. This is the consequence of vast inequalities existing in the composite availability of educational, health services and employment opportunities between Malé and other islands" (ibid).

More recently the Ministry has promoted Population and Development Consolidation (PDC) as part of its Regional Development Policy, hoping that the consolidation of island communities will lead to a "fall in infrastructure expenditure per person, … increase the quality of essential services, … encourage small to large businesses and home industries to develop, [generate] more opportunities for investment, [and] boost the local economy" (MPND 2007, p. 26). In its Seventh National Development Plan 2006–2010 three modes of population consolidation are enumerated: (1) conjoining neighbouring islands (e.g. through land reclamation, bridges or causeway linkages); (2) implementing regular government subsidised ferry services to interlink clusters of islands; (3) sponsoring the "relocation of small populations in remote and small islands to islands with better infrastructure, services and expansion potential" (MPND 2007, p. 27).

With regard to the third model the Ministry stipulated prerequisites, insisting that relocations need the "explicit agreement between the relocating community and potential host community, … should be voluntary and would only be facilitated when the whole community formally lodges the decision to move" (ibid). Moreover, the Ministry offered a number of incentives: "[e]ach relocating family will receive a house for an occupied house and Rf 50,000 (US$ 3,900) as relocation compensation" (ibid). According to several respondents, compensation was also promised for fruit-bearing trees which relocating islanders had to abandon in their "communities of origin" (Q17/Migr/Dest/Nolhivaranfaru/20111229). In light of recent construction cost increases which have seen the unit cost per house rise from US$23,400 in 2005 to US$28,300 in 2007, the Ministry has advocated "to speed up" the migrations and "establish urban areas in the regional divisions of the country that provide an alternative to Malé" (MPND 2007, pp. 22, 27). This demographic decentralisation and urbanisation process was formalised with the passing of the Decentralization Act in October 2010 and intends to relieve congestion pressure in Malé and build up local governance capacities in the regional atolls (SOEM 2011, pp. 3, 15, 28–31).

In two expert interviews four key demographic consolidation phases were identified by the country's Minister of Housing and Environment, Dr Mohamed Aslam, and by the Assistant Executive Director of the Department of National Planning, Mohamed Imad: (1) the Relocation and Housing Model (1980–1998) evolved into the (2) Population and Development Consolidation Model (1998–2004), and was again modified (in the wake of the 2004 Tsunami) to become known (from 2005) as the (3) Safer Island Model.[9] At the time of this research visit the Safer Island Model was in the process of being re-branded into the (4) Resilient Island Model which largely appears to be a continuation of the Safer Island Model concept but with an added focus on grouping islands into "clusters", and a stronger emphasis on human-social adaptation and "soft-systems" community resilience and participation (Q29+30/Exp/Malé/20120102, also MPND 2007, p. 9).

In summary, population consolidation and macro-managed migrations are neither new nor sporadic but have been strategically pursued and sponsored by Maldivian governments for more than two decades, essentially reinforcing a countrywide urbanisation process which is unlikely to dissipate soon (Mohamed 2002; MPND 2006, 2007). In point of fact, it appears that the threat of tsunamis and climate change impacts has further heightened the sense of urgency with which populations are to be concentrated in safer and presumably future-proof urban centres (ADB UN WB 2005, p. 3; MPND 2007; MTAC 2012; Q29+30/Exp/Malé/20120102). According to a notable climate change resilience report published by the Maldives Ministry of Tourism, Arts and Culture with support from UNDP and GEF, a "shift in population was evident following the tsunami of 2004. Climate change will speed up the population displacement" (MTAC 2012, p. 7).

The government's role in coordinating and facilitating inter-island migration appears to be predominantly viewed as a positive migration enabling factor which was gratefully acknowledged on numerous occasions and by various beneficiaries. On the Islands of Hanimaadhoo, Nolhivaranfaru, Huvarafushi, and Dhuvaafaru this researcher was welcomed into such government provided houses (Fig. 3.9).

---

[9]Also known as the Safer Island Development Program (SIDP), the primarily "hard-systems" Safer Island Model "targets to provide the infrastructure necessary to adapt to climate change and to be better prepared for natural disasters. A Safer Island will have better coastal protection, elevated pubic [sic] buildings for vertical evacuation, emergency supplies, appropriate harbour and also more reliable communications systems." (MPND 2007, p. 28).

Fig. 3.9 Government sponsored houses constructed for migrant communities from Faridhoo, Kunburudhoo and Maavaidhoo on the Island of Nolhivaranfaru. Houses were allotted to families by chance when two students from the "host" island drew lots to determine which house would be allocated to which family. This allotment procedure was seen as a "fair and good process". (Q18/Migr/Dest/Nolhivaranfaru/20111229)

Moreover, it also seems that Maldivians want to stay in their country, if at all possible, preferring inter-island migration over international emigration (Q3/Exp/Hanimaadhoo/20121224). The high percentage of internal "lifetime migrants"[10] also appears to corroborate the conjecture that Maldivians are not adverse to mobility per se: while 33% of Maldivians may be classified as "lifetime migrants" (MPND 2006, p. 46), according to the "World Bank Migration and Remittances Factbook 2011" the stock of international emigrants from the Maldives (as a share of population) is a mere 0.6% (Ratha et al. 2011).

Therefore, the question arises whether Maldivians do not emigrate internationally in greater numbers because intrinsically they prefer not to do so, or whether such desires are extrinsically constrained. This question also seems relevant in view of media reports suggesting that Maldivians are seeking to migrate in large numbers to countries like Australia (AVB 2012; Burgess 2012; Doherty 2012; Henley 2008; Schmidle 2009). Hence it seemed useful to ask where migrants would go "if there were no restrictions" (political, financial or otherwise). Five "most preferred destination" categories emerged, namely, (1) the Maldives; (2) Saudi Arabia; (3) Asia/South Asia; (4) Western nations (generally), and (5) Australia (specifically). Answers in response to the hypothetical question highlighted significant restrictions on human movement and seemed to underscore that migrants do not generally consider what they might want in an "ideal world" but rather what seems

[10]Defined by the Ministry of Planning and National Development as people "born outside their island of usual residence [who] lived in that island for more than one year." (MPND 2006, p. 45).

practically attainable under “prevailing circumstances”. Notwithstanding, it appears that most respondents evidently prefer to remain in the Maldives vis-à-vis all other alternatives, whether practically or hypothetically attainable.

Again it needs to be reiterated that while climate change looms large as a future threat with a significant potential to affect future migrations, present-day migrations are primarily influenced by aspirational pursuits, especially education. This particular predilection for education should be recognised as a critical asset.

Policy Recommendations

Given that the study of climate change impacts involves speculative future scenarios that can be hard to project (Luetz 2013, Sect. 2.4), this research has sought to identify so-called “no regrets” policy approaches. According to Brunner and Lynch, “no regrets” policies “make sense regardless of climate change projections” (Brunner and Lynch 2010, p. viii), and according to the Intergovernmental Panel on Climate Change (IPCC), “[n]o regrets options … are by definition … options that have negative net costs, because they generate direct or indirect benefits that are large enough to offset the costs of implementing the options” (Markandya et al. 2001, pp. 474–475). For the purposes of this research, “no regrets policies” are conceptually understood as policy priorities that are broadly beneficial irrespective of what particular scenarios are ultimately realised, and might even make sense in the complete absence of climate change.

While there are a number of issues that would be worth addressing to ameliorate forced or voluntary, present or future migrations in the Maldives, no other issues seem to hold as much preparedness promise as the areas of (1) equitable service provision; (2) expansive education, (3) social integration initiatives; and (4) proactive government planning. Hence the following four recommendations and policy priorities are advised.

First, meet popular aspirational ambitions to support decentralisation and equitable demographic development: There is overwhelming evidence that health care, education and employment are the three key aspirations which are most impacting on migration in the Maldives. All other aspirations were ascribed a subordinate priority by Maldivian respondents. Recognising that the pursuit of health care, education and employment is driving populations to migrate to wherever these essentials are reliably attainable appears to emerge as a compelling opportunity for policy makers seeking to mobilise macro-managed human migrations that are in harmony with decentralised demographic development objectives.

In the simplest of terms, if in response to unabated in-migration to Malé these services are significantly scaled up in Malé itself, more would-be migrants will likely be pulled into the magnetic suction of the country’s already congested capital. Incidentally, this process may have already been triggered by the creation of the artificial Island of Hulhumalé in close proximity to Malé which is already attracting in-migration not just from Malé, but also from the provinces, including Hathifushi

Island (Q24/Migr/Orig/Hathifushi/20111231). Inversely, if essential services are scaled up in regional provincial hubs, further population pressure may be deflected from the compacting conglomerate Malé which may already rank as "the densest city in the world" (MHUD 2008, p. 14; cf also GOV MAL 2009, p. 19). In short, catering to popular aspirations in the provinces may shift and rebalance the country's lopsided centre of gravity, and thereby become a compelling opportunity for policy makers to guide more equitable macro-demographic processes of decentralised development.

Second, promote expansive education: There is overwhelming evidence that "better education" is among the most highly esteemed pursuits in the archipelago. Throughout fieldwork the high level of prioritisation respondents ascribed to education seemed second to no other aspiration. Expressed in simple language, people appear to migrate to wherever high quality education is readily attainable and affordable.

Naturally, education raises options, and inversely, lack of education limits them. Upscaling education may therefore be understood and prioritised as "no regrets" good development practice which will pay dividends irrespective of which climate change scenarios are ultimately realised, and would even appear to make sense in the complete absence of climate change. Upscaling education would invariably enhance prospects of planned emigration by making would-be emigrants from South Asia "desirable as immigrants to other countries" (Belt 2011, p. 81). This proposition can be seen as prudent preparedness practice given that a number of respondents think "there will be some international migration in the future" (Q3/Exp/Hanimaadhoo/20111224). Concurrently and importantly, upscaling education would also contribute expansive domestic hard and soft system expertise needed for the ongoing implementation of in situ adaptation measures (e.g. Contingency Adapted Raised Islands). Hence education appears to make sense with regard to both ex situ and in situ adaptations to climate change. Finally, a number of social problems like loitering, thefts and drug abuse were also attributed to "lack of education among youth" (Q21/Exp/Nolhivaranfaru/20111230; cf GOV MAL 2009). In short, upscaling education holds significant policy maker promise and is recommended as the kingpin for more positive macro-managed migration outcomes. The important point to note is that more expansive education is a wish that respondents have already unequivocally expressed. Hence implementation could be expected to have broad societal uptake and support, and importantly, implementation could begin tomorrow.

Third, implement social integration initiatives to enhance cohesion among consolidating island communities: Despite a strong overall sense that government coordinated migrations are overwhelmingly perceived as more positive than negative, there was a sentiment shared among some respondents that more social programmes should be conducted among migrating or conjoining islander communities "to ensure social cohesion, [and] integration" (Q15+16/Exp/Nolhivaranfaru/20111229).

To overcome social integration challenges respondents suggested initiatives like playing sports, holding football tournaments, "planning a new year's trip to [former island home] Faridhoo, … or inviting people to Eid" (Q21/Exp/Nolhivaranfaru/20111230). Moreover, by increasing government funded inter-atoll and intra-atoll transport links,[11] in addition to existing ferry services to the capital Malé, interconnectivity and mobility could foster more social cohesion because in a geography like the Maldives one "cannot pull the islands together" (Q29/Exp/Malé/20120102). In short, post-migration community integration may stand to benefit significantly from the support of social ("soft") programmes, sports/youth activities, more community consultation (migrants and hosts), as well as pre-migration awareness programmes and post-migration follow-up counselling. Pre-planned job creation should also be integrated into the migration design. In this regard it is important that appropriate strategies give due regard to the development potential of nearby resort islands and be reliant on the availability of local labour rather than expatriate expertise.

Fourth and finally, proactively plan macro-managed migrations where there is reasonable doubt that island communities can persist in perpetuity: It is an unequivocal finding of this research that planned migration (e.g. Nolhivaranfaru) is experienced as inherently more positive than ad hoc migration (e.g. Hathifushi). Importantly, the prospect that storm surge risk will evolve significantly over the course of this century suggests that the archipelago may become more susceptible to flooding events in the future. To incorporate this proposition into island vulnerability assessments would appear to reflect prudent preparedness practice. The two aforementioned migration experiences also highlight two critical points. First, natural disasters and environmental change—of which climate change is a part and to which it contributes—can combine with other problems to swiftly overwhelm a community's collective coping capacity. The resultant situation may be experienced as so frightful and exasperating as to trigger unplanned ad hoc migration responses which are lacking in critical preparation, coordination, and funding support. Second, policy maker foresight and anticipatory preparedness can prompt and support proactive migration in such a way that critical preparations can enable a more benign migration experience where the element of force is more muted. The important point is that critical preparations be made before being needed.

More to the point, forced migration may result where voluntary migration is delayed. Progressive climate change and creeping environmental degradation appear to make anticipatory government coordinated migration planning an increasingly important success factor. As noted above, climate system inertia implies that based on past emissions of greenhouse gases humanity is already committed to 1.3–1.5 °C global warming (Hansen 2008, 2009; UNDP 2007, p. 4; WBGU 2009a, b; Woo et al. 2011; World Bank 2012, 2014). According to a report prepared by the Potsdam

[11]"We've had for a very long time transport between islands and the capital Malé, but we've not had transport inter-atoll and intra-atoll, … so therefore people would have to wait on their island for somebody to come on a casual visit and then go—if the destination of this vessel is where they want to go… Or they would have to hire a boat which becomes very expensive … So since 2008 we've started a nationwide transport ferry system. Every atoll has now the ferry system up and running." (Q29/Exp/Malé/20120102).

Institute for Climate Impact Research and Climate Analytics, "[n]ew estimates for crossing a threshold for irreversible decay of the Greenland ice sheet (which holds ice equivalent to 6 to 7 m of sea level) indicate this could occur when the global average temperature increase exceed roughly 1.5 °C above preindustrial" (World Bank 2012, p. 61, attributed to Robinson et al. 2012).

In this context, the case for learning relevant lessons for "climate migration" from the country's contemporary context of "urbanisation migration" appears to be compelling. Given that the planet is presently on a 4 °C global warming trajectory only underscores the importance of learning the important lessons before need arises and options are foreclosed (Hamilton 2010, pp. 190–208; New et al. 2011; Nicholls et al. 2011; Schellnhuber 2008; Stewart and Elliott 2013; Vince 2009a, b; WBGU 2009a; World Bank 2012, 2014).

Given the long lead time to implement and mainstream some of the above mentioned preparedness priorities it appears essential that the development of appropriate policy instruments is not needlessly delayed. To distil it further, and to conclude, the following preparedness priorities and policy recommendations are advised:

(1) Meet popular aspirational ambitions in regional hubs before options for equitable and decentralised demographic development are foreclosed by unabating in-migration to Malé.
(2) Promote expansive education, free and compulsory for all, before the ambitions for self-actualisation of a whole generation are eclipsed (along with promising consequent options for in situ and ex situ adaptation to climate change).
(3) Mainstream social awareness and community integration initiatives into migration designs before unnecessary social problems evolve which require ad hoc responses or retrospective remediation. Where possible, create jobs before these are needed.
(4) Plan macro-managed migrations wherever there is reasonable doubt that island communities can persist in perpetuity, and importantly, before environmental or climatic changes overwhelm communal coping capacities, trigger ad hoc evacuations, foreclose benign migration scenarios, and/or create unnecessary duress for migrants and/or hosts.

Concluding Synthesis

It is a finding of this research that ongoing urbanisation trends are likely to continue well into the future, implying that more and more people will be living on fewer and fewer inhabited islands, likely artificially adapted, reinforced and/or raised to higher and higher heights. Climate change can be expected to catalyse this trend, making decentralisation an important countermeasure of congestion. It is an unequivocal finding of this research that planned migration is experienced as inherently more positive than ad hoc migration. While natural disasters and environmental change can swiftly overwhelm communal coping capacities, triggering rapid and uncontrolled migration responses which are lacking in critical coordination, preparation and funding support, policy maker foresight and anticipatory preparedness can enable more benign migration processes. Equitable service provision, expansive education, social integration initiatives, and proactive government planning are recommended as essential policy priorities for preparedness informed and more positive migration outcomes. Education is a particularly important success factor in this case study: it holds the dual promise of enhancing options for emigration and contributing critical knowledge for in situ adaptation.

Research Limitations and Future Research

This research is subject to at least three limitations. First, with no Dhivehi speaking proficiency on the part of this researcher and 82% of interviews requiring a Dhivehi interpreter it is conceivable that some information could have been either lost in translation or lack thereof. Second, given the researcher's association with an Australian university—although this did not feature explicitly in interviews—it is conceivable that the responses of some respondents may have been coloured by a possible hope to gain certain benefits or influence ongoing research commitments. Third, research findings are based in part on a comparatively small sample and should be confirmed through other independent research.

As mentioned, extensive planned inter-atoll and inter-island migrations make the Maldives a fertile location for migration research. Two areas for further research appear to be particularly promising: (1) comparative analyses involving multiple study sites; (2) longitudinal studies of social success factors in migration over time (e.g. Black et al. 2011).

Acknowledgements The author wishes to thank Kirsty Andersen for her copy editorial support, and Saffah Faroog for his research assistance in the Maldives. Grateful acknowledgment for relevant Ph.D. research support is also made to John Merson, Eileen Pittaway, Richard Rumsey, Geoff Shepherd, and to the international development organisation World Vision.

References

ADB UN WB—Asian Development Bank, United Nations, & World Bank. (2005). *Tsunami: Impact and recovery. Joint needs assessment world bank-Asian Development Bank-UN System-Asian Development Bank, the United Nations, and the World Bank.* Retrieved November 28, 2012, from http://siteresources.worldbank.org/INTMALDIVES/Resources/mv-na-full-02-14-05.pdf.

Aljazeera. (2012, February 8). Maldives President quits after 'Coup'. *Aljazeera*. Retrieved May 3, 2016, from http://www.aljazeera.com/news/asia/2012/02/20122813513480256.html.

Ananthaswamy, A. (2009). Sea level rise: It's worse than we thought. *New Scientist, 203*(2715), 28–30. Retrieved June 1, 2016, from https://www.newscientist.com/article/mg20327151-300-sea-level-rise-its-worse-than-we-thought/.

Arenstam Gibbons, S. J., & Nicholls, R. J. (2006). Island abandonment and sea-level rise: An historical analog from the Chesapeake Bay, USA. *Global Environmental Change, 16*(1), 40–47. Retrieved April 16, 2016, from http://www.sciencedirect.com/science/article/pii/S0959378005000579.

Aung, T. H., Kaluwin, C., & Lennon, G. W. (1998). *Climate change and sea level: Curriculum modules for the Pacific schools. Part one: Physical science.* Bedford Park, Australia: National Tidal Facility, Flinders University of South Australia.

AVB—Australian Visa Bureau. (2012). *Maldives look to mass Australian migration as solution to rising sea levels*, 11 January 2012. Retrieved December 11, 2012, from http://www.visabureau.com/australia/news/10-01-2012/maldives-look-to-mass-australian-migration-as-solution-to-rising-sea-levels.aspx.

Barnett, J., & Campbell, J. (2010). *Climate change and small island states: Power, knowledge and the South Pacific*. Washington, DC, USA: Earthscan.

Baum, T. (2012). *Migrant workers in the international hotel industry*. International migration paper no. 112. Geneva, Switzerland: International Migration Branch and Sectoral Activities Department, International Labour Organization. Retrieved December 11, 2012, from http://www.ilo.org/public/english/protection/migrant/download/imp/imp112.pdf.

BBC—British Broadcasting Corporation. (2008). Plan for new Maldives homeland. *BBC News*, 10 November 2008. Retrieved November 28, 2012, from http://news.bbc.co.uk/2/hi/south_asia/7719501.stm.

BBC—British Broadcasting Corporation. (2012a). Maldives President Mohamed Nasheed resigns amid unrest: Maldives President Mohamed Nasheed has resigned after weeks of demonstrations and a mutiny by some police officers. *BBC News*, 7 February 2012. Retrieved February 9, 2012, from http://www.bbc.co.uk/news/mobile/world-asia-16922570.

BBC—British Broadcasting Corporation. (2012b). New Maldives leader Waheed Hassan vows calm transition: New Maldives President Mohammed Waheed Hassan has vowed to uphold the rule of law following the dramatic resignation of his predecessor Mohamed Nasheed. *BBC News*, 7 February 2012. Retrieved February 9, 2012, from http://www.bbc.co.uk/news/mobile/world-asia-16932366.

BBC—British Broadcasting Corporation. (2012c). Maldives issues arrest warrant for Mohamed Nasheed: Crowds of supporters have gathered at the house of the former maldives President Mohamed Nasheed, as a warrant was issued for his arrest. *BBC News*, 9 February 2012. Retrieved February 10, 2012, from http://www.bbc.co.uk/news/mobile/world-asia-16959808.

Belt, D. (2011). The coming storm: The people of Bangladesh have much to teach us about how a crowded planet can best adapt to rising sea levels. For them, that future is now. *National Geographic*, May 2011, 58–83.

Berdie, D. R., Anderson, J. F., & Niebuhr, M. A. (1986). *Questionnaires: Design and use* (2nd ed.). Metuchen, USA and London, UK: The Scarecrow Press.

Bettini, G. (2013). Climate barbarians at the gate? A critique of apocalyptic narratives on 'Climate Refugees'. *Geoforum, 45*, 63–72. Retrieved April 17, 2016, from http://www.sciencedirect.com/science/article/pii/S0016718512001947.

Black, R., Bennett, S. R. G., Thomas, S. M., & Beddington, J. R. (2011). Climate change: Migration as adaptation. *Nature, 478*, 447–449.

Blanco, H., Alberti, M., Olshansky, R., Chang, S., Wheeler, S. M., Randolph, J., et al. (2009). Shaken, shrinking, hot, impoverished and informal: Emerging research agendas in planning. *Progress in Planning, 72*, 195–250.

Bluepeace. (2008). *New homeland elsewhere is not the only solution for climate victims of the Maldives*. Bluepeace blog, 30 November 2008. Retrieved December 17, 2012, from http://www.bluepeacemaldives.org/blog/climate-change/new-homeland-not-only-solution-climate-victims-maldives.

Bosley, D. (2012). Inhabited Islands need to be halved: Housing Minister. *Minivan News*, 24 September 2012. Retrieved January 7, 2013, from http://minivannewsarchive.com/tag/housing-minister.

Brunner, R. D., & Lynch, A. H. (2010). *Adaptive governance and climate change*. Boston, USA: American Meteorological Society.

Burgess, J. (2012). *Maldives buying land in Australia as preparation for mass migration*. Oilprice.com, 10 January 2012. Retrieved December 12, 2012, from http://oilprice.com/Latest-Energy-News/World-News/Maldives-Buying-Land-In-Australia-As-Preparation-For-Mass-Migration.html.

Charmaz, K. (2006). *Constructing grounded theory: A practical guide through qualitative analysis*. London, UK: Sage.

Church, J. A., Gregory, J. M., Huybrechts, P., Kuhn, M., Lambeck, K., Nhuan, M. T., et al. (2001). Changes in sea level. In J. T. Houghton, Y. Ding, D. J. Griggs, M. Noguer, P. J. van der Linden, X. Dai, K. Maskell & C. A. Johnson (Eds.), *Climate change 2001: The scientific basis. Contribution of working group I to the third assessment report of the intergovernmental panel on climate change* (pp. 639–693). Cambridge, UK and New York, USA: Cambridge University Press.

Creswell, J. W., & Plano Clark, V. L. (2011). *Designing and conducting mixed methods research* (2nd ed.). London, UK: Sage.

De'ath, G., Lough, J. M., & Fabricius, K. E. (2009). Declining coral calcification on the great barrier reef. *Science, 323*(5910), 116–119. Retrieved April 16, 2016, from http://science.sciencemag.org/content/323/5910/116.

Diamond, J. (2005). *Collapse: How societies choose to fail or succeed*. London, UK, New York, USA: Viking Penguin.

Doherty, B. (2012, January 7). Climate change castaways consider move to Australia: The President of what could be the first country in the world lost to climate change has urged Australia to prepare for a mass wave of climate refugees seeking a new place to live. *The Sydney Morning Herald*, 7 January 2012. Retrieved December 12, 2012, from http://www.smh.com.au/environment/climate-change/climate-change-castaways-consider-move-to-australia-20120106-1pobf.html.

Foresight. (2011). *Migration and global environmental change: Final project report*. London, UK: The Government Office for Science. Retrieved October 21, 2011, from http://www.bis.gov.uk/assets/foresight/docs/migration/11-1116-migration-and-global-environmental-change.pdf.

Frieler, K., Meinshausen, M., Golly, A., Mengel, M., Lebek, K., Donner, S. D., et al. (2013). Limiting global warming to 2 °C is unlikely to save most coral reefs. *Nature Climate Change*, *3*, 165–170. Retrieved April 15, 2016, from http://www.nature.com/nclimate/journal/v3/n2/full/nclimate1674.html.

Godfrey, T. (2007). *Atlas of the Maldives. A reference for travellers, divers and sailors* (5th ed.). Apollo Bay, Australia: Atoll Editions.

GOV MAL—The Government of Maldives. (2009). *"Aneh Dhivehi Raajje". The strategic action plan. National framework for development 2009–2013*. Malé, Republic of Maldives: The President's Office, The Government of Maldives. Retrieved November 27, 2012, from http://planning.gov.mv/en/images/stories/publications/strategic_action_plan/SAP-EN.pdf.

Hamilton, C. (2010). *Requiem for a species: Why we resist the truth about climate change*. Crows Nest, Australia: Allen & Unwin.

Hansen, J. E. (2007). Scientific reticence and sea level rise. *Environmental Research Letters*, *2*(2). Retrieved June 1, 2016, from http://iopscience.iop.org/article/10.1088/1748-9326/2/2/024002/fulltext/.

Hansen, J. (2008). *Global warming twenty years later: Tipping points near*. Briefing before the Select Committee on Energy Independence and Global Warming, U.S. House of Representatives. National Press Club on June 23. Retrieved July 31, 2008–December 1, 2012 from http://www.columbia.edu/~jeh1/2008/TwentyYearsLater_20080623.pdf.

Hansen, J. E. (2009). *Storms of my grandchildren: The truth about the coming climate catastrophe and our last chance to save humanity*. London, UK: Bloomsbury.

Hansen, J., Sato, M., Hearty, P., Ruedy, R., Kelley, M., Masson-Delmotte, V., et al. (2016). Ice melt, sea level rise and superstorms: Evidence from paleoclimate data, climate modeling, and modern observations that 2 C global warming could be dangerous. *Atmospheric Chemistry and Physics, 16*, 3761–3812. Retrieved April 16, 2016, from http://www.atmos-chem-phys.net/16/3761/2016/acp-16-3761-2016.pdf.

Harrabin, R. (2012a). UN climate talks extend kyoto protocol, promise compensation: UN climate talks in doha have closed with a historic shift in principle but few genuine cuts in greenhouse gases. *BBC Science, & Environment*. Retrieved December 10, 2012, from http://m.bbc.co.uk/news/science-environment-20653018.

Harrabin, R. (2012b). Climate compensation row at Doha: Frustration at slow progress of the UN climate talks bubbled over when a spokesman for small island states (AOSIS) rounded on rich nations. *BBC News*. Retrieved January 7, 2013, from http://www.bbc.co.uk/news/science-environment-20613915.

Haveeru Daily. (2011). Government begins relocating Maavaidhoo residents: The government is working on relocating the people of Haa Dhaal Atoll Maavaidhoo to Haa Dhaal Atoll Nolhivaranfaru within 2–3 days. *Haveeru Daily*, 23 January 2011. Retrieved November 19, 2012, from http://www.haveeru.com.mv/english/details/33879.

Henley, J. (2008, November 11). The last days of paradise: The President of the Maldives wants to buy a new home for all 300,000 of his people, to save them from rising sea levels. *The Guardian*. Retrieved December 14, 2012, from http://www.guardian.co.uk/environment/2008/nov/11/climatechange-endangered-habitats-maldives.

Hoegh-Guldberg, O. (2011). Coral reef ecosystems and anthropogenic climate change. *Regional Environmental Change, 11*(1), S215–S227. Retrieved April 16, 2016, from http://link.springer.com/article/10.1007%2Fs10113-010-0189-2#/page-1.

IPCC—Intergovernmental Panel on Climate Change. (1992). *Climate change: The IPCC 1990 and 1992 assessments. IPCC first assessment report overview and policymaker summaries and 1992 IPCC supplement*. Canada: Intergovernmental Panel on Climate Change. Retrieved April 16, 2016, from https://www.ipcc.ch/ipccreports/1992%20IPCC%20Supplement/IPCC_1990_and_1992_Assessments/English/ipcc_90_92_assessments_far_front_matters.pdf.

IPCC—Intergovernmental Panel on Climate Change. (2007). *Climate change 2007: Synthesis report. Contribution of working groups I, II and III to the fourth assessment report of the Intergovernmental Panel on Climate Change*. Geneva, Switzerland: Intergovernmental Panel on Climate Change. Retrieved April 16, 2016, from https://www.ipcc.ch/pdf/assessment-report/ar4/syr/ar4_syr_full_report.pdf.

Kench, P. S., McLean, R. F., Brander, R. W., Nichol, S. L., Smithers, S. G., Ford, M. R., et al. (2006). Geological effects of tsunami on mid-ocean atoll islands: The Maldives before and after the Sumatran tsunami. *Geological Society of America, 34*(3), 177–180.

Lang, O. (2009). Maldives leader in climate change stunt: With fish darting amongst them in a blue lagoon, the Maldivian President and his top team have staged an elaborate stunt to publicise climate change. *BBC News*, 17 October 2009. Retrieved January 4, 2012, from http://news.bbc.co.uk/2/hi/8312320.stm.

Leighton, M. (2012). Population displacement, relocation, and migration. In M. B. Gerrard & K. Fischer Kuh (Eds.), *The law of adaptation to climate change: United States and International Aspects*. Chicago, USA: American Bar Association.

Lubna, H. (2012). Nearly 60 islands hit with water crisis. *ReliefWeb/Minivan News*, 9 April 2012. Retrieved May 24, 2016, from http://reliefweb.int/report/maldives/nearly-60-islands-hit-water-crisis.

Luetz, J. M. (2008). *Planet prepare: Preparing coastal communities in Asia for future catastrophes, asia pacific disaster report*. Bangkok, Thailand: World Vision International. Retrieved April 16, 2016, from http://luetz.com/docs/planet-prepare.pdf.

Luetz, J. M. (2013). *Climate migration: Preparedness informed policy opportunities identified during field research in Bolivia, Bangladesh and Maldives*. Ph.D. dissertation, University of New South Wales. Retrieved May 31, 2016, from http://handle.unsw.edu.au/1959.4/52944.

Markandya, A., Halsnaes, K., Lanza, A., Matsuoka, Y., Maya, S., Pan, J., et al. (2001). Costing methodologies. In T. Banuri, T. Barker, I. Bashmakov, K. Blok, D. Bouille, R. Christ, et al. (Eds.), *Climate change 2001: A report of working group III of the Intergovernmental Panel on Climate Change*. Retrieved May 31, 2016, from http://www.grida.no/climate/ipcc_tar/wg3/pdf/7.pdf.

Marton, F. (1981). Phenomenography: Describing conceptions of the world around us. *Instructional Science, 10*, 177–200.

Masters, T. (2006). *Maldives* (6th ed.). Melbourne, Australia: Lonely Planet.

MEEW NAPA—Ministry of Environment, Energy and Water. (2007). *National Adaptation Programme of Action (NAPA)*. Malé, Republic of Maldives: Ministry of Environment, Energy and Water. Retrieved December 15, 2012, from http://unfccc.int/resource/docs/napa/mdv01.pdf.

MHUD—Ministry of Housing and Urban Development. (2008). *Maldives National Housing Policy May 2008*. Ministry of Housing and Urban Development, Malé, Maldives. Retrieved November 19, 2012, from http://www.mvlaw.gov.mv/pdf/gavaid/minHousing/25.pdf.

Mimura, N. et al. (2007). Small islands. In M. L. Parry, O. F. Canziani, J. P. Palutikof, P. J. van der Linden & C. E. Hanson (Eds.), *Climate change 2007: Impacts, adaptation and vulnerability, contribution of working group II to the fourth assessment report of the Intergovernmental Panel on Climate Change* (pp. 687–716). Cambridge, UK: Cambridge University Press.

Mohamed, K. (2002). *Analytical report, census of the Maldives-2000, Chapter IX: Internal Migration*. Malé, Republic of Maldives: Ministry of Planning and National Development. Retrieved December 11, 2012, from http://www.planning.gov.mv/publications/analytical_report/html/Source/ChapterIX.pdf.

MPND—Ministry of Planning and Development. (2005). Population policy of the Maldives. *Working Document, July 2004*. Malé, Republic of Maldives: Ministry of Planning and Development. Retrieved December 13, 2012, from http://planning.gov.mv/publications/PopulationPolicyMaldives.pdf.

MPND—Ministry of Planning and National Development. (2006). *Analytical report 2006: Population and housing census 2006*. Malé, Republic of Maldives: Ministry of Planning and National Development. Retrieved November 23, 2012, from http://planning.gov.mv/en/images/stories/publications/analysiscd/index.html.

MPND—Ministry of Planning and National Development. (2007). *Seventh national development plan 2006–2010: Creating new opportunities*. Malé, Republic of Maldives: Ministry of Planning and National Development. Retrieved November 19, 2012, from http://www.planning.gov.mv/en/images/stories/ndp/seventh_ndp.pdf.

MTAC—Ministry of Tourism, Arts and Culture. (2012). *Increasing climate change resilience of maldives through adaptation in the tourism sector project: Inception report*. Malé, Republic of Maldives: Project Management Unit, Ministry of Tourism, Arts and Culture. Retrieved December 12, 2012, from http://www.tourism.gov.mv/downloads/tap/2012/reports/Final_InceptionReport_20June2012.pdf.

New, M., Liverman, D., Schroeder, H., & Anderson, K. (2011). Four degrees and beyond: The potential for a global temperature increase of four degrees and its implications. *Philosophical Transactions of the Royal Society, 369*, 6–19. Retrieved April 16, 2016, from http://rsta.royalsocietypublishing.org/content/roypta/369/1934/4.full.pdf.

Nicholls, R. J. (2006). Storm surges in coastal areas. In M. Arnold, R. S. Chen, U. Deichmann, M. Dilley, A. L. Lerner-Lam, R. E. Pullen, et al. (Eds.), *Natural disaster hotspots: Case studies disaster risk management series no. 6*. Washington, DC, USA: The World Bank. Retrieved June 27, 2008, from http://siteresources.worldbank.org/INTDISMGMT/Resources/0821363328.pdf.

Nicholls, R.J., Marinova, N., Lowe, J.A, Brown, S., Vellinga, P., de Gusmão, D., Hinkel, J., et al. (2011). Sea-level rise and its possible impacts given a “Beyond 4 °C World” in the twenty-first century. *Philosophical Transactions of the Royal Society*, *369*, 161–181. Retrieved April 16, 2016, from http://rsta.royalsocietypublishing.org/content/roypta/369/1934/161.full.pdf.

Nunn, P. (2009). Responding to the challenges of climate change in the Pacific Islands: Management and technological imperatives. *Climate Research*, *40*, 211–231. Retrieved April 16, 2016, from http://www.int-res.com/articles/cr_oa/c040p211.pdf.

Punch, K. F. (2014). *Introduction to social research: Quantitative and qualitative approaches* (3rd ed.). London, UK: Sage.

Radhakrishna, R. B. (2007). Tips for developing and testing questionnaires/instruments. *Journal of Extension (JOE)*, *45*(1). Retrieved April 16, 2016, from http://www.joe.org/joe/2007february/tt2.php.

Rahmstorf, S. (2012). Sea-level rise: towards understanding local vulnerability. *Environmental Research Letters*, *7*(2). Retrieved April 16, 2016, from http://iopscience.iop.org/article/10.1088/1748-9326/7/2/021001/pdf.

Rahmstorf, S., Foster, G., & Cazenave, A. (2012). Comparing climate projections to observations up to 2011. *Environmental Research Letters*, *7*(4). Retrieved April 16, 2016, from http://iopscience.iop.org/article/10.1088/1748-9326/7/4/044035/meta.

Ramesh, R. (2008). Paradise almost lost: Maldives seek to buy a new homeland. *The Guardian*, 10 November 2008. Retrieved April 25, 2009, from http://www.guardian.co.uk/environment/2008/nov/10/maldives-climate-change.

Ratha, D., Mohapatra, S., & Silwal, A. (2011). *The migration and remittances factbook 2011*. Maldives: Migration and Remittances Unit, World Bank. Retrieved December 11, 2012, from http://siteresources.worldbank.org/INTPROSPECTS/Resources/334934-1199807908806/Maldives.pdf.

Revkin, A. C. (2008). Maldives considers buying dry land if seas rise. *The New York Times*, 10 November 2008. Retrieved December 17, 2012, from http://www.nytimes.com/2008/11/11/science/earth/11maldives.html.

Robinson, A., Calov, R., & Ganopolski, A. (2012). Multistability and critical thresholds of the greenland ice sheet. *Nature Climate Change*, *2*(4), 429–432. Retrieved April 16, 2016, from http://www.nature.com/nclimate/journal/v2/n6/full/nclimate1449.html.

SAMN—South Asia Media Net. (2011). Government begins relocating residents. *South Asia Media Net*, 24 January 2011. Retrieved November 19, 2012, from http://www.mediawitty.com/test/NewsDetail.aspx?group_id=43&id=13861&folder_id=92&Page_Title=Government%20begins%20relocating%20residents.

Schaeffer, M., Hare, W., Rahmstorf, S., & Vermeer, M. (2012). Long-term sea-level rise implied by 1:5 °C and 2 °C warming levels. *Nature Climate Change*, *2*, 867–870. Retrieved April 16, 2016, from http://climateanalytics.org/files/schaeffer_et_al__2012__long-term_slr.pdf.

Schellnhuber, H. J. (2008). Global warming: Stop worrying, start panicking? In *Proceedings of the National Academy of Sciences (PNAS)*, *105*(38), 14239–14240. Retrieved April 16, 2016, from http://www.pnas.org/content/105/38/14239.full.pdf.

Schmidle, N. (2009). Wanted: A new home for my country. Global warming may drown the Maldives, and the Island Nation’s President is considering relocating the entire population. *The New York Times*, 10 May 2009. Retrieved November 28, 2012, from http://www.nytimes.com/2009/05/10/magazine/10MALDIVES-t.html.

Sedlack, G. R., & Stanley, J. (1992). *Social research: Theory and methods*. Needham Heights, MA, USA: Allyn & Bacon.

Silverman, D. (1993). *Interpreting qualitative data: Methods for analysing talk, text and interaction*. London, UK: Sage.

Slezak, M. (2016). Coral bleaching spreads to Maldives, devastating spectacular reefs. *The Guardian*, 1 June 2016. Retrieved June 1, 2016, from http://www.theguardian.com/environment/2016/jun/01/coral-bleaching-spreads-to-maldives-devastating-spectacular-reefs.

SOEM—State of the Environment Maldives 2011. (2011). Ministry of Environment and Energy, Republic of Maldives. Retrieved December 12, 2012, from http://www.environment.gov.mv/v1/download/386.

Solomon, S., Plattner, G., Knutti, R., & Friedlingsteind, P. (2009). Irreversible climate change due to carbon dioxide emissions. In *Proceedings of the National Academy of Sciences of the United States of America, 106*(6), 1704–1709. Retrieved April 16, 2016, from http://www.pnas.org/content/106/6/1704.full.pdf.

Starke, L., & Mastny, L. (Eds.). (2010). *State of the world 2010—transforming cultures from consumerism to sustainability*. London, UK: Earthscan.

Stewart, H., & Elliott, L. (2013). Nicholas stern: "I got it Wrong on Climate Change—it's Far, Far Worse": Author of 2006 review speaks out on danger to economies as planet absorbs less carbon and is 'On Track' for 4C rise. *The Guardian*, 26 January 2013. Retrieved January 28, 2013, from http://www.guardian.co.uk/environment/2013/jan/27/nicholas-stern-climate-change-davos.

Tesch, R. (1990). *Qualitative research: analysis types and software tools*. New York, USA: Falmer Press.

TPO—The President's Office. (2011). *Press releases: Government to address the existing legal impediments to find a permanent mechanism for mosquito control,* Ref: 2011-141, 19 February 2011. Retrieved November 19, 2012, from http://www.presidencymaldives.gov.mv/?lid=11&dcid=4777.

UNDP—United Nations Development Programme. (2007). *Human development report 2007/2008. Fighting climate change: Human solidarity in a divided world*. New York, USA: United Nations Development Programme. Retrieved April 16, 2016, from http://hdr.undp.org/sites/default/files/reports/268/hdr_20072008_en_complete.pdf.

UNDP—United Nations Development Programme. (2009). *Human development report 2009. Overcoming barriers: Human mobility and development*. New York, USA: United Nations Development Programme.

UNEP—United Nations Environment Programme. (2006). *Marine and coastal eco-systems and human well-being: A synthesis report based on the findings of the millennium ecosystems assessment*. Nairobi, Kenya, Africa: United Nations Environment Programme. Retrieved December 15, 2012, from http://www.unep.org/pdf/Completev6_LR.pdf.

UNHCR—United Nations High Commissioner for Refugees. (2009). *Climate change and statelessness: An overview*. Submitted by UNHCR, supported by the International Organization for Migration (IOM) and the Norwegian Refugee Council (NRC) to the 6th session of the Ad Hoc Working Group on Long-Term Cooperative Action (AWG-LCA 6) under UNFCCC, 1–12 June 2009, Bonn, Germany. Retrieved July 1, 2010, from http://www.unhcr.org/4a1e50082.html.

UNISDR-UNDP—United Nations International Strategy for Disaster Risk Reduction, United Nations Development Programme. (2012). *Disaster Risk Reduction and Climate Change Adaptation in the Pacific: An Institutional and Policy Analysis*, United Nations International Strategy for Disaster Risk Reduction, United Nations Development Programme, Suva, Fiji. Retrieved April 16, 2016, from http://www.unisdr.org/files/26725_26725drrandccainthepacificaninstitu.pdf.

Vince, G. (2009a). Paradise lost: Islanders prepare for the flood. *New Scientist*, No. 2707, pp. 37–39. Retrieved April 16, 2016, from https://www.newscientist.com/article/mg20227071-200-paradise-lost-islanders-prepare-for-the-flood/.

Vince, G. (2009b). Surviving in a warmer world. *New Scientist*, 201(2697), 28–33. Retrieved April 16, 2016, from http://www.sciencedirect.com/science/article/pii/S026240790960587X.

WBGU—German Advisory Council on Global Change. (2009a). *Solving the climate dilemma: The budget approach, German advisory council on global change*. Berlin, Germany: German Advisory Council on Global Change. Retrieved April 16, 2016, from http://www.wbgu.de/fileadmin/templates/dateien/veroeffentlichungen/sondergutachten/sn2009/wbgu_sn2009_en.pdf.

WBGU—German Advisory Council on Global Change. (2009b). *Climate Change: Why 2 °C? Factsheet No 2/2009*. German Advisory Council on Global Change, Berlin, Germany. Retrieved April 28, 2010, from http://wbgu.de/wbgu_factsheet_2_en.pdf.

Woo, S., Hall, B., Yee, A., Primrose, K., Soto, T., Tan, J., et al. (2011). Hit the brakes. *Leadership Network for Climate Change*. Climate Change Simply Explained Teaching Series. Sydney, Australia: UNSW-TV, University of New South Wales. Retrieved from http://tv.unsw.edu.au/video/hit-the-brakes.

World Bank. (2011). *World development indicators (WDI) & Global development finance (GDF)*. Selected data and metadata, The World Bank, exported 23 November 2012, http://data.worldbank.org/country/maldives, http://databank.worldbank.org/ddp/home.do.

World Bank. (2012). *Turn down the heat: Why a 4 °C warmer world must be avoided*. Washington, DC, USA: International Bank for Reconstruction and Development/The World Bank. Retrieved March 15, 2016, from http://documents.worldbank.org/curated/en/2012/11/17097815/turn-down-heat-4C-warmer-world-must-avoided.

World Bank. (2014). *Turn down the heat: Confronting the new climate normal*. Washington, DC, USA: World Bank. Retrieved March 15, 2016, from https://openknowledge.worldbank.org/handle/10986/20595.

Part II
Climate Change Adaptation, Resilience and Hazards

Chapter 4
Climate Change and Extreme Weather Events: The Mental Health Impact

David N. Sattler

Introduction

Climate change has the potential to increase the strength, intensity, and frequency of extreme weather events that threaten lives and property, mental health and well-being, and the livelihood of millions of people (United Nations 2014). The mental health consequences of climate change and extreme weather events associated with climate change are so severe that the Task Force on the Interface between Psychology and Global Climate Change states that "Addressing climate change is arguably one of the most pressing tasks facing our planet and its inhabitants" (American Psychological Association 2009, p. 6). This chapter considers how exposure to extreme weather events may influence preparedness for new threats and mental health and psychological functioning during recovery, as well as prospects for projects that educate about extreme weather events.[1]

In recent years, climate change has been associated with Cyclone Winston and Super Typhoon Haiyan (two of the strongest storms in recorded history to have made landfall in Fiji and the Philippines, respectively); heatwaves, fires, and extended periods of drought in Australia, Russia, Canada, the Amazon, and southern and western Africa, and flooding in United Kingdom, Bangladesh, and other South Asian countries (Johnson 2016; Union of Concerned Scientists 2016; World Bank 2013). In the wake of research conducted by the author of this chapter

[1]The term extreme weather event is used to refer to any natural disaster whose strength can be intensified by climate change, such as a hurricane, flood, drought, fire, and heatwave. The terms hurricane, cyclone, and typhoon refer to the same weather phenomena; use of the term varies depending on the location of the storm.

D.N. Sattler (✉)
Department of Psychology, Western Washington University, Bellingham, WA 98225-9172, USA
e-mail: David.Sattler@wwu.edu

W. Leal Filho (ed.), *Climate Change Adaptation in Pacific Countries*, Climate Change Management, DOI 10.1007/978-3-319-50094-2_4

examining psychological functioning among persons who experienced Cyclone Winston in Fiji and Super Typhoon Haiyan in the Philippines, the need for a chapter that considers the mental health consequences of exposure to extreme weather events associated with climate change became apparent.

Experiences During an Extreme Weather Event

Experiences with extreme weather events can influence the way people perceive, prepare for, and respond to subsequent extreme weather threats. This section considers psychological responses based on exposure to and experiences during extreme weather events, cognitive appraisal, and individual difference variables.

Exposure to Extreme Weather Events. The characteristics of extreme weather events associated with climate change can have a profound influence on the nature of the experience and psychological reactions. Extreme weather events vary in strength, intensity, duration, predictability, degree to which they inflict damage on property and community infrastructure, the number of people whose lives are threatened, and the number of people who are impacted and displaced as a result of damage or destruction to their home and community. They may occur with little or no advanced warning and can overwhelm the ability of individuals and communities to mitigate damage. These characteristics can influence the amount of time people and communities have to prepare and to take actions to reduce exposure to life threat (e.g., evacuate to shelters or another location) and to minimize losses to property (Sattler et al. 2000).

People who are not able to evacuate for extreme weather events often are at increased risk of life threat and injury (Sattler 2001; Shukla 2013). The sheer force of wind can cause extensive damage to and destruction of homes and property. Flying debris, felled power lines that continue to transmit electricity, and rising flood waters are examples of dangerous conditions that place people at increased risk for loss of life or injury. For certain disasters, such as a heatwave or drought, threat to one's life or livelihood may continue for many years. For example, in Australia, a relentless drought severely impacted the ability of farming communities to grow crops for years, and created enduring psychological distress and coping challenges (Berry et al. 2011; World Bank 2013).

The threat of loss of life or injury to oneself and to loved ones and others during the event is associated with psychological distress, including posttraumatic stress (Sattler et al. 2002, 2014; Zwiebach et al. 2010). Experiencing or witnessing an event that results in life threat or serious injury to oneself, a close family member, or friend is a significant risk factor for developing posttraumatic stress disorder, which involves "clinically significant distress or impairment in the individual's social interactions, capacity to work or other important areas of functioning" (American Psychiatric Association 2013). Experiencing an extreme weather event can have a profound negative influence on one's sense of control, predictability, safety, and trust (Sattler et al. 1995; Weisath 1993).

Experience with Extreme Weather Events and Cognitive Appraisal. Cognitive appraisal concerns the way in which people interpret and give meaning to their experience. For example, when people are threatened by an extreme weather event, their prior experiences with similar events can influence perceptions of risk and actions to prepare for or respond to the event. Consider the case of Charleston, South Carolina, which experienced extensive damage as a result of Hurricane Hugo, a strong category four storm. Four and seven years later, two hurricanes again threatened Charleston. People who had direct experience with Hurricane Hugo and who perceived a greater threat associated with each new hurricane took greater measures to prepare (Sattler et al. 2000). Similar findings have been reported with people who have experienced floods, such that people who had direct experience with a flood perceived greater threat associated with a new flood threat than those with vicarious flood experience (Kellens et al. 2013). Other research shows that the degree to which people anticipate life threat and risk of damage to property influence actions to prepare (Lindell and Perry 2012). As such, people without direct personal experience relevant to a new extreme weather event may be at increased risk of being unprepared and to experience life threat, posttraumatic stress disorder, and property loss.

The availability heuristic helps explain the influence of direct experience on perceptions of threat and preventive actions. Heuristics are mental shortcuts people use to make decisions, and the availability heuristic states that information that comes to mind quickly and is easier to recall has greater influence on decisions than information that is more difficult to recall. Because experiences with an extreme weather event may come to mind more quickly and may be associated with a stronger emotional response, it may be more likely to influence behavior than vicarious information concerning the experiences of others (Sattler et al. 1995).

In certain situations, direct experience with an extreme weather event may lead to decisions that place people at risk. For example, direct experience might create an optimistic bias that influences people to discount the importance of preparing for a new threat. For example, people who had experienced a flood whose waters came only as high as their driveways (and did not damage their homes) incorrectly believed that the water level of a future flood threat would be no higher than the prior flood level, and they did not evacuate for a new flood threat (Lindell and Perry 1992). Another example is when people who experience a weak hurricane incorrectly assume that a future hurricane will cause similar damage, even if it is rated as stronger. Some people may assume that the intensity of all hurricanes is essentially the same, especially when they are not aware how hurricanes are classified (Sattler and Marshall 2002). This is a significant concern given that climate change has the potential to increase the strength of hurricanes, and people may not appreciate the destructive potential of stronger storms. For example, in the Philippines and Fiji, few people had prior experience with storms the magnitude of Super Typhoon Haiyan or Cyclone Winston, which were two of the most powerful storms to make landfall. If people overestimate the importance of their own prior experiences, especially when they do not have complete or appropriate information about the new disaster threat and the new threat is of greater magnitude, then they may be at increased risk of under preparation (cf. Viscusi and Zeckhauser 2015).

When people do not associate climate change with their extreme weather experience, they may be less likely to take actions to minimize behaviors that contribute to it. Although daily choices and actions can contribute to climate change, people often receive little or no feedback regarding how their actions (e.g., driving a car, purchasing an appliance, using electricity in the home, food consumed) influence climate change. This is so, in part, because (a) people receive little to no information about how much carbon dioxide was produced as a result of their actions (e.g., emissions as a result of heating a home or driving a car), and (b) the consequences of climate change may be felt more profoundly in a different location (Gardner and Stern 2002).

Individual Difference Variables. Individual difference variables such as personality traits, gender, age, and socioeconomic status can influence how people respond to extreme weather events.

Sense of coherence is a

> …global orientation that expresses the extent to which one has a pervasive, enduring, though dynamic feeling of confidence that (a) the stimuli deriving from one's internal and external environments are structured, predictable, and explicable; (b) the resources are available to one to meet the demands posed by these stimuli; and (c) these demands are challenges, worthy of investment and engagement (Antonovsky 1987, p. 19).

Sense of coherence is associated with a strong sense of mastery, hardiness, and optimism. People with higher levels of sense of coherence may be more likely to view the world as predictable and explicable. In the wake of a devastating hurricane, sense of coherence was negatively associated with psychological distress and positively associated with providing help to others (Kaiser et al. 1996). Further, individuals who view the world as predictable may be more likely to prepare for an extreme weather event (Sattler et al. 2000).

Locus of control concerns the degree to which people believe their outcomes are influenced by their own choices or fate. People with an internal locus of control are more likely to believe their actions control their outcomes, whereas those with an external locus of control are more likely to believe that outside forces, such as luck or fate, influence one's outcomes. Locus of control is associated with preparation for a hurricane: those with an internal locus of control are more likely to prepare for a hurricane than those with an external locus of control (Sattler et al. 2000). During a disaster threat, taking protective action, finding a safe location in one's community, or evacuating the area can increase the sense of control and reduce anticipatory anxiety (cf. Tally et al. 2013). Research is needed to examine how culture and community experiences with natural disasters influence preparedness (Sattler et al. 2016).

Experiences During Recovery

The intensity and size of Super Typhoon Haiyan and Cyclone Winston were associated with climate change. On 8–9 November 2013, Super Typhoon Haiyan devastated coastal areas of the Philippines, and was the second deadliest typhoon to strike the country: 6,293 persons lost their lives and 28,689 persons were injured (Mullen 2013). Property damage estimates were close to US $1 billion, 890,895 families (or 4,095,280 million persons) were left homeless, and 1,140,332 homes were damaged or destroyed. It destroyed significant portions of the ecosystem, including 33 million coconut trees and crops (National Disaster Risk Reduction and Management Council 2014). Maximum wind speed was estimated at 196 miles per hour.

On February 20, 2016, Cyclone Winston devastated communities on many islands throughout Fiji. Maximum wind speed was estimated at 185 miles per hour (Di Liberto 2016; Yulsman 2016). Just under half of the population of Fiji (40%) was impacted by the cyclone, and 43 persons lost their lives. The cyclone damaged or destroyed 28,000 houses, and nearly all of the buildings were leveled in some communities and villages (Yulsman 2016). Property damage estimates exceeded $470 million, or close to 10% of Fiji's gross domestic product (Varandani 2016). A significant portion of the country depends on subsistence production, especially in rural areas on outer islands. The cyclone destroyed significant portions of the ecosystem, including essential crops.

Extreme weather events such as Super Typhoon Haiyan and Cyclone Winston create significant challenges when people lose access to resources essential for survival, such as food, water, shelter, and clothing (Chan and Rhodes 2014; Hobfoll 2012; Sattler et al. 2014). These events can compromise the ability of communities to provide basic services, such as electricity, telephone, potable water, and waste disposal. Businesses may be closed for an indefinite period of time, resulting in loss of access to goods and services for the community and loss of employment and income. Roadways may be impassable when they are damaged or covered in debris. As a result of transportation and communication difficulties, recovery crews may face extended delays in reaching people and providing aid, and people in the affected communities may not be able to provide aid to others. When recovery personnel are not able to promptly access the area, those who are injured or need medical attention may have little or no access to medical services. Loss of access to essential resources is a significant source of stress (Hobfoll 2012).

Loss of Resources and Social Support. Conservation of resources stress theory provides a useful framework for understanding how exposure to an extreme weather event and life threat can result in psychological stress. The theory asserts that people utilize four types of resources: object (e.g., home, car, furniture), condition (e.g., marital/employment status), personal characteristic (e.g., age, locus of control, self-esteem), and energy (e.g., money, insurance). By maintaining a variety of resources, people can enhance resiliency and their ability to cope with stressors. According to the theory, psychological stress occurs when resources in one or more of these categories is threatened or lost (Hobfoll 2012).

Conservation of resources stress theory views object, condition, personal characteristic, and energy resources as interconnected. The lack or loss of resources in one area can affect resources in another area and create a resource loss spiral, wherein resource loss in one area propagates loss across other areas (Hobfoll 2012). For example, following a hurricane, the loss of a person's house (an object resource) can be accompanied by depletion of personal wealth (an energy resource). Consequently, self-esteem (a personal characteristic resource) may suffer when the person cannot fulfill prior roles (a condition resource), which can place a strain on social relationships. Continued loss of resources across areas can negatively impact psychological functioning and well-being. Survivors may report sleeping disturbances, anxiety, depression, substance abuse and antisocial behavior, relationship difficulties, and feeling of helplessness and pessimism (American Psychological Association 2009; Sattler et al. 2002; Shukla 2013).

Loss of resources is a significant predictor of posttraumatic stress symptoms. Posttraumatic stress symptoms include dissociative reactions (e.g., emotional detachment), re-experiencing the event (e.g., recurrent thoughts and/or dreams), avoiding reminders of the event, and anxiety or arousal (e.g., difficulty sleeping, exaggerated startle response). Hurricane Katrina, one of the strongest hurricanes to make landfall in the United States, created unprecedented damage in the Gulf Coast states. Kessler et al. (2008) report that more than one-quarter (26%) of their sample met the criteria for posttraumatic stress disorder, and nearly half (49%) reported symptoms associated an anxiety-mood disorder.

Four weeks after Cyclone Winston made landfall in Fiji, more than three-quarters (77%) of persons sampled in coastal communities on the main island reported posttraumatic stress disorder symptomatology. Resource loss was associated with posttraumatic stress symptoms (Sattler et al. 2016).

Six weeks after Super Typhoon Haiyan devastated coastal areas of the Philippines, more than three-quarters (76%) of persons sampled reported posttraumatic stress symptomatology. Home damage and distress due to prior disasters, life threat during the typhoon, resource loss due to the typhoon, health problems, and low social support predicted posttraumatic stress symptoms (Sattler et al. 2016).

One month after Hurricane Georges devastated areas of the Dominican Republic, Puerto Rico, the U. S. Virgin Islands, and the United States, personal characteristic and energy resource loss, perceived life threat, prior traumatic event exposure, stressful life events, and low social support predicted stress disorder symptoms at each location (Sattler et al. 2002).

Resource loss spirals are especially problematic when recovery is drawn out and people have difficulties returning their lives to normal (Hobfoll 2012). Secondary stressors are related to resource loss spirals, and they commonly occur during the recovery period. Secondary stressors involve stressful life events, strains, and hassles that develop during recovery. Examples include employment difficulties, extended financial difficulties, living situation, relocating to a new area, and a delay in one's life returning to normal. Communities that depend on agricultural production for sustenance and livelihood may experience significant resource loss spirals (Berry et al. 2011).

People who are able to maintain, enhance, or acquire resources after a disaster are more likely to experience positive outcomes and to reduce the potential for resource loss spirals (Hobfoll 2012). For example, those who experience an extreme weather event and who learn effective coping skills may develop an enhanced sense of self-efficacy (a personal characteristic resource). Enhanced self-efficacy may help people to better maintain their social roles (a condition resource) while pursuing resource replenishment (e.g., object and energy resources). Property insurance can help people protect object resources and reduce the risk of resource loss in other areas.

Many mental health problems may resolve within the first few weeks or months, some may continue for years, and new problems may develop, especially when the disruption to one's life and community are not resolved in a timely manner (Hobfoll 2012). It can take years for communities to rebuild when extreme weather events damage and destroy property on a massive scale.

Posttraumatic Growth. The hardships imposed by extreme weather events may present opportunities for personal growth. Posttraumatic growth involves reappraising and reassessing one's values and life priorities following exposure to a traumatic or life threatening event, and reinterpreting the event by giving it new meaning (Groleau et al. 2012). Survivors may develop new coping skills, a heightened sense of self-efficacy, and new appreciation for others, especially during the first few months following a disaster when people often report making friends and receiving and providing assistance to others. These and other positive outcomes help counter negative experiences (Groleau et al. 2012; Linley and Joseph 2004; Sattler et al. 2014).

Six weeks after Super Typhoon Haiyan struck the Phillipines, the majority of people sampled reported posttraumatic growth (Sattler et al. 2016). Four weeks after Cyclone Winston make landfall in Fiji, the majority of persons sampled reported posttraumatic growth (Sattler et al. 2016).

In the weeks and months after an extreme weather event, posttraumatic growth may reflect a coping strategy wherein people focus on the positive aspects of their life that they can control, and reassess their life priorities and attitudes about the situations. They also may engage in social comparison wherein they compare themselves and their situation to others who are less fortunate in order to reassess their situation more favorably (Updegraff and Taylor 2000). This process may help people feel more optimistic and counter negative feelings (Linley and Joseph 2004). Seeing positives can help people control their emotions during a seemingly uncontrollable situation (Sattler et al. 2014). Research is needed to explore how posttraumatic growth may change over time during the recovery process, and how cultural variables may influence its development.

Main Lessons

The work discussed above offers guidance for preparation and intervention and recovery efforts. Exposure to an extreme weather event, life threat, resource loss, resource loss spirals, and secondary stresses can negatively affect mental health and psychological functioning. Resource loss spirals can become pervasive and extensive. The enduring economic impact can have significant impact on the ability of individuals and communities to function (Padhy et al. 2015).

In the aftermath of an extreme weather event, it is essential that prompt action is taken to meet the basic survival needs of survivors, to prevent or minimize the development of resource loss spirals and secondary stressors, and to promote normalcy within families and communities. Demands on coping may intensify when recovery is prolonged or when an extreme weather event continues for an extended time, as in the case of some heatwaves and droughts (Padhy et al. 2015; Shelby and Tredinnick 1995). Coping resources are especially challenged when a location experiences multiple hurricane threats within the same season.

In the wake of three recent storms linked to climate change—Hurricane Katrina in the United States, Super Typhoon Haiyan in the Philippines, and Cyclone Winston in Fiji—a sizable percentage of persons reported posttraumatic stress symptomatology. Posttraumatic stress symptoms were associated with life threat and loss of resources. Living in a community ravaged by the event can present constant reminders of the terror experienced and the immense challenges, and people may have new fears about their safety (e.g., vulnerability to new disasters, looting, meeting basic survival needs). As such, people may have difficulty maintaining a positive outlook and question the degree to which they can cope with the situation. Similar findings have been observed following other natural disasters (Chan and Rhodes 2014; Hobfoll 2012).

The mental health needs of persons in affected communities can overwhelm mental health service providers. In many communities, the number of people needing services can greatly outnumber the number of mental health professionals available, and many people may not have insurance or funding for services. For example, Wang et al. (2008) report that "Many Hurricane Katrina survivors with mental disorders experienced unmet treatment needs, including frequent disruptions of existing care and widespread failure to initiate treatment for new-onset disorders" (p. 34). Research is needed to examine how to most efficiently provide mental health services on a large scale to individuals living in communities ravaged by disaster.

Programs that teach stress management techniques and coping strategies can help people process and resolve issues and enhance personal characteristic and energy resources. These programs should consider cultural variables, including local knowledge concerning disaster response and community functioning. Community officials and others should facilitate self-help efforts, such that people work together in their neighborhoods and communities to help one another. Self-help efforts can help reestablish feelings of control and self-worth (that may have been compromised as a result of the event), counter feelings of helplessness, and strengthen and renew social support networks (Sattler et al. 2006). Research is needed to examine the effectiveness of these programs.

Strong winds and heavy rain associated with hurricanes can lead to crop failure. In the wake of Cyclone Winston in Fiji, there was discussion concerning the types of crops that may be able to withstand extensive wind and rain, and how to prepare the crops in order to minimize damage. For example, prior to a hurricane, farmers may prepare cassava plants by chopping their tops and prepare banana plants by cutting the leaves; doing so may increase the chance that the plant will survive. Having diverse and resilient crops may be essential for self-sufficiency and sustainability in many communities (Thaman 1982; Thaman et al. 2010).

After Cyclone Winston in Fiji and Super Typhoon Haiyan in the Philippines, there were concerns about the degree to which people were informed about the storm and understood the magnitude of the situation. Research is needed to examine how best to notify citizens in advance of an impending extreme weather event (Lindell and Perry 2012; Sattler and Marshall 2002).

Limitations

This chapter presents select experiences and mental health issues associated with exposure to extreme weather events. Research is needed to examine cultural expression of distress, to evaluate appropriate treatment for psychological reactions across cultures, and to assess these treatment approaches (Marsella 2010; Sattler et al. 2014). Research also is needed to examine how cultural knowledge concerning extreme weather events influences recognition of and preparation for the event and coping responses.

Prospects for Projects that Educate About Extreme Weather Events

The studies reviewed in this chapter highlight the need to provide education concerning how to prepare, minimize damage and destruction, and promote mental health. Education may be provided in schools, through the media, educational centers, and other methods, including museums (e.g., Louisiana State Museum, the Pacific Tsunami Museum in Hawaii, and earthquake museums in Japan). The author of this chapter developed the International Tsunami Museum in Thailand after the Indian Ocean Tsunami—the fourth deadliest disaster in history—

devastated coastal communities. The museum provided information concerning how to recognize a tsunami threat and how to evacuate, the conditions that create a tsunami, the environmental causes of the tsunami, warning signs, the world response to assist with the recovery, and actions to take in the event of a future tsunami. Each exhibit conveyed hope and resilience and showed the vitality of the human spirit. Visitor surveys indicated that the museum effectively educated visitors about these topics, decreased fears, and helped survivors reestablish feelings of control, predictability, safety, and trust (Sattler 2010). To promote education in schools along the coast in southern Thailand, an educational video based on information in the museum was produced by the present author and it was distributed at no cost to schools along the coast (Sattler et al. 2014). This approach can be adapted to provide information about the connection between climate change and extreme weather events and address the specific needs of the community.

For example, after Cyclone Winston damaged crops in Fiji, there was discussion concerning sustainable agriculture. Exhibits and information may illustrate the value of "non-perishable or more resistant staples such as taro, yam and sweet potato, or to wild yams, or other terrestrial and marine foodstuffs" for communities in the Pacific, in the event that other staples are destroyed during an extreme weather event (Thaman 1982). Exhibits also could show how people have provided assistance to one another, including traditional trading networks used by Pacific Islanders to distribute food and materials to rebuild, diversify crop production, and provide social support.

Finally, museums, educational centers, and schools could develop ways to educate people how daily choices and actions can contribute to climate change. Exhibits could offer visitors immediate feedback about how actions such as driving a car, using an appliance or television, using electricity contribute, or eating various food items contribute carbon dioxide to the atmosphere. Research is needed to develop and assess such educational programs.

References

American Psychological Association. (2009). *Psychology and global climate change: A report of the American Psychological Association task force on the interface between psychology and global climate change*. Washington, DC: American Psychological Association.

American Psychiatric Association. (2013). Posttraumatic stress disorder. http://www.dsm5.org/Documents/PTSD%20Fact%20Sheet.pdf.

Antonovsky, A. (1987). *Unraveling the mystery of health*. San Francisco: Jossey-Bass.

Berry, H. L., Hogan, A., Owen, J., Rickwood, D., & Frager, L. (2011). Climate change and farmer's mental health: Risks and response. *Asia-Pacific Journal of Public Health, 23*, 1295–1325.

Chan, C. S., & Rhodes, J. E. (2014). Measuring exposure in Hurricane Katrina: A meta-analysis and an integrative data analysis. *PLoS ONE, 9*(4), e92899.

Di Liberto, T. (2016, February 23). Tropical cyclone Winston causes devastation in Fiji, a tropical paradise. Climate.gov. Retrieved April 4, 2016, from https://www.climate.gov/news-features/event-tracker/tropical-cyclone-winston-causes-devastation-fiji-tropical-paradise.

Gardner, G., & Stern, P. (2002). *Environmental problems and human behavior* (2nd ed.). Boston: Pearson Custom Publishing.

Godwin, E. E., Foster, V. A., & Keefe, E. P. (2012). Hurricane Katrina families: Social class and the family in trauma recovery. *The Family Journal: Counseling and Therapy for Couples and Families, 21*, 15–27.

Groleau, J. M., Calhoun, L. G., Cann, A., & Tedeschi, R. G. (2012). The role of centrality of events in posttraumatic distress and posttraumatic growth. In *Psychological trauma: Theory, research, practice, and policy*. Advance Online Publication.

Hobfoll, S. E. (2012). Conservation of resources and disaster in cultural context: The caravans and passageways for resources. *Psychiatry, 73*, 227–233.

Johnson, I. (2016, April 27). UK flooding driven by soaring temperatures in Greenland, study says. *The independent*. http://www.independent.co.uk/news/science/greenland-temperature-global-warming-climate-change-floods-britain-uk-a7002186.html.

Kaiser, C. F., Sattler, D. N., Bellack, D. R., & Dersin, J. (1996). A conservation of resources approach to a natural disaster: Sense of coherence and psychological distress. *Journal of Social Behavior and Personality, 11*, 459–476.

Kellens, W., Terpstra, T., & De Maeyer, P. (2013). Perception and communication of flood risks: A systemic review of empirical research. *Risk Analysis, 33*, 24–49.

Kessler, R. C., Galea, S., Gruber, M. J., Sampson, N. A., Ursano, R. J., & Wessely, S. (2008). Trends in mental illness and suicidality after Hurricane Katrina. *Molecular Psychiatry, 13*, 374–384.

Lindell, M. K., & Perry, R. W. (1992). *Behavioral foundations of community emergency planning*. Washington, DC: Hemisphere Publishing.

Lindell, M. K., & Perry, R. W. (2012). The protective action decision model: Theoretical modifications and additional evidence. *Risk Analysis, 32*, 616–632.

Linley, P. A., & Joseph, S. (2004). Positive change following trauma and adversity: A review. *Journal of Traumatic Stress, 17*, 11–21.

Marsella, A. J. (2010). Ethnocultural aspects of PTSD: An overview of concepts, issues, and treatments. *Traumatology, 16*, 17–26.

Mullen, J. (2013, November 8). Super Typhoon Haiyan, one of the strongest storms ever, hits central Philippines. CNN. Retrieved March 15, 2014, from http://www.cnn.com/2013/11/07/world/asia/philippines-typhoon-haiyan.

National Disaster Risk Reduction and Management Council (2014, April 3). Effects of Typhoon "Yolanda" (Haiyan), SitRep No. 108, Quezon City, Philippines. Retrieved November 25, 2014, from http://www.ndrrmc.gov.ph/attachments/article/1329/Effects_of_Typhoon_YOLANDA_(HAIYAN)_SitRep_No_108_03APR2014.pdf.

Padhy, S. K., Sarkar, S., Panigrahi, M., & Paul, S. (2015). Mental health effects of climate change. *Indian Journal of Occupational and Environmental Medicine, 19*, 3–7.

Sattler, D. N. (2001). Repeated exposure to hurricanes and willingness to evacuate: Implications from the hurricane floyd experience. In J. R. Maiolo, J. C. Whitehead, M. McGee, L. King, & J. C. Johnson (Eds.), *In the aftermath of hurricane floyd: Recovery in the coastal plain* (pp. 69–76). Wilmington, NC: Coastal Carolina Press.

Sattler, D. N. (2010, March). *International Tsunami Museum in Thailand*. Invited talk presented at the International Forum on Telling Live Lessons from Disasters Symposium, Kobe, Japan.

Sattler, D. N., & Marshall, A. (2002). Hurricane preparedness: Improving television hurricane watch and warning graphics. *International Journal of Mass Emergencies and Disasters, 20*, 41–49.

Sattler, D. N., Adams, M., & Watts, B. (1995a). Effects of personal experience on judgments about natural disasters. *Journal of Social Behavior and Personality, 10*, 891–898.

Sattler, D. N., Sattler, J. M., Kaiser, C., Hamby, B. A., Adams, M., Love, L., et al. (1995b). Hurricane Andrew: Psychological distress among shelter victims. *International Journal of Stress Management, 2*, 133–143.

Sattler, D. N., Kaiser, C. F., & Hittner, J. (2000). Disaster preparedness: Relationship between prior experience, personal characteristics, and psychological distress. *Journal of Applied Social Psychology, 30*, 1398–1420.

Sattler, D. N., Preston, A., Kaiser, C. F., Olivera, V. E., Valdez, J., & Schlueter, S. (2002). Hurricane Georges: A cross-national study examining preparedness, resource loss, and psychological distress in the U. S. Virgin Islands, Puerto Rico, Dominican Republic, and the United States. *Journal of Traumatic Stress, 15*, 339–350.

Sattler, D. N., Glower de Alvarado, A. M., Blandon de Castro, N., Van Male, R., Zetino, A. M., & Vega, R. (2006). El Salvador earthquakes: Relationships among acute stress disorder symptoms, traumatic event exposure, and resource loss. *Journal of Traumatic Stress, 19*, 879–893.

Sattler, D. N., Assanangkornchai, S., Moller, A., Kesavatana-Dohrs, W., & Graham, J. M. (2014). Indian Ocean Tsunami: Relationships among posttraumatic stress, posttraumatic growth, resource loss, and social support at three and fifteen months. *Journal of Trauma and Dissociation: Systemic Trauma Special Issue, 15*, 219–239.

Sattler, D. N., Atienza, R., & Graham, J. (2016). Climate change risk perceptions among typhoon survivors: Relationships among resource loss, posttraumatic stress, coping, and posttraumatic growth. Manuscript Submitted for Publication.

Sattler, D. N., Claramita, M., & Muskavage, B. (2016). Disasters in Indonesia: Relationships among resource loss, posttraumatic stress, somatic problems, and posttraumatic growth. Manuscript Submitted for Publication.

Sattler, D. N., Whippy, A., Johnson, J., Vucago, N., & Graham, J. (2016, July). *Cyclone Winston in Fiji: Climate change risk perceptions, posttraumatic stress, coping, and resilience among survivors*. Paper presented at the Symposium on Climate Change Adaptation in the Pacific Region, Lautoka, Fiji.

Shelby, J. S., & Tredinnick, M. G. (1995). Crisis intervention with survivors of natural disaster: Lessons from Hurricane Andrew. *Journal of Counseling and Development, 73*, 491–497.

Shukla, J. (2013). *Extreme weather events and mental health: Tackling the psychosocial challenge* (p. 127365). Article: ISRN Public Health.

Tally, S., Levack, A., Sarkin, A. J., Gilmer, T., & Groessl, E. G. (2013). The impact of the San Diego wildfires on a general mental health population residing in evacuation areas. *Administration Policy and Mental Health, 40*, 348–354.

Thaman, R. (1982). Hurricane Isaac and Tonga: A natural or cultural disaster? *Review, 3*(8), 22–35.

Thaman, R. R., Puia, T., Tongabaea, W., Namona, A., & Fong, T. (2010). Marine biodiversity and ethnobiodiversity of Bellona (Mungiki) Island, Solomon Islands. *Singapore Journal of Tropical Geography, 31*, 70–84.

Union of Concerned Scientists. (2016). Heavy flooding and global warming: Is there a connection? http://www.ucsusa.org/global_warming/science_and_impacts/impacts/heavy-flooding-and-global-warming.html#.VxWbjE32aMk.

United Nations. (2014, November 1). *Intergovernmental panel on climate change, Climate change 2014 synthesis report*. http://ipcc.ch.

Updegraff, J. A., & Taylor, S. E. (2000). From vulnerability to growth: Positive and negative effects of stressful life events. In J. H. Harvey & E. D. Miller (Eds.), *Loss and trauma: General and relationship perspectives* (pp. 3–28). Philadelphia, PA: Brunner.

Varandani, S. (2016, March 8). Cyclone Winston: Fiji's estimated cost of damages exceeds $470M, 10% of the island nation's total GDP. *International Business Times*. Retrieved March 29, 2016, from http://www.ibtimes.com/cyclone-winston-fijis-estimated-cost-damages-exceeds-470m-10-island-nations-total-gdp-2332151.

Viscusi, W., & Zeckhauser, R. (2015). The relative weights of direct and indirect experiences in the formation of environmental risk beliefs. *Risk Analysis, 35*, 318–331.

Wang, P. S., Gruber, M. J., Powers, R. E., Schoenbaum, M., Speier, A. H., Wells, K. B., et al. (2008). Disruption of existing mental health treatments and failure to initiate new treatment after Hurricane Katrina. *American Journal of Psychiatry, 165*, 34–41.

Weisath, L. (1993). Disasters: Psychological and psychiatric aspects. In L. Goldberger & S. Breznitz (Eds.), *Handbook of stress: Theoretical and clinical aspects* (2nd ed., pp. 591–616). New York: Free Press.

World Bank. (2013). Turn down the heat: Climate extremes, regional impacts, and the case for resilience. *International Bank for Reconstruction and Development*. http://www-wds.worldbank.org/external/default/WDSContentServer/WDSP/IB/2015/07/17/090224b0828c0f20/1_0/Rendered/PDF/Turn0down0the00lience000full0report.pdf.

Yulsman, T. (2016, February 22). How Winston became Earth's strongest Southern hemisphere storm in recorded history. *Discovery Magazine*. Retrieved March 30, 2016, from http://blogs.discovermagazine.com/imageo/2016/02/22/how-winston-became-strongest-southern-hemisphere-storm-on-record/#.VvsINU32aot.

Zwiebach, L., Rhodes, J., & Roemer, L. (2010). Resource loss, resource gain, and mental health among survivors of Hurricane Katrina. *Journal of Traumatic Stress, 23*, 751–758.

Chapter 5
Preparing for Better Livelihoods, Health and Well-Being—A Key to Climate Change Adaptation

Peni Hausia Havea, Sarah L. Hemstock and Helene Jacot Des Combes

Introduction

Climate-change adaptation and mitigation encompass a broad range of human policies and activities primarily intended to improve the sustainable development capacity of communities whose livelihoods, health and well-being are under threat from climate change in order to improve future resilience (Conway and Mustelin 2014; Georgeson et al. 2016; Smajgl et al. 2015; Tanaka et al. 2015; Convertino et al. 2013; Ipcc 2014; Ferrario et al. 2014). In 2013, this project investigated five coastal communities in Tongatapu, Tonga: Kanokupolu, ‘Ahau, Tukutonga, Popua and Manuka (see Fig. 5.1 on p. xx), who faced substantial negative impacts of climate change such as rising sea levels, temperature rise, rainfall pattern variations, extreme weather events (EWEs), droughts and seasonal shift (Havea et al. 2016).

Community perceptions revealed that climate change has affected livelihoods in the following ways: decreasing crop yields, reduced land productivity and horticultural production; coastal erosion, collapse of marine resources and loss of traditional fishing practices; damage to infrastructure, settlements and the built environment, including loss of property; disruption of food production and water supply; employment impacts and family income revenue disorder (especially for those who relied on their jobs and small-scale business to earn money); impact on

P.H. Havea (✉) · H.J.D. Combes
PaCE-SD, The University of the South Pacific, Suva, Fiji
e-mail: ilaisiaimoana@yahoo.com

H.J.D. Combes
e-mail: descombe_h@usp.ac.fj

S.L. Hemstock
The Pacific Community, 3rd Floor, Lotus Building, SPC, Nabua, Fiji
e-mail: sarahh@spc.int

W. Leal Filho (ed.), *Climate Change Adaptation in Pacific Countries*,
Climate Change Management, DOI 10.1007/978-3-319-50094-2_5

Fig. 5.1 Location of the studied communities and sample size. Map of Tongatapu Island: Kanokupolu, 'Ahau, Tukutonga, Popua and Manuka Created by Peni Hausia Havea (2015)

training and education (loss of school days, seminars or training and damaged school property); and financial impact on government investment since the government cannot fund all adaptation projects to reduce the impact of climate change and these communities depend largely on international donors (Havea et al. 2016).

As for the perceived impacts of climate change on human health, participants indirectly linked several diseases to climate change: non-communicable disease (NCD)—asthma, influenza, pneumonia, shortness of breath, inter alia, and communicable disease (CD)—STIs (sexually transmitted infections), typhoid fever, leptospirosis, dengue fever and tinea. However, consequences for mental health and human well-being were also identified. The following perceived impacts of climate change on well-being were revealed: worry, stress and anxiety, physical impacts (e.g. coastal erosion and death) and spiritual impacts (e.g. church services cancelled). Because climate change affected Tongans aged 15–75 livelihoods, health, and well-being negatively, in Kanokupolu, ‘Ahau, Tukutonga, Popua and Manuka, an integrated mixed method approach has been used, to model these perceived impacts and the adaptation measures (Havea et al. 2016).

In this dynamic model, the study explored climate change impacts and adaptation. It was developed from two types of data: quantitative data and qualitative data, to contribute to the eliminations of gaps and a lack of published research in this area. The data was modelled using an analytical approach: ‘dynamic modelling’, to help solve impact of climate change on livelihood, health and well-being of people in these communities (QSR International 2014; Wong 2008; Welsh 2002; Wiltshier 2011; Bazeley 2007).

The model was built by carefully examining the dynamics that linked impacts and perceived impacts to adaptation grounded in the data. The result of this noding analysis was an ‘impact and adaptation model’, which enabled the emergence and identification of important lessons learnt from perceptions of community experiences of climate change and adaptation. By sharing this information and model of adaptation from Pacific experiences, it is expected that the model can be replicated and contextualized for other communities.

Method

Concurrent Convergence Parallel Triangulation Design

This study used an integrated mixed method design that was model generating. Data from both quantitative and qualitative studies were converged and then merged. For the quantitative study, information was collected on the impact of climate change on livelihoods, health and well-being among Tongans age 15–75 via a self-administered questionnaire. The survey was administered by local villagers, who were trained by the researcher in the local language in order to facilitate the data collection processes. This ensured information and perception bias were minimized.

Concurrent with the quantitative study, qualitative studies were also implemented. Information was gathered via: in-depth interviews (IDI), information interview (II) and Focus Group Discussion (FGD). Documents from the government, NGOs, international and regional organisations were also collected for supplementary information. After the completion of the data collection, data were then translated into English and analysed using a Computer-Assisted Quantitative and Qualitative Data Analysis Software Packaging (CAQQDASP), by the researcher for validity and reliability (Havea et al. 2016; Creswell and Plano Clark 2007, 2011).

This type of mixed method design is called “Concurrent Convergence Parallel Triangulation Design” because data is collected over the same timeframe (concurrent), and merged using convergence design, then it is equally weighted (parallel) and multiple methods were employed to examine same issues (triangulation). All efforts have been put into minimizing bias in this study. The result is that a model was formulated: ‘impact and adaptation’, that would help these communities make more effective choices in their sustainable efforts in future (Havea et al. 2016; Creswell and Plano Clark 2007, 2011).

Results

Sample Characteristic

A random sampling framework from the Department of the Statistics of the Government of Tonga (GOT) was used to recruit 460 participants aged 15–75 from five coastal communities in Tongatapu, Tonga: Kanokupolu—provided 8.7% (40), ‘Ahau—11.3% (52), Tukutonga—16.7% (77), Popua—53.5% (246) and Manuka—9.8% (45), (see Fig. 5.1). By gender and age, the study recruited more women (53%) than men (47%). The studied population were adults with a median age of 25–34 (15.7%), (see Fig. 5.2). This study has also benefitted greatly from the National Capacity Self-Assessment (NCSA) project to review the Multilateral Environmental Agreement (MEA) on three Rio conventions: United Nation Convention on Biological Diversity, United Nation Framework Convention on Climate Change and

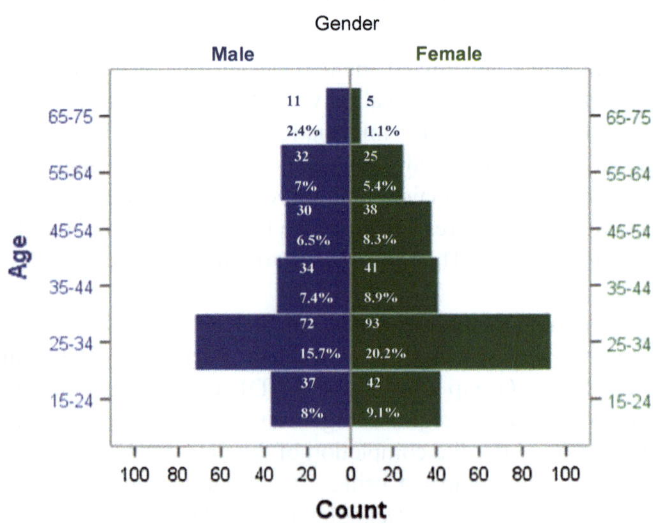

Fig. 5.2 Sample population pyramid: gender and age

United Nation Convention to Combat Desertification, using 35 cases of directors, managers, finance analysts, environmentalists, urban planners, project coordinators, legal advisors, inter alia in Tonga (Havea et al. 2016; Havea 2014).

Themes that Emerged in Survey, Interviews and FGD

Several themes emerged from careful review of the transcripts and documents. Five major themes were identified and described for their climate change impact, mitigation and adaptation importance. These themes were related to the participants perceived negative impact on their livelihoods, health and well-being and adaptation. These five themes are: (1) mitigation policies, (2) modulating factors, (3) impacting factors, (4) impact on livelihood, health and well-being and (5) climate change adaptation. The themes were used to construct the impact and adaptation model. Based on the theory and ideas that were grounded in the data, such as survey, interviews and FGD, codes were generated into themes and then label as project items and linked them into one another (QSR International 2014; Welsh 2002; Wiltshier 2011; Wong 2008; Zamawe 2015; Ozkan 2004; Bazeley 2007). The result of that is described in the discussion below.

Discussion

Climate Change Impact and Adaptation Model

The integrated, mixed methods approach, allowed the construction of this “impact and adaptation model” in a single study. This model is divided into five main categories: mitigation policies, modulating factors, impact, impact on livelihoods, health and well-being and climate change adaptation. Each category is mandated to prepare these communities for a better future (see Fig. 5.3). Firstly, in mitigation policies, this paper addressed human-induced activities and natural variability influences climate for these communities by focussing on reducing carbon footprint, reducing deforestation, forest preservation, improving waste management, renewable energy use and improving energy efficiency, and air pollution control. Secondly, in modulation factors, issues related to the contributing factors that would make these communities more vulnerable and highly at risk to be affected by climate change impact factors were addressed. By addressing these modulating influences that affect these communities, is possible to ascertain the key factors that make adaptation behaviours and activities more sustainable for them in the future.

These modulating factors include, but are not limited to the climate change impact, pre-existing livelihood, health and well-being conditions and care available, country infrastructure, the rate of deforestation, access to renewable energy and energy efficiency, environmental pollution and population growth and density.

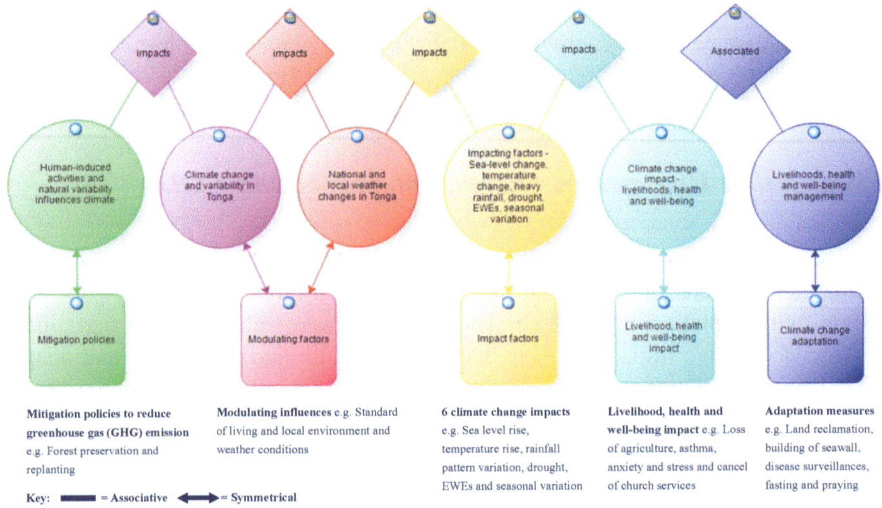

Fig. 5.3 Impact and adaptation model

Thirdly, for the climate change impact, this paper focussed particularly on sea level rise, temperature rise, EWEs, drought, rainfall pattern variations and seasonal shift. Furthermore, these six impacts were used to measure the effects of climate change on livelihoods, health and well-being. For livelihood impacts, different sectors affected negatively by climate change in these communities were identified as: livestock, women group activities (e.g. weaving & koka'anga (tapa making), traditional textiles), agriculture, employment, coastal erosion, sand mining, small-scale businesses, deforestation, education, remittances and fisheries.

For the impacts on health the model identifies a myriad of perceived climate-change related diseases: STIs, typhoid fever, boils, heart condition, general malaise, depression, anxiety, cancer, diabetes, cough, folliculitis, common cold, measles, filariasis, epilepsy, skeletal pain, asthma, bronchitis, chest pain, abdominal pain, anaphylaxis, leptospirosis, menorrhagia, gastroenteritis, pneumonia, migraine, influenza, shortness of breath, emesis, urticaria, fever, rhinorrhoea, poliomyelitis, back pain, sore throat, CVA (cerebral vascular accident), CVD (cerebral vascular disease), cramps, pruritus, hypertension, conjunctivitis and tinea. The well-being impacts, cover mental, physical and spiritual impacts. The mental impacts correspond to, health impact, communication impact and increasing the level of worry. The physical impacts include, livelihoods impact, governmental impact, hygienic impact, injuries, dying (e.g. the end of life and transition to death—semi-conscious to completely unconscious) and death. Then disruption of the church services and religious activities accounted for the spiritual impact.

The last category of the model was the climate change adaptation. The adaptation was divided into 3 main parts: livelihood adaptation, health adaptation and well-being adaptation. The additional feature of adaptation is the policy proposal.

Based on the sample, for the livelihood adaptation the following were recommended, but may not be limited to:

- Continue supporting of the community-based policing programme
- Ensure food maintenance of crop and vegetable based adaptation
- Establishment of a community-based multipurpose centre hall
- Improve community sanitation adaptation programme
- Improve road map to cater for communities during evacuation and emergency
- Initiate proposal for beach resorts with hotels and motels
- Initiate a proposal for national park adaptation programme
- Increase marine-based conservation special management area (SMA)
- Increase information sharing and public awareness
- Installation of cyclone and tsunami warning siren
- Link waste management act to climate change adaptation
- Locate community-based high point spot for relocation in advanced
- Maintenance and repairing of main road
- Nationalise building and maintenance of seawall programme
- Proposal for community-based landfill and reclamation programme
- Reinforce policy on prevention and control burning of rubbish
- Reinforce policy on national environmental protection
- Reinforce policy on illegal fishing
- Reinforce policy on illegal sand mining
- Revisit chemical imported usage protocol for Tonga
- Revisit community-based fishing boat and programme
- Revisit evacuation and emergency plan to suit community-based needs
- Revisit recovery plan to include financial aid for affected communities
- Support community-based fund raising initiatives
- Community participant in government decision making
- Community-based reforestation initiatives programme of mangroves and other plant species
- District officer proposal allocated for Tukutonga
- Electricity connection proposal for Tukutonga
- House repairing codes guidelines reinforcement
- Land property ownership right review for Tukutonga and Popua
- Large water tank aid programme
- Policy proposal on climigration and relocation further inland
- Small-scale home farming initiatives.

For the health adaptation the following were recommended but may not be limited to: increase of community-based physical activity programme on public health promotion, encourage people to utilise modern medical facility and use of rain protection while on duty. For the well-being adaptation the following were recommended: nationalise fasting day for worshipping god about climate change, and nationalise message on climate change and adaptation through church. Then, on policy proposal, this paper recommended the following policies:

- Establishment of a community-based adaption multipurpose centre
- Evacuation plan enhancement—road mapping architecture will improve with emergency disaster kits valid for a day
- Initiate a policy for climigration relocation for Tonga
- Land ownership review for Tukutonga
- Mark the global and national climate change campaign day in Tonga
- Nationalise survey on livelihood, health and well-being into the census data
- Policy to support community-based initiatives on landfill and reclamation programme
- Policy to support community-based seawall construction and maintenance
- Policy to support community-based reforestation initiatives of mangroves and other plant species
- Reinforcement of policy on illegal sand mining and fishing
- Reinforcement of policy on prevention and control burning of rubbish in the community
- Review of policy on uninhabited land tenure for Popua according to the government protocol
- Strengthening community-based policing programme
- Tsunami and cyclone siren installed into the high vulnerable and most at risk communities.

In support of these adaptation approaches, here is a cross-section sample of the transcripts that had guided this paper in this model. A high official from the Ministry of Agriculture in Tonga, with the details about deforestation and reforestation initiatives, stated explicitly: "Another problem is that there are too many deforestations; that is the impact of climate change on commercial agriculture. We don't have any reforestation programs to address this issue in Tonga." And another man from Manuka proposed the same policy by stating: "I think the only policy I would like to propose for the people of this community is to ban them from cutting off the trees on the beach." Then another man from Kanokupolu clearly stated his opinion by saying: "But when we talk about climate change, it is clearly driven by human activities. This means that human destroyed the trees that were support and prevent sea from coming in land and cause coastal erosion. Today, however, we have tried to build seawalls to prevent sea level rise from inundation."

Then another woman from Kanokupolu backed this up by saying: "And sea level rise is one of the major problems that affect us here in this community. Mainly because the rising of sea level is above the land benchmark in addition to the fact that we are located in the narrowest part of Tongatapu. This impact of sea level rise eroded the land that usually serve the purpose of agriculture in the past. As a result, our land is getting smaller the same way it diminishes our sources of livelihoods, especially the eastern side of Kanokupolu. In the past, these areas have been used for bush allotment ('api tukuhau) but as sea level rise affected these areas, we now shift to the western side of Kanokupolu to do agriculture to sustain our livelihoods, maintain health and restore well-being."

Preparing Kanokupolu, 'Ahau, Tukutonga, Popua and Manuka to Adapt

To prepare these communities for better livelihoods, health and well-being in the future, this paper concluded five most important climate change adaptation actions: (1) land reclamation, (2) reforestation, (3) building of seawall, (4) building of a multipurpose centre, (5) preaching about climate change and nationalising a national fasting day for Tonga.

Limitations

This study had some limitations in building this model. Firstly, the translation from Tongan into English. The collection of data in communities was conducted in Tongan. To minimise the risk of varied interpretations affecting the model, the researcher completed all translations and had interpretations varied by dual language speakers in the communities studied. Second, the model relied on the participant's perception of climate change. To prevent perception bias from affecting the model, the study used parallel triangulation design.

Audience Benefit from the Model

This model will be useful to any affected communities in Tonga and the region to plan adaptation for the future, the governments, NGOs, international and regional organizations as well as those who are planning adaptation projects in the region. The graduate students who have some or advance experienced with both quantitative and qualitative research will find this model, useful for designing their first modelling as well. Writers of mixed method research, however, may see this model as state-of-the-art ideas.

Conclusion

Lesson Learnt

In conclusion, four important lessons learnt from the dynamic modelling of livelihoods, health and well-being. They are listed as follow: A key to climate change adaptation, model production, god scriptures fulfilled, and community-based adaptation initiative.

Better Livelihood, Health and Well-Being—A Key to Climate Change Adaptation

This study is the first to modelling perceived climate change impact and adaptation in a single study in Tonga. It brought together information from five communities on specific mitigation and adaptation options, including environmentally sound technologies and infrastructure, sustainable livelihoods, health behaviour, lifestyle choices and well-being. It describes common contributing factors and constraints, and policy approaches, finance and technology on which effective response measures depend, in order to give these communities, the best possible options. This model integrates responses and links mitigation and adaptation with other societal objectives so that these communities will become prosperous in their lives in the future.

Most importantly, as an indirect benefit of this study, some of these mitigation and adaptation approaches have been implemented in Tonga recently. The result is that the people are now more aware about the impact of climate change on their livelihoods, health and well-being and how to adapt to these impacts. For instance, the installation of an E-Disaster Early Warning System that detects tsunamis and cyclones in Kanokupolu, and the rejuvenation of the Ancient Sia Heulupe Heritage site and the construction of a new community Evacuation Centre for Popua, inter alia, shows Tonga's commitment to Resilient Tonga 2035 is real (Matangi Tonga Online 2016a, 2015, 2014, 2016b). Because this model supported the community development goal for Tonga, this paper argued that it should be adapted by the GOT, NGOs, international and regional organisation in order to build Tonga's sustainable development capacity, to become a more resilient nation in the future.

Model Production

In terms of modelling, another important lesson learnt, is that the model can be replicated and reproduced by any stakeholders with an interested in modelling. In order to be effective in modelling, researcher must sketch or created the project items right from the beginning of the study—the research design, so that later on, model can help to show links or relationships between concepts, ideas, theories or items in the project (QSR International 2014; Welsh 2002; Wiltshier 2011; Wong 2008).

God Scriptures Fulfilled

The study also learnt these communities have prioritised fulfilling the Christian God's scriptures as the most reliable adaptation strategy. This is a very interesting finding regarding Tonga, because the sample—of 460—were all Christians, regardless of their denominations. Tonga is classified as a Christian nation, and the Christian faith plays an important role in the Tongan culture. The model has identified religious doctrine is perceived to play a crucial role in adaptation approaches as well. This could be very important for Tongan responses to climate change impacts because church politics are very powerful tools (Havea et al. 2016).

Community-Based Adaptation

Finally, adaptation should be community-based driven like in Kanokupolu, where the community have raised their own finance and built their own seawall, which seems to be efficient and sustainable.

Future Research

To reach a more definitive conclusion, future work must test out this model—see how cost-effective and sustainable it would be for the region—and develop a better way to implement this programme in another small Pacific island nation or low-lying atoll with the same risk and vulnerabilities as Tonga. Then, there are new dimensions open up for health research and well-being research. This could be correlation studies, psychological impact or spiritual impact studies inter alia.

Acknowledgements This paper begins by thanking the people of Kanokupolu, 'Ahau, Tukutonga, Popua and Manuka for taking their time out of their busy schedule in order to fully participate in this study. Special thank, also goes, to the GOT. Moreover, the project would like to thank personally to PaCE-SD, who monitored closely, and make it a reality. Finally, this study held a deep indebted to the Research Office at USP, who funded this project right from the beginning to successfully completion.

References

Bazeley, P. (2007). *Qualitative data analysis with Nvivo*. London: SAGE.

Convertino, M., Foran, C. M., Keisler, J. M., Scarlett, L., Loschiavo, A., Kiker, G. A., et al. (2013). Enhanced adaptive management: Integrating decision analysis, scenario analysis and environmental modeling for the everglades. *Scientific Reports, 3,* 2922.

Conway, D., & Mustelin, J. (2014). Strategies for improving adaptation practice in developing countries. *Nature Climate Change, 4,* 339–342.

Creswell, J. W., & Plano Clark, V. L. (2007). *Designing and conducting mixed methods research.* Thousand Oaks, USA: SAGE.

Creswell, J. W., & Plano Clark, V. L. (2011). *Designing and conducting mixed methods research.* Thousand Oaks, USA: SAGE.

Ferrario, F., Beck, M. W., Storlazzi, C. D., Micheli, F., Shepard, C. C., & Airoldi, L. (2014). The effectiveness of coral reefs for coastal hazard risk reduction and adaptation. *Nature Communications, 5.* doi:10.1038/ncomms4794.

Georgeson, L., Maslin, M., Poessinouw, M. & Howard, S. (2016). Adaptation responses to climate change differ between global megacities. *Nature Climate Change,* Advance online publication. doi:10.1038/nclimate2944.

Havea, P. H. (2014). *Tonga NCSA capacity development programme and implementation action plan, 2014: A mixed methods review.* Fiji: UN-MCO UNDP.

Havea, P. H., De Combes, H. J., Hemstock, S. L., Gee, K., Han, D., Khan, M. G. M. et al. (2016). *Climate change impact on livelihood, health and well-being: A mixed method approach.* Ph.D. thesis, The University of the South Pacific.

Ipcc. (2014). *Climate change 2014: Synthesis report. Contribution of working groups I, II and III to the fifth assessment report of the intergovernmental panel on climate change (Core writing team).* In R. K. Pachauri, & L. A. Meyer (Eds.). Geneva: IPCC.

Matangi Tonga Online. (2014). *Early warning tsunami system for vulnerable villages.* Tonga: Matangi Tonga Online. Retrieved April 15, 2016, from http://matangitonga.to/2014/03/22/early-warning-tsunami-system-vulnerable-villages.

Matangi Tonga Online. (2015). *PM says govt will "cement" Sia Heulupe heritage site.* Tonga: Matangi Tonga Online. Retrieved April 15, 2016, from http://matangitonga.to/2015/09/17/pm-says-govt-will-cement-sia-heulupe-heritage-site.

Matangi Tonga Online. (2016a). *Cyclone resilient evacuation centre for Popua.* Tonga: Matangi Tonga Online. Retrieved April 15, 2016, from http://matangitonga.to/2016/02/24/cyclone-resilient-evacuation-centre-popua.

Matangi Tonga Online. (2016b). *Tonga policy aims to build climate change resilience by 2035.* Tonga: Matangi Tonga Online. Retrieved April 17, 2016, from http://matangitonga.to/2016/03/29/tonga-policy-aims-build-climate-change-resilience-2035.

Ozkan, B. C. (2004). Using NVivo to analyze qualitative classroom data on constructivist learning environments. *The Qualitative Report, 9,* 589–603.

Qsr International. (2014). *Nvivo 10 for windows getting started.* Australia: QSR International Pty Ltd.

Smajgl, A., Toan, T. Q., Nhan, D. K., Ward, J., Trung, N. H., Tri, L. Q., et al. (2015). Responding to rising sea levels in the Mekong Delta. *Nature Climate Change, 5,* 167–174.

Tanaka, A., Takahashi, K., Masutomi, Y., Hanasaki, N., Hijioka, Y., Shiogama, H., et al. (2015). Adaptation pathways of global wheat production: Importance of strategic adaptation to climate change. *Scientific Reports, 5,* 14312.

Welsh, E. (2002). Dealing with data: Using NVivo in the qualitative data analysis process. *Forum Qualitative Social Research Sozialforschung,* 3.

Wiltshier, F. (2011). Researching with NVivo 8. *Forum Qualitative Social Research Sozialforschung,* 12.

Wong, L. P. (2008). Data analysis in qualitative research: A brief guide to using Nvivo. *Journal of the Academy of Family Physicians of Malaysia, 3,* 14–20.

Zamawe, F. C. (2015). The implication of using NVivo software in qualitative data analysis: Evidence-based reflections. *Malawi Medical Journal, 27,* 13–15.

Author Biographies

Mr Peni Hausia Havea is a PhD candidate in climate change who holds a Master degree in Public Health majoring in International Health from the University of Melbourne, Australia. He has been awarded a Regional PhD Scholarship by the Research Office at USP (University of the South Pacific).

Dr Sarah L Hemstock holds a PhD in bioenergy systems modeling and is an author and adviser to the Alofa Tuvalu "Small Is Beautiful" project—recognised by UNESCO as one of its "Decade of Achievement Projects". Over the past 20 years, she has worked on community and domestic bioenergy schemes and agricultural extension projects with communities in Sub-Saharan Africa, Europe, India, South America, and the Pacific Islands Region. She also worked on national and regional energy. In 2010, she was made Government of Tuvalu Honorary Ambassador—Officer for Environmental Science. She was the Team Leader for EU-GCCA Project at PaCE-SD. Currently, she is the Team Leader for the European Union PacTVET Project at the Economic Development Division, The Pacific Community; Visiting Fellow at Nottingham Trent University and Adjunct Fellow at the University of the South Pacific.

Dr Helene Jacot Des Combes obtained her PhD on "reconstruction of the paleoceanography and paleoproductivity variations in the NW Indian Ocean during the last 300 kyr: the geochemical response compared to the biological record", at the University of Lille, France, in 1998. From 2000 to 2008, she worked as a scientist at the Alfred Wegener Institute for Polar and Marine Research in Bremerhaven, Germany. She joined PaCE-SD as a Research Fellow. Currently, she is a Senior Lecturer in Climate Change Adaptation at USP. She is currently developing a training needs and gap analysis on TVET in 15 Pacific island countries for the 6 million Euro European Union Pacific Technical Vocational Education and Training on Sustainable Energy and Climate Change Adaptation Project (EU PacTVET). She is the Course Coordinator for the Postgraduate Diploma course on disaster risk reduction and also delivers guest lectures on the ocean response to climate change.

Chapter 6
Climate-Friendly Adaptation Strategies for the Displaced Atoll Population in Yap

Murukesan V. Krishnapillai

Introduction

Ethnic communities of the oceanic nation of the Federated States of Micronesia (FSM) are among the most economically disadvantaged, socially backward and environmentally vulnerable groups in the western tropical Pacific. Yap is the westernmost State in the FSM and consists of Yap Proper, a group of four conjoined islands, and 14 atoll islets that are inhabited by traditional communities who are dependent on fishing, agroforestry, groundwater and rainfall. Environmental problems associated with climate variability, sea level rise, coastal flooding, loss of biodiversity, saltwater intrusion and lack of freshwater, soil degradation, and problems related to energy converge to place the Micronesian Islands, especially the atoll islets and other coastal settings, at the forefront of climate change (Fletcher and Richmond 2010). The climate change effects are disproportionately borne by the island communities. The recent report by the Intergovernmental Panel on Climate Change (IPCC 2014) came as no surprise to these communities, who have been living with increasing severe impacts of climate change amplified by regional El Niño Southern Oscillation (ENSO) phenomena for years.

The climate-poverty puzzle is one of the intractable problems limiting the development of island communities and the uptake of agricultural innovations. Impacts of climate change add to the problem and hinder the efforts to successfully achieve development goals. With an Environmental Vulnerability Index score of 392, the FSM is currently one of the highly vulnerable Small Island Developing States (SIDS) in the Pacific. Communities living in these islands are already experiencing higher temperatures, shifts in rainfall patterns, rising sea levels, and changes in frequency and intensity of extreme climatic events. Further changes are

M.V. Krishnapillai (✉)
Cooperative Research and Extension, College of Micronesia-FSM,
Yap Campus, Colonia, Yap 96943, Federated States of Micronesia
e-mail: muru@comfsm.fm

W. Leal Filho (ed.), *Climate Change Adaptation in Pacific Countries*,
Climate Change Management, DOI 10.1007/978-3-319-50094-2_6

expected long into the future as a result of climate change associated with global warming and regional ENSO phenomena. As a consequence, changes of this magnitude are having a profound impact directly on the livelihoods of island communities.

The extension intervention was initiated with the main objective to develop the adaptive capacity and enhance livelihood opportunities of a group of displaced atoll communities on a barren landscape. The paper presents the results of more than 10 years of restoration efforts in regreening the degraded landscape to bring fresh promise to the displaced atoll populations. While the importance of volcanic soil management and mosaic[1] restoration are emphasized, it was done in recognition of full range of mutually-reinforcing activities (i.e. economic and community development, capacity building, education, etc.) that helped the poor and most vulnerable atoll communities adapt to climate change. It demonstrates the significance of a traditional agroforestry system in regreening a degraded landscape as a sustainable food production system.

Climate Stressors for Micronesia

The climate of the FSM varies considerably from year to year due to the regional ENSO phenomena. Recent research shows that the frequency of extreme El Niño is expected to double due to climate change, with the average frequency increasing from once every 20 years to once per decade (Cai et al. 2014). Various climate models indicate that El Niño and La Niña events will continue to occur and have a significant impact on inter-annual variability in the region. Some of the impacts of ENSO on rainfall (e.g. floods) may intensify in a warmer climate due to increased atmospheric moisture (Seager et al. 2012). Global warming is also expected to enhance average rainfall along the equator, and new research suggests it will also enhance El Niño-driven drying in the western tropical Pacific and El Niño-driven increases in rainfall over the central and eastern tropical Pacific (Power et al. 2013). Climate projections based on the analysis of about 26 new Global Climate Models (GCMs) in the Coupled Model Intercomparison Project Phase 5 (CMIP5) database show that for all emissions scenarios, temperatures will continue to rise in the FSM, as will sea level and ocean acidification (Australia Bureau of Meteorology and CSIRO 2014; IPCC 2014).

Yap's weather, in particular, is influenced by a number of factors including the paired Hadley cells and Walker circulation and ENSO phenomenon (Chowdhury et al. 2010). Under El Niño conditions the islands typically experience drought and

[1]Mosaic restoration integrates trees into mixed-use landscapes, such as agricultural lands and settlements. Trees support people by improving water quality, increasing soil fertility, and boosting other ecosystem services. This type of restoration works best in deforested or degraded landscapes with moderate population density.

under La Niña conditions the islands experience higher than normal rainfall. Studies indicate that this seasonal climatology of extreme events in the western Pacific is variable in spatial and temporal scales. Yap islands display considerably higher seasonal extremes than the other islands in the region for 20 year return period due to typhoon-related storm surges. Decadal observations reveal mean sea level in the western Pacific rose alarmingly (3.1 mm/year) from 1988 to 2008. This is in close correlation with the faster rate of rise as observed by IPCC (2015). The small size of Yap and the neighboring atoll islets, their lower elevations and extensive coastal areas, their remoteness and limited financial resources, and poor economic and social decisions thus contribute to great ecosystem and human vulnerability to disasters (Shea et al. 2001).

While sea level rise alone is causing adverse impacts, the combinations of extreme events have a profound impact on the livelihoods of communities on atolls. Climate stressors of various magnitudes threaten the life-support systems of atoll communities essentially in three ways: (i) impact on land security—the very physical presence of land on which to live and sustain livelihoods, (ii) threat on livelihood security (food security) where productivity of subsistence agriculture reduced or lost, and (iii) declining habitat security as atoll environment becomes less habitable. Failure of these life-support systems resulting from climate shocks and disasters propel vulnerable atoll communities into poverty traps. Low adaptive capacity eventually forces them to migrate to highlands on Yap Proper in search of better living opportunities.

The past two decades have seen tremendous influx of atoll communities to Yap Proper. Although such migrations[2] were classified as "sojourning" and recorded in the past (Nelson 1976), today such movements are largely permanent and are synonymous with environmental migration. Atoll communities on Yap Proper reside principally on four settlements namely, Gargey, Daboch, Ruu and Makiy. Gargey settlement (9°33′24″N, 138°08′15″E), the project site in discussion here, is situated on Gargey-Tomil plateau on Yap Proper (Fig. 6.1). The area was barren until 2004 when environmental migrants began colonizing especially after the devastation from Typhoon Sudal.

[2]Migration typically describes a process of population movement—forced or voluntary—across an international border or within a state and encompassing any kind of movement of people, whatever the movement's length, composition, and causes (IOM 2011; Kilot 2004). For those movement for survival due to imminent or acute environmental disaster, the term displacement is more appropriate (IOM 2011). No internationally accepted term exists for people moving for environmental reasons. In an effort to capture the complexity and breadth of the phenomenon, the International Organization for Migration (IOM) has advanced the following working definition of *environmental migrants* (IOM 2008). *Environmental migrants* are persons or groups of persons who, predominantly for reasons of sudden or progressive changes in the environment that adversely affect their lives or living conditions, are obliged to leave their homes or choose to do so, either temporarily or permanently, and who move either within their country or abroad.

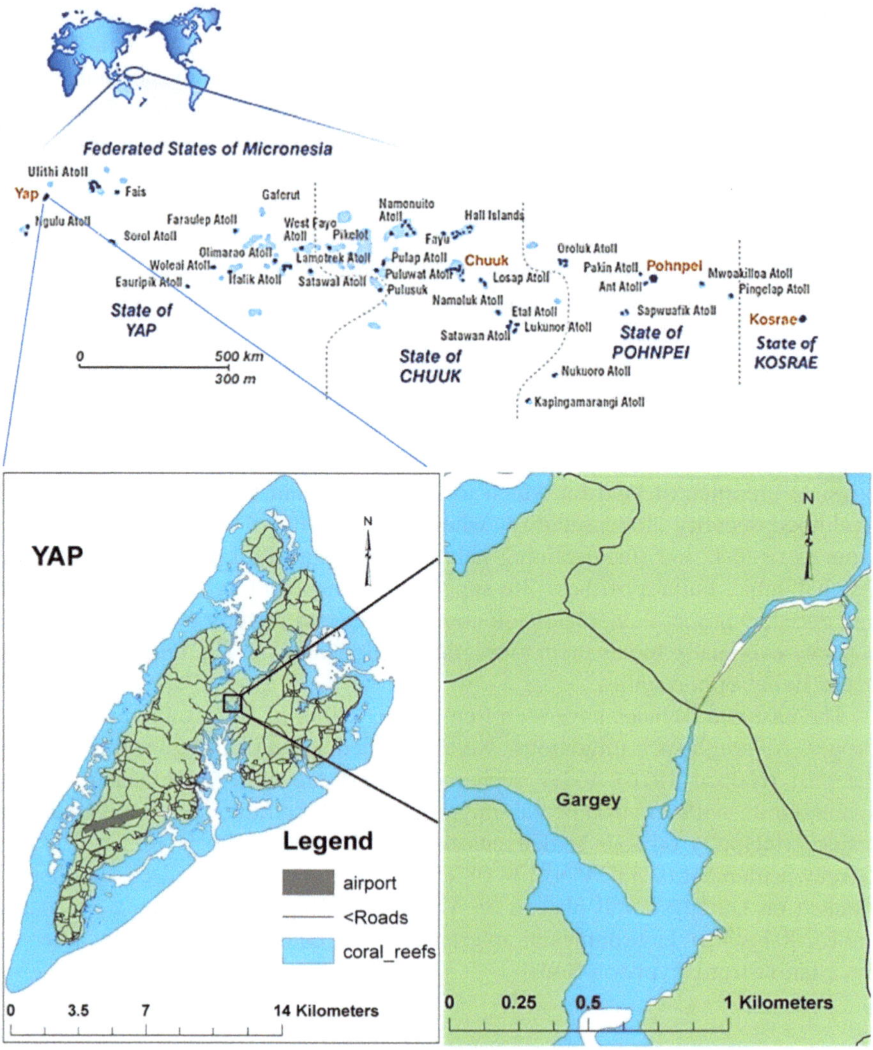

Fig. 6.1 Location map

Cultivating Success Through Context-Specific Climate-Smart Adaptation Strategies

Climate change induced migration threatens food security of atoll migrants by affecting its four dimensions—availability, accessibility, stability and utilization. Hence, the core of this extension intervention was to enable atoll communities to meet food security challenges while reducing climate related risks. Adaptation through traditional agroforestry systems with a focus on improved resilience of food production systems and the local communities depending upon them was of paramount importance. This was primarily achieved through restoration of degraded lands by utilizing sustainable volcanic soil management practices and sustainable production intensification through mosaic restoration approaches and climate-smart agriculture[3] strategies. The emphasis was to educate the communities with scientific knowledge and provide technical assistance to meet nutritional requirements through vegetable gardening activities. In addition to volcanic soil management strategies (composting, mulching, liming etc.), communities were also trained in alternative vegetable gardening activities such as SPIN farming, raised bed gardening, micro-gardens and container gardening.

Soils on Yap's volcanic landscapes developed in rocks that probably formed during the Miocene but the soils may have formed in the Pleistocene (Johnson et al. 1960). That timeframe allows ample time for deep rock weathering in Yap's humid tropical environment. However, because of heavy rainfall (110–130 in./year) and warm climate, many of Yap's soils have been depleted of nutrients through leaching. Leaching of nutrients causes a residual buildup of iron and aluminum in many of the soils giving them a reddish color (when oxidized). The soils highlighted in this study are degraded, dominated either by ferns or grasses (open savannah) (Fig. 6.2). The amount of nutrients and the ability to hold on to nutrients are very low in these soils (Fig. 6.3). The persistent physical, chemical and biological limitations found in this degraded land create barriers to natural forest regeneration. Building depleted topsoil is accomplished through additions of organic matter in the form of compost and mulch. Addition of crushed coral at the rate of about 40 lbs. per 100-foot row raised pH while lowering aluminum content and provided needed calcium for the crops. Additionally, pH also increases the negative charge on these variable charged soils and thereby increases the cation-exchange capacity. Understanding the status and condition of soil and their properties thus facilitates atoll communities to adopt sustainable volcanic soil management practices that also contributes to climate-smart land use. Over the

[3]FAO (2010) has developed and promotes a unified approach, known as Climate-Smart Agriculture (CSA), to developing the technical, policy and investment conditions to support its member countries achieve food security under climate change. It integrates the three dimensions of sustainable development (economic, social and environmental) by jointly addressing food security and climate challenges. It is composed of three interlinked pillars:

1. Sustainably increasing agricultural productivity and incomes;
2. Adapting and building resilience to climate change; and
3. Reducing and/or removing greenhouse gases emissions, where possible.

This approach aims to strengthen livelihoods and food security, especially of smallholders, by improving the management and use of natural resources and adopting appropriate methods and technologies for the production, processing and marketing of agricultural goods. A key component is the integrated landscape approach that follows the principles of ecosystem management and sustainable land and water use. CSA is not a single specific agricultural technology or practice that is universally applied. It is site-specific and identifies appropriate agricultural production technologies and practices.

Fig. 6.2 The vegetation of Gagil soil series is mostly ferns that tolerates very low soil fertility and high soluble aluminum

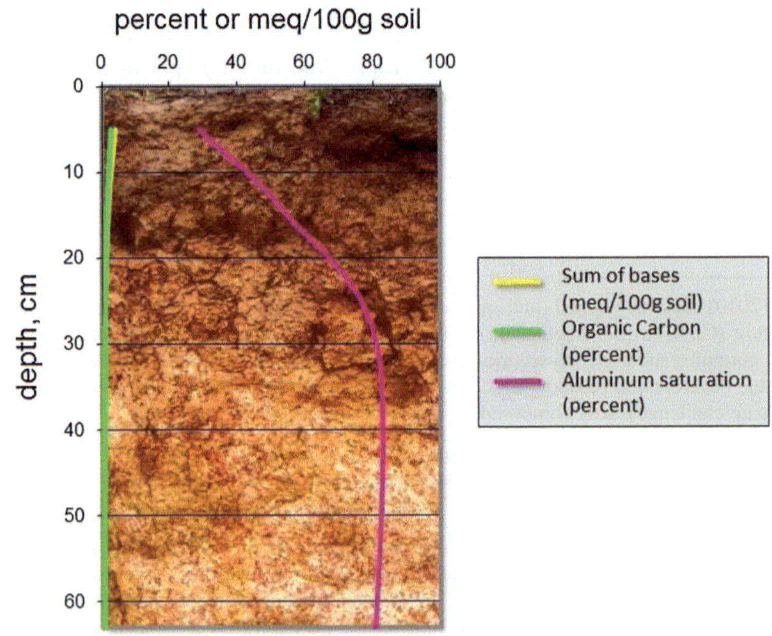

Fig. 6.3 Key quality parameters of Gargey soil series. Note very low levels of soil organic carbon and up to 80% soluble aluminum in about 20–30 cm depth

years, vegetable gardening outreach activities were gradually complemented and progressed toward restoration agroforestry activities.

Degraded land restoration efforts thus far aimed at supporting communities as they strive to increase and sustain the benefits from the marginal lands. These efforts also aimed to strengthen the resilience of degraded landscape and to keep future management options open. The various components of the restoration efforts were not new. They combined adaptive management, participatory methods and traditional methods to create a flexible and creative approach to the use of various indigenous trees and crops in the degraded landscape. It involved communities' efforts to improve ecological functioning of the degraded landscape and the well-being of the communities residing there. After several crop cycles, the traditional and science-based knowledge along with human ingenuity helped communities to successfully re-establish a productive agroforestry system (Figs. 6.4 and 6.5). The multiple products that come from the complementary mixtures are available at different time intervals, use space more effectively and utilize nutrients and other garden inputs more efficiently. Agroforestry adoption at the local setting was improved by complementing it with local practices, atoll communities' cultural norms and traditions.

The mosaic restoration efforts at Gargey settlement using traditional agroforestry practices, smallholder agriculture, and home gardens have some distinct advantages

Fig. 6.4 Early establishment of crops at Gargey settlement. Crops like pineapple, sugarcane, cassava tolerate elevated levels of soluble aluminum in the soil (Photo credit: Dr. Bob Gavenda)

Fig. 6.5 Present day biodiverse garden at Gargey settlement consists of an array of indigenous trees and traditional crops

over similar practices usually employed in restoration efforts. It plays a significant role in providing ecological corridors and habitats for many indigenous trees and crops aiding biodiversity conservation. It was not only a compromise option for rehabilitating degraded lands which de jure belongs to local government but also a method of developing livelihood opportunities in a barren land under the de facto control of displaced atoll communities. More importantly, integration of trees with smallholder agriculture, home gardens, and other cultural land uses fulfills the 'double filter' requirements as recommended in forest landscape restoration (FLR) efforts.[4] By blending their traditional knowledge with food security requirements, over the years, atoll migrants have developed a sustainable, bio-culturally diverse 'food forest' to suit the new terrain characteristics. Today, this 'food forest' consists of over 40 different food crops/trees and several

[4]The term forest landscape restoration (FLR) has been widely recognized as an important means of not only restoring ecological integrity at scale but also generating additional local-to-global benefits by boosting livelihoods, economies, food and fuel production, water security and climate change adaptation and mitigation. It combines adaptive management, participatory techniques and new and not-so-new technologies to create a flexible and creative approach to the use of trees in degraded landscapes. FLR initiative implies that any restoration efforts should improve the ecological functioning of a landscape and the well-being of the human communities that reside in that landscape. It is a *forward-looking* and *dynamic* approach, focusing on strengthening the resilience of landscapes and creating future options to adjust and further optimize ecosystem goods and services as societal needs change or new challenges arise (IUCN and WRI 2014).

varieties/cultivars in many layers: an over-story (for example: breadfruit, mango, coconut, chestnut, betelnut), a middle canopy (for example: banana, noni, citrus, papaya) and understory of crops like sweet potato, taro, yam, cassava, pineapple and a variety of vegetables. Fundamentally it supports the basic principles of Sustainable Land Management—increased land productivity, improved livelihoods, and improved ecosystems. This multifunctional land use system ensures economic viability and improves biodiversity, ecosystem services and carbon capture, benefiting local communities and society as a whole (Krishnapillai and Gavenda 2014).

Coping with chronically variable yields of food crops is critical for survival of vulnerable populations in marginal environments. Addressing climate challenges requires specific climate-smart agriculture practices that suit local conditions. Realizing the importance of nutrition and food security in the enhanced action on adaptation, one of the key activities was to empower communities to build sustainable food production systems. Over the years, a range of climate-smart alternate crop production systems were tested with high success rate. Sustainable intensification of vegetable production using methods such as raised bed gardening, container home gardening, micro gardens and small plot intensive (SPIN) farming ensures nutrition and food security requirements of the communities (Figs. 6.6 and 6.7).

Fig. 6.6 Small Plot Intensive (SPIN) farming is a nontechnical, organic-based, easy to understand and inexpensive to implement vegetable farming system designed specifically for small land bases. It reduces two big barriers for limited resource farming communities—land and capital. It is geared toward making significant income from farming on a limited space

Fig. 6.7 Raised beds are freestanding garden beds constructed above the natural terrain. It is a proven method for growing a variety of crops to bypass otherwise challenging soil conditions. Raised bed gardens improve growing conditions for plants by lifting their roots above poor soil and gives greater control over soil quality and nutrition

Impacts of restoration agroforestry systems at Gargey settlement on a 12-year temporal scale are encouraging (Table 6.1). The analysis based on household surveys reveal that displaced atoll communities readily adopted agroforestry practices with both short term and long term benefits from multistoried systems. Multistoried agroforestry systems provide a variety of agricultural products and income sources. The multiple products that come from these complementary mixtures are available at different time intervals, utilize space more effectively, and use nutrients and other farm inputs more efficiently. These diverse combinations buffer communities from the risk of crop failure or unanticipated risks. More importantly, these systems are location-specific and promote recreational, educational and cultural options.

Table 6.1 Impacts of restoration agroforestry system at Gargey settlement

Benefits	Impacts	Communities/community level
Production	+++	Crop diversification
	++	Combined yields
	++	Provide products year round
	+++	Improved food security
Economic	+	Limited income, mostly subsistence level production
	++	Improved livelihood and well being
Ecological	++	Improved soil cover
	+++	Reduced soil erosion
	++	Favorable changes in microclimatic conditions
	+++	Improved biological activity
	++	Improved organic carbon content (above ground)
	++	Improved soil structure
	++	Enhanced biodiversity
	++	Increased resilience to climate change
	++	Arresting land degradation
Sociocultural	++	Improved conservation
	+++	Multipurpose trees, meeting various needs
	++	Community cohesion strengthening
	++	Aesthetic value
	+++	Culturally important crops
	++	Protecting natural resources
	+++	Source of traditional medicine

+ slightly positive, ++ positive, +++ very positive

Discussion

Agroforestry is a sustainable land management system that increases the overall yield of the land, combines the production of crops, including tree crops, and forest plants and/or livestock simultaneously or sequentially, on the same unit of land, and applies management practices that are compatible with the cultural practices of the local population (Nair 1989). It is one of few land use strategies that promises such synergies between food security and climate change adaptation (Mbow et al. 2014). It also helps in water cycle regulation (Mendez et al. 2012; Smith and Olsen 2010; Vermeulen et al. 2012) and biodiversity conservation (Ravindranath 2007), all of which are integral aspects of climate-smart agriculture (FAO 2012). Agroforestry options provide a means for restoring, diversifying and increasing the sustainability of smallholder farming systems. Tree-based systems have some obvious advantages for maintaining production during varying climate regimes and different soil types. Diversifying the production systems by incorporating economically and traditionally important trees such as mango, breadfruit, noni, coconut, chestnut, soursop etc. helped to buffer against food insecurity associated with climatic variability.

On nutrient depleted soils, the long-term prospects of systems purely based on annual food crops are grim and a transition into tree-based farming offers better prospects. Moreover, establishing agroforestry on a land without tree cover is identified as one of the most promising strategies to raise food production (Bayala et al. 2011; Garrity et al. 2010).

For many Micronesian islands, agroforestry is a sustainable land use system, and an integral component of the traditional subsistence system which provided the people with many of the necessities of life. They developed a range of agroforestry systems capable of sustainable food production in widely differing ecosystems on high and low islands and atolls. Polyculture and the cultivar diversity (which minimized the impacts of seasonality and varietal failure) in the mixed agroforest, wetland taro fields, intermittent tree gardening, and the kitchen garden provided the islanders with a variety and perhaps surplus of foods throughout the year. The inherent diversity of agroforestry systems leads to a potentially broad crop portfolio, allowing farmers to reduce the market risks of single-crop systems, distribute labor more evenly over the year, and derive income from different crops over short-, medium and long-term time horizons after planting (Manner 2008).

Agroforestry is increasingly recognized as a sustainable land use that enhances the ability of the communities to adapt to climate change because of the multiple benefits it delivers including food provision, supplementary income, in addition to environmental services (Lasco et al. 2011; Schoeneberger et al. 2012; Syampungani et al. 2010; Verchot et al. 2007). Trees help to buffer subsistence farmers against environmental extremes by providing shade and acting as alternate sources of food during the period of drought. The immediate economic benefits of agroforestry are felt more at the local village level. Observations made during the present study are consistent with other studies on the multifunctional role of trees by sustaining production during different seasons (Sayer et al. 2013; Tscharntke et al. 2012; Ziegler et al. 2012).

Challenges and Constraints

This study demonstrates the transformative powers of CSA on a carbon-neutral land. The climate-smart approach enabled needed changes of crop production on a degraded land given its necessity to address food security requirements in a changing climate. However, it has some enduring challenges and constraints:

- Currently this extension intervention is limited to Gargey settlement where a moderate population of displaced atoll communities live. Although communities have adopted agroforestry practices as a viable option along with other livelihood activities, small size of landholdings and infertile soils are still impediment to consistent productivity. Maintaining an appropriate level of soil organic matter and biological cycling of nutrients is challenging in tropical humid conditions.

- Although this extension intervention is based on CSA concept, it has a different perspective on a scale at which it was implemented. It is more to be seen as an adaptive strategy on a 'farmscape' than providing ecosystem services on a landscape.
- Communities' adoption of a wide range of culturally appropriate trees and traditional crops was exclusively to meet food security requirements and, therefore, complex management requirements or supportive functions of agroforestry have not been thoroughly explored. Exploring market opportunities for multipurpose tree products (for example, noni) will have significant economic, social and environmental benefits.
- While the interventions help the communities to achieve their short term food security requirements while building a lasting and environment friendly solution, the complex nature of tree/traditional crops systems is less understood. There is a scope for implementing innovative agroforestry practices such as fodder banks, improved fallows, windbreaks/shelterbelts, fertilizer tree systems etc.
- Policy, incentives and government support are essential for the successful adoption of agroforestry practices. Although agroforestry is one of the thematic areas of the National Biodiversity Strategy Action Plan of the FSM (Federated States of Micronesia 2002), action plans are mostly disjointed or nonexistent. With climate change high on government's agenda, agroforestry warrants a high profile in the FSM and raising its profile in the national policy domains is critical.

Conclusion

This study described the potential of agroforestry system in addressing food security and livelihood needs of displaced atoll population in Yap Island. This agroforestry system reflected diversity in terms of the multiple benefits from trees and crops integrated in agriculture systems.

Agroforestry products improved the resilience of displaced population particularly by improving crop production, ecosystems services, and household income and nutrition security. Restoration of degraded lands is slow and challenging particularly in areas like Gargey settlement that have been fully deforested and where severe soil erosion has occurred. Creation of a more beneficial, multi-functional land system involving indigenous trees along with shade-tolerant crops and medicinal and ornamental plants was a viable alternative. This approach ensured economic viability, improved biodiversity, and ecosystem services benefiting local communities and islanders as a whole.

Given the reliance of the poor on environmental services for their livelihoods, a central element of adaptation approach in the present case study was restoration activities. By protecting and enhancing the natural services that support livelihoods, vulnerable communities maintain local safety nets and expand the range of options for coping with disruptive shocks and trends. This combination of a secured natural resource base, reduced exposure to natural hazards, and diversified livelihood activities increase community resilience to future threats, including climate change. In fact, a mosaic restoration approach has the advantage of meeting immediate development needs while contributing to longer-term capacity development that creates a basis for reducing future vulnerabilities.

Being a small island grappling with the challenges of food insecurity, climate change, land degradation and rural poverty, regreening offers a path forward, especially in degraded areas. The transformation of degraded landscapes—restoring productivity and increasing resilience through the widespread adoption of agroforestry and sustainable land management practices delivers food, climate, and livelihood benefits. A return to agroforestry modeled on traditional systems holds promise for displaced communities who are benefiting from multifaceted values of agroforestry.

Future Prospects

Although this work is being carried out in Yap, it applies to similar situations in small islands where displaced population are found in marginal areas. This significantly add to the world knowledge of how we approach the climate change-induced and climate change-forced migrants in different locations. Restoration agroforestry has great potential for regreening degraded lands in a less expensive and participatory way, creating basis for improved livelihoods, water provision and sustainable food production. Realizing this potential, work is currently advancing under the umbrella of USAID's Pacific American Climate Fund (PACAM)[5] to scale up the multiple benefits of climate-smart agriculture among three additional displaced community settlements on Yap Proper. Climate-smart adaptation strategy is a "multiple benefit" approach as it typically builds climate resilience alongside other benefits.

Table 6.2 lists some of the climate-smart adaptation interventions with multiple-benefits being implemented.

Climate risk management focused on community-based adaptation through effective climate-smart agriculture strategies will enhance the adaptive capacity by improving food, water and nutrition security, livelihood development activities, and sustainability. The methods for coping with climate variability through this project will inevitably pave the way for better climate change resilience tomorrow.

[5]The Pacific-American Climate Fund (PACAM) is a 5-year USAID grant facility that provides support to assist Pacific island communities to adapt to the negative impacts of climate change and provide co-benefits or solutions to other development challenges, such as livelihood enhancement, improved health, food security, water availability, ecosystem conservation and improved and improved governance.

Table 6.2 Climate-smart adaptation interventions with multiple-benefits

Improved soil management for climate-smart agriculture	Scaling up sustainable land management practices to improve soil functions, volcanic soil management, improving soil structure with organic matter, managing soil organic matter, improving soil water storage, integrated nutrient management, controlling soil erosion, restoring degraded soils for climate change resilience, rotating and diversifying crops, minimizing topsoil disturbance, increasing the use of perennial crops
Increased awareness on sustainable water use	Adopting a range of rainwater harvesting techniques such as use of low-cost 'Bob bags,' communal water catchments, promoting water conservation awareness programs, enhancing soil moisture retention capacity, changing crop pattern and diversification, minimizing water losses from field, improving water use efficiency
Increased livelihood opportunities and gender-focused climate-smart agriculture practices	Rehabilitating degraded lands to grow traditional crops, using traditional knowledge and community-generated innovations suitable for adaption, practicing alternative crop production practices such as container gardening, raised bed gardening, small plot intensive gardening etc., using intensive high-nutrition vegetable production systems, strengthening food security systems, providing gender-focused training on adaptation topics, providing support to men and women in climate-smart agriculture methods, providing training on value-addition and marketing

Acknowledgements Financial support from USDA-NIFA (Award #: 2006-51200-03594), USDA-OAO (Award #: 59-2501-10-008), and Hatch and Smith-Lever funds from the USDA—NIFA through College of Micronesia Land Grant Program and Cooperative Research and Extension, College of Micronesia-FSM, USAID—Pacific American Climate Fund (Award #: PACAM-14-0005) is gratefully acknowledged. NRCS-PIA Office in Guam is thanked for technical support with soil analysis. Author wish to thank the Executive Director, College of Micronesia Land Grant Program, Vice President, Cooperative Research and Extension, College of Micronesia-FSM, the Chief of Party, USAID Pacific American Climate Fund for encouragement and support. Thanks are also due to atoll communities in Gargey settlement for their active participation, enthusiasm and enduring support.

References

Australian Bureau of Meteorology & CSIRO. (2014). *Climate variability, extremes and change in the Western Tropical Pacific: New science and updated country reports*. Pacific-Australia Climate Change Science and Adaptation Planning Program Technical Report. Melbourne, Australia: Australian Bureau of Meteorology and Commonwealth Scientific and Industrial Research Organization.

Bayala, J., Kalinganire, A., Tchoundjeu, Z., Sinclair, F. & Garrity, D. (2011). Conservation agriculture with trees in the West African Sahel—a review. World Agroforestry Centre Occasional Paper No. 14. Nairobi, 72 pp.

Cai, W., Borlace, S., Lengaigne, M., van Rensch, P., Collins, M., Vecchi, G., et al. (2014). Increasing frequency of extreme El Niño events due to greenhouse warming. *Nature Climate Change, 4*, 111–116.

Chowdhury, M. R., Chu, P. S., Zhao, X., Schroeder, T. A., & Marra, J. J. (2010). Sea level extremes in the U.S.-Affiliated Pacific Islands—a coastal hazard scenario to aid in decision analyses. *Journal of Coastal Conservation, 14*, 53–62.

FAO. (2010). *"Climate-Smart" agriculture: Policies, practices and financing for food security, adaptation and mitigation*. The Hague Conference on Agriculture, Food Security and Climate Change. Hague, October 31–November 5, 2010.

FAO. (2012). *Greening the economy with climate-smart agriculture*. Background Paper for the Second Global Conference on Agriculture, Food Security and Climate Change. Rome, Italy: FAO.

Federated States of Micronesia. (2002). *The Federated States of Micronesia national biodiversity strategy and action plan* (p. 66). Palikir, Pohnpei: FSM Department of Economic Affairs.

Fletcher, C. H., & Richmond, B. M. (2010). *Climate change in the Federated States of Micronesia; food and water security, climate risk management, and adaptive strategies* (28 pp). Report, University of Hawaii, Honolulu.

Garrity, D. P., Akinnifesi, F. K., Ajayi, O. C., Weldesemayat, S. G., Mowo, J. G., Kalinganire, A., et al. (2010). Evergreen agriculture: A robust approach to sustainable food security in Africa. *Food Security, 2*, 197–214.

IOM. (2008). *World migration report 2008: Managing labour mobility in the evolving global economy* (562 pp.). Geneva: International Organization for Migration.

IOM. (2011). *Glossary on migration, international migration law series no. 25* (p. 114). Geneva: International Organization for Migration.

IPCC. (2014). *Climate change 2014: Impacts, adaptation and vulnerability. Fifth assessment report*. http://www.ipcc.ch/report/ar5/wg2/.

IPCC. (2015). *The IPCC's fifth assessment report—what's in it for small island developing states*? http://cdkn.org/wp-content/uploads/2014/08/IPCC-AR5-Whats-in-it-for-SIDS_WEB.pdf.

IUCN & WRI. (2014). *A guide to the restoration opportunities assessment methodology (ROAM): Assessing forest landscape restoration opportunities at the national or sub-national level* (125 pp.). Working Paper (Road-test edition). Gland, Switzerland: IUCN.

Johnson, C. G., Alvis, R. J., & Hetzler, R. L. (1960). *Military geology of Yap Islands, Caroline Islands*. Prepared under the direction of the Chief of Engineers, U.S. Army by the Intelligence Division, Office of the Engineer, Headquarters United States Army Pacific with personnel of the United States Geological Survey.

Kliot, N. (2004). Environmentally induced population movements: Their complex sources and consequences, a critical review. In J. D. Unruh, M. S. Krol, & N. Kliot (Eds.), *Environmental change and its implications for population migration* (pp. 69–99). Dortrecht: Kluwer.

Krishnapillai, M., & Gavenda, R. (2014). From barren land to *biodiverse home gardens. Farming Matters, 30*(1), 26–28.

Lasco, R. D., Habito, M. S., Delfino, R. J. P., Pulhin, F. B., & Concepcion, R. N. (2011). *Climate change adaptation guidebook for smallholder farmers in southeast Asia* (65 pp.). Philippines: World Agroforestry Centre.

Manner, H. I. (2008). Directions for long-term research in traditional agricultural systems of Micronesia and the Pacific Islands. *Micronesica, 40*, 63–86.

Mbow, C., et al. (2014). Agroforestry solutions to address food security and climate change challenges in Africa. *Current Opinion in Environmental Sustainability, 6*, 61–67.

Mendez, V. E., Castro-Tanzi, S., Goodall, K., Morris, K. S., Bacon, C. M., Laderach, P., et al. (2012). Livelihood and environmental trade-offs of climate mitigation in smallholder coffee agroforestry systems. In E. Wollenberg, M. L. Tapio-Bostrom, M. Grieg-Gran, & A. Nihart (Eds.), *Climate change mitigation and agriculture* (pp. 370–381). ICRAF-CIAT: London-New York.

Nair, P. K. R. (1989). *Agroforestry systems in the tropics*. Dordrecht: Kluwer Academic Publishers, The Netherlands. 664 pp.

Nelson, J. M. (1976). Sojourners Versus New Urbanites: Causes and consequences of temporary Versus permanent cityward migration in developing countries. *Economic Development and Cultural Change, 24*, 721–757.

Power, S., Delage, F., Chung, C., Kociuba, G., & Keay, K. (2013). Robust twenty-first-century projections of El Niño and related precipitation variability. *Nature, 502*, 541–545.

Ravindranath, N. H. (2007). Mitigation and adaptation synergy in forest sector. *Mitigation Adaptation Strategies for Global Change, 12*, 843–853.

Sayer, J., Sunderland, T., Ghazoul, J., Pfund, J.-L., Sheil, D., Meijaardb, E., et al. (2013). Ten principles for a landscape approach to reconciling agriculture, conservation, and other competing land uses. *Proceedings of the National Academy of Sciences, USA, 110*, 8349–8356.

Schoeneberger, M., Bentrup, G., de Gooijer, H., Soolanayakanahally, R., Sauer, T., Brandle, J., et al. (2012). Branching out: Agroforestry as a climate change mitigation and adaptation tool for agriculture. *Journal of Soil and Water Conservation, 67*, 128–136.

Seager, R., Naik, N., & Vogel, L. (2012). Does global warming cause intensified interannual hydroclimate variability? *Journal of Climate, 25*, 3355–3372.

Shea, E. L., et al. (2001). Preparing for a changing climate: The potential consequences of climate variability and change. A report of the Pacific Islands Regional Assessment Group for U.S. Global Change Research Program, 8 pp.

Smith, P., & Olesen, J. E. (2010). Synergies between the mitigation of, and adaptation to, climate change in agriculture. *The Journal of Agricultural Science, 148*, 543–552.

Syampungani, S., Chirwa, P. W., Akkinifesi, F. K., & Ayayi, O. C. (2010). The potential of using agroforestry as a win-win solution to climate change mitigation and adaptation and meeting food security challenges in Southern Africa. *Agricultural Journal, 5*, 80–88.

Tscharntke, T., Clough, Y., Wanger, T. C., Jackson, L., Motzke, I., Perfecto, I., et al. (2012). Global food security, biodiversity conservation and the future of agricultural intensification. *Biological Conservation, 151*, 53–59.

Verchot, L. V., Noordwijk, M. V., Kandji, S., Tomich, T., Ong, C., Albrecht, A., et al. (2007). Climate change: Linking adaptation and mitigation through agroforestry. *Mitigation and Adaptation Strategies for Global Change, 12*, 901–918.

Vermeulen, S. J., Campbell, B. M., & Ingram, J. S. I. (2012). Climate change and food systems. *Annual Review of Environment and Resources, 37*, 195–222.

Ziegler, A. D., Phelps, J., QI Yuen, J., Webb, E. L., Lawrence, D., Fox, J. M., & et al. (2012). Carbon outcomes of major land-cover transitions in SE Asia: Great uncertainties and REDD+ policy implications. *Global Change Biology, 18*, 3087–3099.

Chapter 7
Integrating Disaster Risk Reduction and Climate Change Adaptation in Vanuatu: The Art and Practice of Building Resilience to Hazards

Astrid Vachette

Introduction

This paper[1] contributes to the current international discussions on the need to integrate Climate Change Adaptation (CCA) and Disaster Risk Reduction (DRR) into a comprehensive approach to better, and more sustainably, build resilience to hazards (e.g. Thomalla et al. 2006; Venton and La Trobe 2008; Djalante 2013; UN 2015a, b; UNFCCC 2015 etc.). Building resilience of communities relies on their "ability […] to resist, absorb, accommodate, and recover from the consequence of a hazard event or of climate change in a timely and efficient manner, including through the preservation and restoration of [their] essential basic structure and function" (UNISDR 2009, p. 24). The development of such ability is complex and depends on continuous, consistent, complementary and reciprocal relationships between the agendas, institutions, policies and practices of the different sectors of DRR, CCA, Sustainable Development and Disaster Management (DM). Although these interlinked sectors merge through their impacts on the overriding process of resilience-building, they need to remain independent to enable the development of appropriate and effective structures and expertise in each sector. Therefore it is important to find a complementary balance between integration for a comprehensive approach, and sectoral fragmentation enabling the development of specialised capital. To achieve such a goal, vulnerable small islands (facing local resource limitations, high exposure to hazards and climate change, as well as complex cultural, political, environmental and economic diversity) need to establish complex

[1]This paper presents results that are part of the author's Ph.D. research on Networked Disaster Governance in Vanuatu.

A. Vachette (✉)
Centre for Disaster Studies, College of Science and Engineering, James Cook University, Townsville, QLD 4811, Australia
e-mail: astrid.vachette@my.jcu.edu.au

W. Leal Filho (ed.), *Climate Change Adaptation in Pacific Countries*, Climate Change Management, DOI 10.1007/978-3-319-50094-2_7

governance systems to enable reciprocal positive impacts between the development of expertise and resource at a specialised level (e.g. climate change adaptation, disaster management, gender protection etc.), and the development of a comprehensive resilience-building approach. Despite the general consensus on the need to developed integrated and cooperative governance systems, empirical research on networked cooperation for climate change and disaster risks, and its impact, remains rare (Kinnear et al. 2013). The analysis of networked governance systems in practice is essential to increase understanding on these systems and develop more effective strategies to enable this balance between integration for a comprehensive approach, and fragmentation for expertise building and appropriate decision-making. Hence, the purpose of this paper is to utilise the highly at-risk Small Island Developing State of Vanuatu, where a networked governance system was developed, as a case study to better understand the potential of such a system to more effectively build resilience by developing a complex and extensive integrated approach. It analyses how the different mechanisms in place within the Vanuatu Climate Change and Disaster Governance system focus on promoting cooperation across CCA, DRR, DM and resilience-related Sustainable Development, and the impact on the resilience-building process. It was found that, to achieve this, all formal and informal components of the governance system (its structure, leadership and processes) consider and foster the development of an integrated approach through the promotion of cross-sectoral networking.

Methodology

This paper is based on a mixed-methodology, using Social Network Analysis (SNA) and Comparative Qualitative Analysis, complemented by an extensive literature and policy review. SNA is a research tool used for mapping relationship structures and exchange patterns between and among members of a network (Scott 1988; Chung et al. 2005; Hossain and Kuti 2010). SNA is increasingly utilised in the sectors of DRR and DM (e.g. Kapucu 2005; Comfort and Haase 2006; Varda et al. 2009; Hossain and Kuti 2010) and CCA (e.g. Bharwani et al. 2013; Kinnear et al. 2013; Corlew et al. 2013). SNAs are critical tools to collect and analyse data, not only on how a network is structured but also why, and which information is shared within this system (Borgatti 2005; Chung et al. 2005; Hossain and Kuti 2010).

Data were collected through surveys, interviews and participant observation of key stakeholders, across sectors and levels, involved in the process of resilience-building, in routine times and during the emergency period following Cyclone Pam (March–June 2015). Data on routine times were collected from 90 respondents, resulting in detailed information on a governance system, referred to as the Vanuatu-Networked-System, comprising 54 networks and 260 stakeholders directly involved in the processes of CCA, DRR and/or DM in Vanuatu. Collection of data during the emergency period following Cyclone Pam was more difficult and focused on individual interviews and group observations (operation briefs, cluster meetings etc.), in order to analyse during disaster preparedness, response and early recovery the implementation and impact of the climate change and disaster governance system developed in routine times.

Key methods developed through SNA concepts were utilised for this research to assess networking patterns. Network members' centrality and influence (based on connection density, position in the whole network and position compared to the other members) were measured using the software UCINET 6 (Borgatti et al. 2002) and maps were drawn with the software Netdraw (Borgatti 2002). These measures and maps were a source of essential numerical data and visual representation of the different patterns of cross-sectoral cooperation and integration (structure, leadership and processes). A comparative analysis of the data collected through SNA, and the formal and institutional cross-sectoral decision-making as stated in policies, plans and strategies built a comprehensive understanding of the contrast and complementarity between formal and informal structure and processes of networked governance. Furthermore, a complementary analysis of the data on cross-sectoral cooperation before Cyclone Pam and during the emergency period, using SNA and Qualitative Comparative Analysis, increased the understanding of the impact of pro-active integration on the level of preparedness to face, cope with and respond to an extreme event. The rationale behind this mixed-methodology was to capture and analyse the complementarity between the formal (cross-sectoral institutions, policies and strategies) and informal (respondents' perceptions, individual network, supportive customs etc.) spheres that make interconnections between CCA and DRR possible and effective.

A Networking System Relying on CCA and DRR Interconnections to Build Resilience

Institutionalised Networking Structure

The Vanuatu-Networked-System analysed in this paper represents the governance system in which the 260 stakeholders 'captured' by the SNA cooperate to develop and utilise their expertise, capacities and resources to participate in the shared process of resilience-building. The Vanuatu-Networked-System is developed from a set of 54 complementary institutionalised networks. These networks cover key dimensions essential to consider in the process of resilience-building: meteorology and geo-hazards, climate change, disaster management, people in situations of vulnerability, and project effectiveness.

Three of these 54 networks, referred to as the umbrella networks, have a particular position within the Vanuatu-Networked-System enabling them to oversee the development of an integrated and inclusive vision, while supporting expertise, capacity and resource building at the sectoral level. The first umbrella network, the

Vanuatu National Advisory Board for Climate Change and Disaster Risk Reduction (referred to as the NAB), is government-led. The NAB's purpose is to supervise the consistency and appropriateness of government and non-government CCA and DRR policies, agendas, strategies, projects and decision-making in the country (NAB 2012). Observations of recurrent and ineffective overlaps and gaps between CCA and DRR strategies triggered the establishment of the NAB. The emergence of the NAB addressed the divide of advisory institutions for policy-making, by replacing two prior bodies, the National Task Force on Disaster Risk Reduction and the National Advisory Committee on Climate Change. The configuration of the NAB committee strongly supports integration of CCA, DRR and resilience-related Sustainable Development. It is composed of government members from all Ministries and representatives of Non-government Organisations (NGOs) and civil society groups, as well as regular visitors from the private sector and academia, involved in CCA, DRR and development projects. Among other duties, the NAB is in charge of reviewing, advising, endorsing and monitoring project proposals, as well as Information, Communication and Education materials. These endorsement processes aim particularly to promote cross-sectoral cooperation to develop and conduct more effective projects.

The second umbrella network, the Vanuatu Humanitarian Team (VHT), is non-government-led. It is composed of government and non-government humanitarian stakeholders involved in disaster preparedness, risk reduction, response and recovery in the country. The VHT supports the coordination of and interconnection among and between the Vanuatu clusters (loosely based on the international cluster approach), which are networks regrouping humanitarian actors based on their field of expertise (Food Security and Agriculture; Protection and Gender; Logistics; Water, Sanitation and Hygiene; Education; Health and Nutrition, and Shelter). The VHT, and the Vanuatu clusters are key platforms to share and disseminate information from one sector to another one, for harmonised, coordinated and effective action before, during and after a disaster, continuously bridging the sectors of DRR, CCA, DM and Sustainable Development.

The third umbrella network, the Vanuatu Climate Action Network (VCAN) is non-government led. It is primarily composed of NGO representatives, but also has strong connection with government agencies, academia and donors, in order to facilitate expertise building and sharing among all stakeholders involved more or less directly with building resilience to climate change in the country. Although focused on climate change, all members reported the network outcomes on all-hazard resilience and general community development.

The NAB, VHT and VCAN are well interlinked, each of them having representatives on the board of the two other umbrella networks. Simultaneously, the three umbrella networks are strongly connected with 51 other networks involved in resilience-building that were captured by the SNA. These satellite networks evolve within specific sectors at different levels of governance. They are key sources of expertise building that are optimised by the umbrella networks, which disseminate this expertise across the whole Vanuatu-Networked-System with the objective of developing a comprehensive vision at the national level.

These 54 networks present differences in objectives, configurations and function:

- Sectoral/integrated;
- Social/binding;
- Small/medium/large;
- Government/non-government/inclusive;
- Participant governed/lead-organisation governed/network administrative organisation (leadership forms theorised by Provan and Kenis 2007);
- Local/national/regional/international/across-levels;
- Flexible/rigid in member commitment.

This diversity builds a whole system that provides stakeholders with a wide choice of platforms to share their information, expertise and resources adapted to their individual capacities, goals and development. These networks are strongly interconnected, with pathways for sharing through their common members. More particularly, the strong social connections of the umbrella networks among the set of networks through their common members illustrate the umbrella networks' potential (and responsibility) in dissemination of capital across the Vanuatu-Networked-System, as shown on Fig. 7.1.

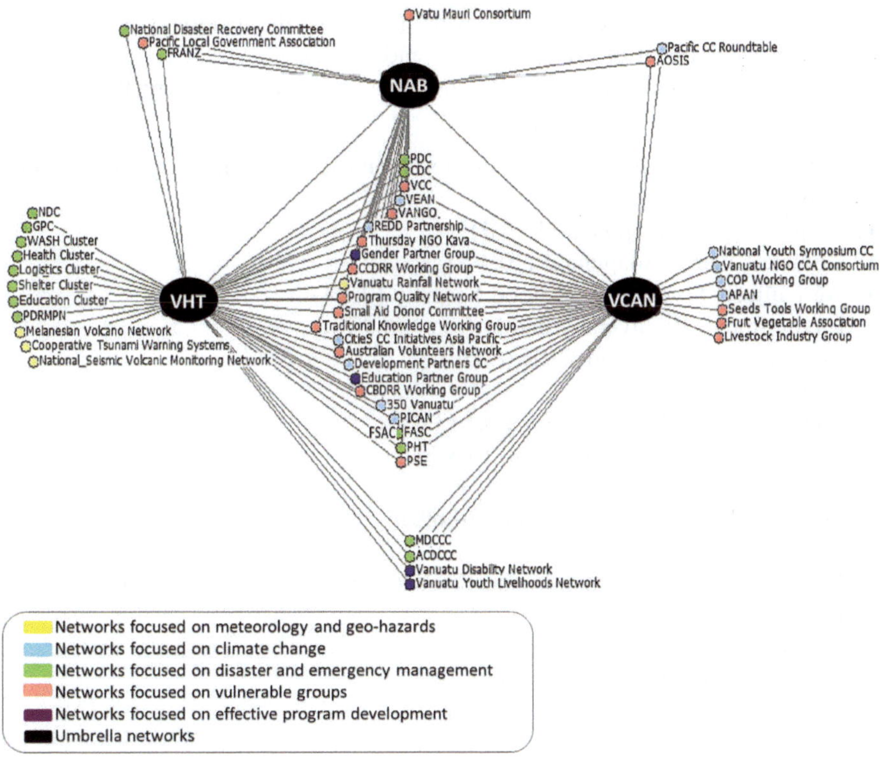

Fig. 7.1 Networking pathways between the three main networks of Vanuatu and the other networks involved in the different dimensions of resilience-building

These active links between the NAB and 53 networks in which non-government stakeholders (and more particularly locals) actively participate, were considered by most government and non-government respondents to be key assets for effective decision-making, because they are based on the sectoral expertise needed in the practice and the analysis of overlaps between projects focused on climate change, disaster risk, and/or development. This extensive set of networks, which are independent but interconnected, represents significant governance capital for effective resilience-building.

Social Networking Processes

Supported by the existence of these 54 networks, the Vanuatu-Networked-System is maintained by extensive and continuous social networking among the network members. The 260 stakeholders captured in the SNA represent more than 80 organisations (government agencies 35%, NGOs 36%, Regional organisations and regional United Nations offices 12%, Research Institutes 5%, donors 6%, private sector 4% and Civil Society groups 2%), from the six provinces of Vanuatu, as well as the regional and international levels (such as the Pacific Community or Australian Department of Foreign Affairs and Trade).

These stakeholders covered 20 domains of expertise directly affecting resilience-building:

- Disaster Risk Reduction and Disaster Management (14.2% *of the Vanuatu-Networked-System members*)
- Community Development (13.5%)
- Climate Change Adaptation (10.4%)
- Agriculture, Livestock and Food Security (9.2%)
- Environment, Biodiversity and Forestry (6.9%)
- Integration of CCA and DRR (6.1%)
- Education (5%)
- Meteorology and Climate Sciences (3.8%)
- Fisheries and Marine Resources (3.8%)
- Water Management (3.5%)
- Policy and Public Administration (4.6%)
- Gender (3.1%)
- Communication, Information and Knowledge Management (3.1%)
- Planning (2.7%)
- Program Management (2.7%)
- Monitoring and Evaluation (1.6%)
- Health (1.1%)
- Energy (0.8%)

- Traditional Knowledge and Culture (0.8%)
- Logistics (0.8%)
- Other (2.3%)

Networking patterns between these Vanuatu-Networked-System members demonstrate the strong cooperation across sectors, with 79.6% of the ties captured by the SNA (417 ties between the 260 stakeholders) happening across sectors, as shown on Fig. 7.2.

As displayed in Fig. 7.2, in the process of resilience-building, the stakeholders whose main expertise is 'CCA', 'DRR/DM', 'CCA and DRR Integration', 'Community Development' or 'Policy and Public Administration' are particularly well connected with each other and with the other domains of expertise existing in the system. Figure 7.2 illustrates the potential of the system for integration of decision-making and its implementation at the individual and common levels. However, based on respondents, the first incentives for cross-sectoral cooperation among respondents are not the development of a comprehensive approach but are (1) trusting relationships built over the years and with experience of collaboration, and (2) having access to more resources. More than half of the respondents (and almost all NGO respondents) spontaneously stressed the role played by informal networking within the Vanuatu-Networked-System to address these two incentives.

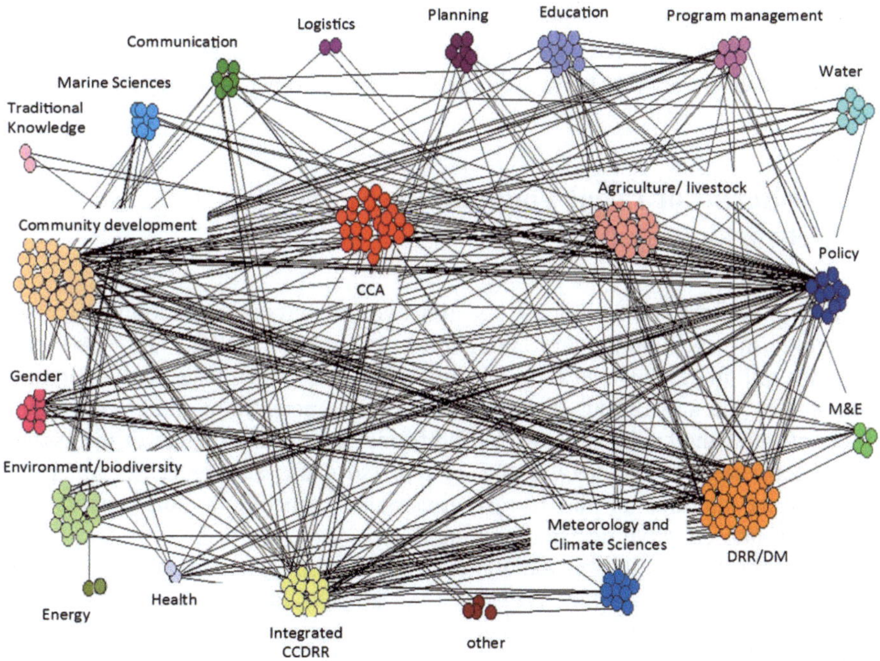

Fig. 7.2 Social networking across the 260 Vanuatu-Networked-System members featuring cross-sectoral cooperation for resilience-building. *Source* Adapted from Vachette (2015)

Indeed, informal networking occurring within the Vanuatu-Networked-System results in continuous, long-term and trusting relations supporting the development of conscious and unconscious understanding and aspirations for integration among climate change and disaster risk stakeholders to build resilience.

Echoing the need for long-term integration, the SNA highlighted that the current CCA progress in the country relied on non-national expertise, with 66.5% of the CCA experts involved in Vanuatu being from other countries compared to only 32.5% for DRR/DM experts. This trend can be explained by the fact that CCA integration into general resilience-building agendas is a new priority in Vanuatu, while the need to link DRR/DM and day-to-day Development was already recognised in the very first national disaster plan, the Disaster Preparedness Plan for Vanuatu released in 1986. This challenge for national CCA expertise, however, may not remain long, since the sector benefits from the established cross-sectoral structure developed over the years by the DRR and development sectors. More particularly, the VCAN coordinator highlighted that the development of the climate-focused umbrella network had been significantly facilitated by the integrated and inclusive networking structures and processes already developed by the members of the disaster-focused umbrella network VHT. Also, respondents highlighted that the current strong mobilisation of the Vanuatu-Networked-System to promote capacity building in CCA among staff and communities results in the rapid development of expertise building strongly based on training and workshops, compared with the slow-developing disaster risk expertise mainly built on practice, experience and simulations over the years. This two-way process for capacity building in CCA and DRR was considered by government and non-government respondents as an asset to build a comprehensive approach to resilience.

Establishment of Legal Foundations Supporting the Networking System

The set of networks and their dense social networking processes undoubtedly play a key role in facilitating the integration of CCA, DRR and DM into resilience-building. However, the sustainability and effectiveness of integration promoted by the networking system depend on the recognition and empowerment of the informal and social structure by the formal and legal structure of the system. The effectiveness of the networking system (structure and processes) therefore relies on the aptitude of the formal governance structure to turn unconscious integration into effective decision-making.

Addressing this challenge, the Vanuatu-Networked-System is based on extensive legal foundations for DRR, CCA and resilience-related Sustainable Development, with specialised policies and regulatory frameworks, such as the National Disaster Act (2000), national support plans for different hazards (e.g.

tsunami or cyclone), Provincial Disaster Plans, National Sustainable Development Plan (2016–2030) etc.

The development of the Vanuatu legal background for DRR, CCA and Sustainable Development conforms to international and regional guidelines. Hence, in accordance with the integration objectives of the recent Sendai Framework, the 2030 Agenda for Sustainable Development and the Paris Agreement in particular, the Government of Vanuatu developed a set of policies and regulatory frameworks strongly promoting integration of DRR, CCA, DM and resilience-related Sustainable Development into the diverse national agendas, strategies and programs.

Key strategies, such as the 'Planning Long Acting Short' agendas 2009–2012 and 2013–2016 or the 'Priorities and Action Agenda' 2006–2015 explicitly recognise the priority to integrate DRR and CCA considerations in development plans, and vice versa. For instance, the National Adaptation Program for Action (2009–2015) was developed as a guideline to mainstream climate change considerations across national strategies. More particularly, the National Climate Change and Disaster Risk Reduction Policy (2015–2030) strengthens the credibility, legitimacy and sustainability of the integration process evident in the country.

Hence, all legal frameworks related to DRR, CCA and/or resilience-related Sustainable Development developed strategies to integrate these sectors. All these documents recognise the need for inclusive action with a strong empowerment of non-government stakeholders, and promotion of government–non-government cooperation to better achieve this integration. This integrated and inclusive approach brings about the legal background to officially recognise the impact of institutionalised networks. More particularly, the umbrella networks are explicitly referred to in legal frameworks (e.g. the NDMO Standard Operating Procedures or the Vanuatu Climate Change and Disaster Risk Reduction Policy 2016–2030) as key tools to be utilised by stakeholders in the process of resilience-building to ensure integration and inclusiveness.

In accordance with the legal background and to facilitate formal integration of DRR, CCA and resilience-related Sustainable Development, the Government of Vanuatu established a single leading institution: the Ministry for Climate Change, Meteorology, Geo-Hazards, Environment, Energy and Disaster Management. The Ministry for Climate Change is an emblem of the national commitment to integrated resilience-building. The NAB is a strategic component of the activities of the Ministry for Climate Change, promoting integrated and inclusive social networking and decision-making to pursue the Ministry agenda.

The reciprocal impacts of DRR, CCA, DM and Community Development are illustrated by the strong ties between two departments: the Vanuatu Meteorology and Geo-Hazards Department (VMGD) and the National Disaster Management Office (NDMO). The VMGD and NDMO recognise the potential of integrated actions to respectively develop scientific information on all hazards, and to prepare communities to face them. The two departments have developed a strong network with non-government stakeholders (in particular with the VHT and VCAN members), facilitating inclusive and integrated decision-making within the Ministry for Climate Change.

Likewise, the well-regarded Community Disaster Committees and Provincial Disaster Committees, considered as key assets for implementation of national strategies at the local and provincial levels evolved to become respectively Community Disaster and Climate Change Committees and Provincial Disaster and Climate Change Committees. This evolution and the configuration of these committees strongly promote integrated activities, with government representatives from key departments (e.g. Agriculture, Forestry or Education), and non-government actors involved in community protection and development (e.g. NGOs or Civil Society Groups).

This formal legal and institutional background illustrates the formal national efforts to achieve better integration of response to, and preparation for, climate change and disaster risks. The development of this background is strongly and continuously facilitated by the networking structure (the structure composed of the diverse networks and the continuous cross-sectoral social networking). In support, the formal background officially recognises and guides the process of integrated resilience-building, making the whole system more credible, legitimate and sustainable.

Building Capacities to Process CCA and DRR Information and Expertise into Knowledge for Resilience-Building

The sustainability of the system (formal and informal) described in the previous sections strongly depends on the long-term capacities of the diverse stakeholders to share information and expertise on climate change and disaster risks, and to process them into knowledge utilised in decision-making for resilience-building. To address this need, a set of tools was developed within the Vanuatu-Networked-Governance to build the capacities of the diverse stakeholders to identify, understand and use expertise in CCA and DRR across all sectors.

The first dimension to consider in the process of integration is the set of tools to share information and expertise across sectors. Network meetings were considered as one of the main platforms to share information and expertise. At the level of smaller networks, this sharing process is essential to build strong expertise, which is then disseminated across all sectors during umbrella network meetings. This highlights the reliance of face-to-face interaction between stakeholders within and across sectors to share information.

Tools based on modern technology are increasingly being used in the archipelago as a complement to the more traditional sharing platforms. One of the most valuable modern tools is a national integrated database, the NAB Portal, on which stakeholders are requested to publish any information and intellectual resources related to climate change and disaster risks in Vanuatu. The NAB Portal is linked with the Pacific Climate Change Portal, increasing the potential for information and expertise sharing in the sectors of CCA, DRR and resilience-building. Respondents highlighted that the NAB Portal was particularly useful for sharing information not valuable to their own sectoral network, but potentially useful to other sectors. The NAB Portal also significantly increases the transparency and visibility of the programs and projects conducted in the country covering all aspects of disaster, climate change and resilience. Many respondents reported that they often find projects from different sectors that could be useful for their own, or projects to which they could add value by integrating their sectoral input in the development of goals. Therefore, the NAB Portal significantly improves the process of integration, and participates in the development of a more comprehensive approach for resilience-building in general.

A key challenge to cross-sectoral sharing, however, is the inconsistency of terminology and concepts between the different sectors of CCA, DRR, DM and resilience-related Sustainable Development. Indeed, the different sectors often use different expressions for the same overriding goals. For instance, 'food security', and 'sustainable agriculture' cover the same range of activities from the point of view of communities. The differences between the two themes may not always be understood by local staff and communities, leading to confusion concerning the complementarity of the different sectors in the overriding process of resilience-building. Furthermore, these terminology differences can critically hinder communication among the diverse stakeholders. More critically, this inconsistency sometimes occurs within the sectors between the different organisations, increasing the difficulty to conduct inter-sectoral sharing, limiting the development of sectoral expertise to be shared across the Vanuatu-Networked-System. To address these issues, the VMGD and NDMO, with the support of ad hoc and already established networks, work on the standardisation of climate, climate change and disaster risk terminology to be used by all Vanuatu-Networked-System members.

The second dimension to consider after information and expertise sharing is the capacity of individuals to identify the value of this information and expertise, and to integrate them into their own activities in resilience-building. A key obstacle to integrated CCA, DRR and resilience-building in the Vanuatu-Networked-System was considered to be the limited individual capacities of national and local stakeholders to use their own capacities developed within one sector for the benefit of other sectors. The challenge of turnover strengthened these difficulties. Indeed, due to the specific work context of Vanuatu, the rate of staff turnover is very high, leading actors to work consecutively in government and non-government organisations, in very diverse sectors. Although this developed the potential of these individuals to have a comprehensive perspective on resilience-building needs, their awareness of this potential is lacking. Workshops, training and meetings often conducted by the umbrella networks are particular assets to increase the understanding of the potential ties between the different sectors to achieve the overriding goal of resilience. More particularly, educating youth to understand the integration process is a main priority within the Vanuatu-Networked-System. For instance, key organisations and networks involved in CCA and DRR cooperated to integrate climate change and disaster subjects in the school curriculum. Likewise, the VHT and VCAN members cooperated to develop questions for the Climate Zone competition in secondary schools around the topics of climate change and disaster risks. This competition aimed to increase youth understanding of climate change and disaster matters, and illustrated the capacity of the future Vanuatu leaders to integrate climate change and disaster risks into their everyday life perspective.

Cyclone Pam Preparedness and Management: An Example of Effective Resilience-Building Based on the Integration of Climate Change and Disaster Risks

Integrating CCA, DRR and Resilience-Related Sustainable Development for More Effective DM

The proactive integrated networking process strongly builds capacities of staff and communities to address climate change and disaster risks. Building capacities to face geological and climate hazards (but also political instability, and economic, environmental and social development issues) falls under different perspectives, while similarly affecting the general level of resilience. Furthermore, preparedness, response and recovery of small versus major, slow versus rapid onset disasters may significantly vary in strategies but rely on a similar background approach. The balance between sectoral fragmentation and integration provided by the Vanuatu-Network-System (its structure, leadership and processes) enables actors and communities to build capacities to face all types of hazards.

The occurrence of Category 5 Cyclone Pam demonstrated the positive impact of cross-sectoral relationships on resilience-building. Undoubtedly, the principal positive outcomes of proactive relationship-building was the timely mobilisation of stakeholders and resources to lead preparedness and response to the cyclone. This was facilitated by the acquaintance, understanding and trust between these stakeholders. As said by a long-term expatriate, "this coordination [was] not an accident", but an outcome of the networking system (Van Rooyen 2015, p. 1). The existence of the Vanuatu clusters before Cyclone Pam provided strong, established platforms for humanitarian actors to promptly gather around their domain of expertise to better coordinate sectoral operations (e.g. food security, health or education). The established role of co-coordinator of the operations of the NDMO and the VHT enabled fluent cooperation between government and non-government stakeholders in each sector, with strong pre-determined collaboration pathways between Ministries, NGOs and donors involved in the same areas (e.g. the Department of Women's Affairs, the Council of Women, UN Women, CARE, the Vanuatu Christian Council and the sectoral networks involved in gender protection, sharing the goal of helping women to cope with Cyclone Pam emergency).

Short-term international actors reported that the cooperative intensity to build shared expertise at each sector level in Vanuatu was impressive.

These international respondents stressed even more, the particular capacity of the system in place to promote the development of a comprehensive approach to capacity building for resilience-building. The recognition of the value of government–non-government and cross-sectoral cooperation value among leaders also raised the appreciation of international actors for the system in place. Short-term humanitarian actors reported that compared to their experience in other countries, the pooling of resources in Vanuatu across sectors, types of organisations and levels was particularly impressive, and seemed "natural". This was a direct outcome of the proactive efforts to build a governance system focused on inclusive and integrated decision-making and implementation.

The strong established structure and ties within the Vanuatu-Networked-System, however, made it more difficult for some short-term internationals to intervene directly within the system in place, leading to the emergence of cooperation difficulties between Ni-Vanuatu, long-term expatriates and short-term humanitarian actors. These difficulties were partly due to the continuous (and relatively still unconscious) integration processes occurring in the country. The dichotomy between response and recovery periods was unclear for locals, who often started long-term recovery (building shelters for the next hazard) while first response was just starting. This highlighted the need to reinforce the role played by the formal background of the Vanuatu-Networked-System to build better understanding and clarity of the informal integration and inclusiveness processes.

Despite remaining challenges, the development of the Vanuatu-Networked-System, based on a comprehensive structure using all expertise and resources from CCA, DRR and resilience-related Sustainable Development, undoubtedly built capacities of staff and communities to prepare, adapt and respond to Cyclone Pam. Conversely, the management of Cyclone Pam contributed to building resilience for climate change, future disasters and resilience-related Sustainable Development. Management of the Cyclone Pam response highlighted key arenas to work on to build resilience (e.g. development of Build-Back-Better guidelines, empowerment of certain institutions and networks such as the Community Disaster and Climate Change Committees, or better integration of traditional knowledge and modern sciences). These arenas of improvement would have significant impacts on the general level of resilience for day-to-day lives.

The Particular Case of the Vanuatu Food Security and Agriculture Cluster

The particular analysis of the impact of the Vanuatu Food Security and Agriculture Cluster (FSAC) activities on integrated DM-DRR-CCA highlights the links between hazard, climate change and resilience-building. The FSAC includes diverse members, such as government agencies, NGOs, private sector, development partners and faith-based organisations. The high dependence of local communities on agriculture, livestock and fisheries for subsistence in routine times and the lack of sustainable food preservation methods increase community vulnerability to weather fluctuations and hazards. This situation makes food security a key general priority in CCA and DRR agendas, and climate change and fundamental disaster dynamics to consider in the development of sustainable agriculture strategies.

The food security situation was precarious following Cyclone Pam, after the destruction of 69% of food crops, 16% of the forestry, 9% of the livestock and 6% of the fisheries, critically affecting the level of resilience of communities (RRU 2015a). Food security was a key priority, and aid was essential in this sector. Therefore, the network of the Vanuatu FSAC was activated to conduct response and early-recovery. Established late in 2012, the Vanuatu FSAC is co-led by the Department of Agriculture and Rural Development, Vanuatu SPC/GIZ Coping with Climate Change in the Pacific Islands Programme, and the Food and Agriculture Organisation of the United Nations (UN OCHA 2015; RRU 2015b). This heterogeneous leadership illustrates the strong relationship between disaster management, climate change and long-term agriculture sustainability.

Subsistence continuity was supported by the simultaneous distribution of food aid (such as rice, cans and pasta to substitute lost staples), seeds and tools for land preparation within the first days following Cyclone Pam (UN OCHA 2015; Gero and Thiessen 2015). This strategy aimed that seeds should have started to produce by the time food distribution ceased at the end of the emergency period (Bolis 2015). Despite these efforts, Cyclone Pam affected long-term food security; it is expected that 10 years will be needed for the agriculture sector to fully recover (RRU 2015a). This situation not only affected the nutritional subsistence of the population but also the whole economy of the country, representing in routine times 25% of the national Gross Domestic Product (RRU 2015a). Agriculture rehabilitation was (and still is) therefore a key priority to build resilience to any future small and major, slow and rapid onset hazard.

The Vanuatu FSAC demonstrated the impacts of the networking structure and processes to prepare for hazards and climate change in long-term resilience building through the prompt establishment of the Risk and Resilience Unit (RRU) by the Vanuatu FSAC members before the end of the emergency period. The RRU was developed to be a long-term network falling under the authority of the Ministry of Agriculture to extend the impact of integrated and inclusive cooperation occurring within the FSAC for the specific purpose of agricultural recovery, while pursuing the development of general capacities to face future hazards and climate change.

The network members showed their capacities to develop a comprehensive, flexible and effective agenda, particularly with their simultaneous mobilisation early in Cyclone Pam recovery and preparation for 2015 El-Niño season. This simultaneous mobilisation was fully enabled by the proactive development of relations among the network members through the diverse institutionalised networks, social networking pathways and formal foundations promoting continuous integrated cooperation within the Vanuatu-Networked-System. The tools, such as SMS, radio or leaflet drops, utilised to disseminate key information on preparedness, response and recovery had already been cooperatively developed by government and non-government CCA, DRR and DM actors.

The FSAC and RRU activities were challenged during the response and recovery periods by remaining coordination challenges, mainly due to the outsiders' confusion concerning the differences between the two networks. The strong focus on cross-sectoral cooperation held by these networks in particular and the whole Vanuatu-Networked-System in general, addresses this challenge by developing all stakeholders' awareness of the interlinkages between the separated agendas of long-term recovery from Cyclone Pam, agriculture rehabilitation, climate change adaptation and disaster preparedness conducted within the FSAC and RRU. The significant proactive relationship-building between FSAC members involved in CCA, DM, DRR and/or resilience-related Sustainable Development resulted in building strong potential for the Vanuatu-Networked-System to facilitate transition between routine times and emergency period, as well as to build-resilience in general.

Conclusion

Undoubtedly, the Vanuatu-Networked-System still faces challenges to conduct and achieve integration for climate change, disaster risks and sustainable development (such as remaining competition among stakeholders, or resource scarcity to conduct cross-sectoral activities). This paper builds understanding on remaining challenges, but more importantly on the strong potential of a networked governance system to promote integration of DRR, CCA, DM and Sustainable Development to effectively conduct the overriding process of resilience-building. The effectiveness of the Vanuatu-Networked-System relies on the systematic integration process of CCA, DRR and resilience-related Sustainable Development into all formal and informal governance components, supporting effective decision-making and practice. Based on SNA as a methodology and method to assess the patterns and potential of a networked governance system, the findings of this paper outline the necessity to integrate cross-sectoral strategies with all dimensions of governance to ensure consistency, sustainability and credibility of the process. In the case of Vanuatu, this holistic approach is reflected by the complementarity of:

- a profuse set of more or less formal cross-sectoral networks (enabling all stakeholders to participate in the process based on their own capacities, resources, types, objectives etc., and facilitating dissemination of expertise and resources across the whole system);
- a government-led umbrella network and two non-government-led umbrella networks (facilitating and overseeing the process of integrated and cooperative decision-making and practices related to Disaster Risk Reduction, Disaster

Management and Climate Change Adaptation conducted across the whole system);
- dense cross-sectoral social networking (continuously building relationships and exchange pathways for knowledge and resources developed intra- and cross-sectors);
- a formal background (building an endorsed, consistent, stable and legitimate common vision and strategy propitious to cross-sectoral actions), and
- capacity-building mechanisms (enabling stakeholders to build stronger intra- and cross-sectoral understanding, expertise and knowledge to facilitate the implementation of integrated strategies for resilience-building).

Preparedness and response to Cyclone Pam, which optimised the complementarity of the sectoral and cross-sectoral networking processes, highlighted the potential of this whole system to facilitate appropriate decision-making and to be implemented. This governance system makes the Vanuatu-Networked-System a remarkable example of the potential of Pacific Small Island Developing States to build their own resilience.

The findings of this paper, however, face some limitations. Due to the high rate of staff turnover in the country and the difficulties to conduct SNA during emergency periods, data could not be collected among the same sample of respondents before and after Cyclone Pam limiting the assessment of networking impacts on the transition from routine to emergency times. Also, due to time and resource constraints, data was mainly collected on Efate, the main island of Vanuatu. Although, very diverse stakeholders meet there to make decisions, and ties were found with more remote areas, the actual impact of decision-making and its implementation in the different provinces on the integration process could not be fully covered. Further research in the different provinces of Vanuatu could complement findings from this paper to build a more comprehensive dataset on networked governance systems in geographic scattered contexts. Further research in other contexts (other Pacific islands, other developing country or other at-risk communities) could also complement this paper by comparing the Vanuatu system with other networked governance systems to analyse their differences and similarities in successes and difficulties to address resilience-building challenges.

Acknowledgements The author would like to thank the significant financial and technical support of the Centre for Disaster Studies of James Cook University, the Vanuatu SPC/GIZ Coping with Climate Change in the Pacific Island Region (CCCPIR) and the Vanuatu National Advisory Board for Climate Change and Disaster Risk Reduction (NAB) in the pursuit of the Ph.D. research of which this paper is part.

References

Bharwani, S., Downing, T. E., Varela-Ortega, C., Blanco, I., Esteve, P., Carmona, G., et al. (2013). *Social network analysis: Decision support methods for adaptation.* MEDIATION Project, Briefing Note 8. Funded by the EC'S 7FWP.

Bolis, A. (2015, July 28). Vanuatu reconstruction moves ahead in the aftermath of Cyclone Pam. *The Guardian.* Retrieved July 30, 2015, from http://www.theguardian.com/world/2015/jul/28/vanuatu-cyclone-pam-el-nino-reconstruction.

Borgatti, S. P. (2005). Centrality and network flow. *Social Networks, 27*, 55–71.

Borgatti, S. P. (2002). *Netdraw network visualization.* Harvard, MA: Analytic Technologies.

Borgatti, S. P., Everett, M. G., & Freeman, L. C. (2002). *Ucinet 6 for Windows: Software for social network analysis.* Harvard, MA: Analytic Technologies.

Chung, K. K., Hossain, L. & David, J. (2005). Exploring sociocentric and egocentric approaches for social network analysis. In *Proceedings of the 2nd International Conference on Knowledge Management in Asia Pacific* (pp. 1–8).

Comfort, L. K., & Haase, T. W. (2006). Communication, coherence, and collective action. The impact of Hurricane Katrina on communications infrastructure. *Public Works Management and Policy, 10*(4), 328–343.

Corlew, L. K., Keener, V. & Finucane, M. (2013, January 22). *Pacific regional integrated sciences and assessments (RISA) network analysis of climate change professionals in the pacific.* Poster presented at the Pacific Islands Climate Services Forum in Suva, Fiji. Retrieved April 24, 2016, from http://www.pacificrisa.org/wp-content/uploads/2013/05/Suva-Poster.pdf.

Djalante, R. (2013). Identifying drivers, barriers and opportunities for integrating disaster risk reduction and climate change adaptation in Indonesia: An analysis based on the earth system governance framework. In W. Leal Filho (Ed.), *Climate change and disaster risk management.* Berlin: Springer.

Gero, A. & Thiessen, J. (2015, March 23). Beyond cyclone Pam: Identifying Vanuatu's strengths and needs. *The Conversation.* Retrieved April 02, 2015 from http://theconversation.com/beyond-cyclone-pam-identifying-vanuatus-strengths-and-needs-38913.

Hossain, L., & Kuti, M. (2010). Disaster response preparedness coordination through social networks. *Disasters, 34*(3), 755–786.

Kapucu, N. (2005). Interorganizational coordination in dynamic context: Networks in emergency response management. *Connections, 26*(2), 33–48.

Kinnear, S., Patison, K., Mann, J., Malone, E., & Ross, V. (2013). *Network governance and climate change adaptation: Collaborative responses to the Queensland floods.* Gold Coast: National Climate Change Adaptation Research Facility.

NAB. (2012, December). NAB terms of reference. Retrieved October 10, 2013 from http://www.nab.vu/nab-terms-reference.

Provan, K. G., & Kenis, P. (2007). Modes of network governance: Structure, management, and effectiveness. *Journal of Public Administration Research and Theory, 18*(2), 229–252.

RRU. (2015a). Medium and long term recovery and rehabilitation strategy 2015–2017. Risk and Resilience Unit (RRU) of the Ministry of Agriculture, Livestock, Forestry, Fisheries, Biosecurity, May 2015.

RRU. (2015b). Key lessons from the agriculture response to severe tropical cyclone Pam. Lessons Learned Workshop. Risk and Resilience Unit (RRU) of the Ministry of Agriculture, Livestock, Forestry, Fisheries, Biosecurity, June 2015.

Scott, J. (1988). Social network analysis. *Sociology, 22*(1), 109–127.

Thomalla, F., Downing, T., Spanger-Siegfried, E., Han, G., & Rockström, J. (2006). Reducing hazard vulnerability: Towards a common approach between disaster risk reduction and climate change adaptation. *Disasters, 30*(1), 39–48.

UN. (2015a). Sendai framework for disaster risk reduction 2015–2030. Resolution adopted by the General Assembly on 3 June 2015. United Nations Resolution A/RES/69/283.

UN. (2015b). Transforming our world: The 2030 agenda for sustainable development. resolution adopted by the general assembly on 25 September 2015. United Nations Resolution A/RES/70/1.
UN OCHA. (2015). Flash Appeal. Emergency response plan for Vanuatu. Tropical Cyclone Pam, March–June 2015.
UNFCCC. (2015). Adoption of the Paris Agreement. Proposal by the President. Draft decision on 12 December 2015. United Nations Framework Convention on Climate Change Resolution FCCC/CP/2015/L.9.
UNISDR. (2009). UNISDR Terminology on Disaster Risk Reduction. UNISDR, Geneva, Switzerland.
Vachette, A. (2015). The little handbook of disaster and climate change networked governance structure in Vanuatu. Centre for Disaster Studies of James Cook University, Secretariat of the Pacific Community & Deutsche Gesellshaft für Internationale Zusammenarbeit GmbH.
Van Rooyen, C. C. (2015). Vanuatu: A brave nation responds to the storm. Devpolicy Blog. From the Development Policy Centre, 7 April 2015. Retrieved April 15, 2016, from http://devpolicy.org/vanuatu-a-brave-nation-responds-to-the-storm-20150407/.
Varda, D. M., Forgette, R., Banks, D., & Contractor, N. (2009). Social network methodology in the study of disasters: Issues and insights prompted by post-Katrina research. *Population Research and Policy Review, 28*(1), 11–29.
Venton, P., & La Trobe, S. (2008). *Linking climate change adaptation and disaster risk reduction*. London: Tearfund.

Author Biography

Astrid Vachette is a social scientist at the Centre for Disaster Studies (JCU, Australia) specialising in organisational cooperation for emergency management, disaster risk reduction, climate change adapation and sustainable development. She has a MSSc in International Humanitarian Action and NGO Management for Extreme Environments (UPEC, France) and a PhD in Environmental Science and Management (JCU, Australia) on Networked Disaster Governance in Vanuatu.

Chapter 8
Response of *Marginopora vertebralis* (*Foraminifera*) from Laucala Bay, Fiji, to Changing Ocean pH

Roselyn Naidu, Pamela Hallock, Jonathan Erez and Matakite Maata

Introduction

Climate change and human activities have increased the release and accumulation of greenhouse gases, such as CO_2, into the atmosphere (Gattuso and Lavigne 2009). Emissions of CO_2 into the atmosphere over the past two centuries, combined with changes in land use and terrestrial vegetation, have resulted in increasing concentration of dissolved CO_2 in the ocean. The resulting impact on marine carbonate chemistry has been called a phenomenon called ocean acidification (e.g., Feely et al. 2004; Caldeira and Wickett 2005; Cao and Caldeira 2008; Gattuso and Hansson 2011). As CO_2 dissolves in seawater (H_2O), it combines with water molecules to form HCO_3^- and H^+ ions, the latter increasing the acidity of seawater and decreasing pH. The world oceans have already taken up about 50% of the anthropogenic CO_2 released (Sabine 2004). The recent global mean pH is 8.1, which is 0.1 pH units below the pre-industrial value of the open surface ocean (Sabine 2004). The International Panel on Climate Change (IPCC) predicted future scenarios, indicating that the atmospheric CO_2 will increase to concentrations as much as 970 ppm by the year 2100 (IPCC 2013). The continuous increase in CO_2 concentrations likely will result in a pH decrease of 0.46 units by 2100 (Caldeira

R. Naidu (✉)
UniFiji-USAID PACAM Project, The University of Fiji, Saweni Campus, Lautoka, Fiji
e-mail: RoselynN@unifiji.ac.fj

P. Hallock
College of Marine Science, University of South Florida, St. Petersburg, FL 33701, USA

J. Erez
Freddy and Nadine Hermann Institute of Earth Sciences, The Hebrew University of Jerusalem, Edmond J. Safra Campus, Givat Ram, 91904 Jerusalem, Israel

M. Maata
School of Biological and Chemical Sciences, University of South Pacific, Laucala Campus, Suva, Fiji

W. Leal Filho (ed.), *Climate Change Adaptation in Pacific Countries*,
Climate Change Management, DOI 10.1007/978-3-319-50094-2_8

and Wickett 2005). Reduced carbonate ion concentrations and saturation state will require greater energy expenditure for organisms to calcify (Erez 2003). An increasing number of field and laboratory studies are indicating negative impacts on marine calcifying organisms, such as corals and foraminifera (Kleypas et al. 1999; Riebesell et al. 2000; Doo et al. 2014 and references therein).

The most abundant shelled organisms, and second-most abundant carbonate-precipitating organisms (after coccolithophorids), in the world's oceans are the Foraminifera. Their shells are commonly known as "tests". The most abundant foraminifers in the upper oceans precipitate calcium carbonate for their test formation by seawater vacuolization (Erez 2003). According to Erez (2003), the calcification is manifested by the formation of organic matrix where the granules provide Ca^{2+} for the first calcium carbonate crystals. These crystals precipitate over previously formed chambers, providing the overall shape of the foram. The bulk of the calcification process continues with secondary lamination involving vacuolization of seawater and is responsible for bulk of skeletal deposition. The secondary calcification occurs in a delimited space created by the pseudopodia (Erez 2003).

Based on the absence of calcareous benthic foraminifera in acidic seawater associated with natural seafloor venting of carbon dioxide in a variety of shallow-water locations, Uthicke et al. (2013) concluded that, if ocean acidification continues at predicted rates, many benthic foraminifera will be extinct by 2100. However, Pettit et al. (2013) did not find a significant response in the benthic foraminiferal assemblages to reduced pCO_2 in deeper shelf and slope settings.

Similarly, laboratory experiments examining responses of calcareous foraminifera to lower pH and elevated pCO_2 have shown significant differences among different taxa and even within the same species (Doo et al. 2014). The majority of studies have shown that pH lower than 7.8 can have deleterious effects on calcareous benthic foraminifera (Kuroyanagi et al. 2009; Dias et al. 2010; Fujita et al. 2011; Vogel and Uthicke 2012; McIntyre–Wressnig et al. 2013; Knorr et al. 2015). These effects include reduced calcification (Kuroyanagi et al. 2009; Moy et al. 2009; Haynert et al. 2011) and decreased growth rates (Manno et al. 2012; Reymond et al. 2012), indicating the decline in carbonate sediment production by foraminifera as ocean acidification proceeds (Dias et al. 2010; Knorr et al. 2015). However, a few studies have reported increased rates of calcification at responded to increased pCO_2 (e.g., Fujita et al. 2011; Vogel and Uthicke 2012).

The present study investigated the effects of ocean acidification on the growth of *Marginopora vertebralis* Quoy and Gaimard (1830). Larger benthic foraminifers, such as *M. vertebralis,* produce calcite skeletons and are major contributors to sand production on reef flats that nourish the beaches in Pacific Islands. Our working hypothesis was that *M. vertebralis* are sensitive to changes in seawater carbonate chemistry such that reduced pH should affect their ability to form calcareous tests. *Marginopora vertebralis* lives in the shallowest reef-associated environments (i.e., the intertidal and shallow subtidal zone) and is an important producer of sediment in Fijian beaches (Sharma 2007); its demise could have serious consequences for beach stability, especially under rising sea level.

Methods

Sample Collection

The samples of *M. vertebralis* were collected from Makuluva and Nukubuco reefs on the Suva Barrier Reef (Fig. 8.1). Sharma (2007) identified a total of 68 different species of benthic foraminifers in the Nukubuco Reef flat in Laucala Bay, including abundant *M. vertebralis*. Moreover, Sharma cultured living specimens to study growth, abundance and reproduction of marine benthic foraminifers.

The location of the collection site was determined using GPS data. Salinity and temperature of the sample collection area were assessed using an YSI-85 m. The map of Viti Levu is enlarged in Fig. 8.1c to show the research site. During low tide

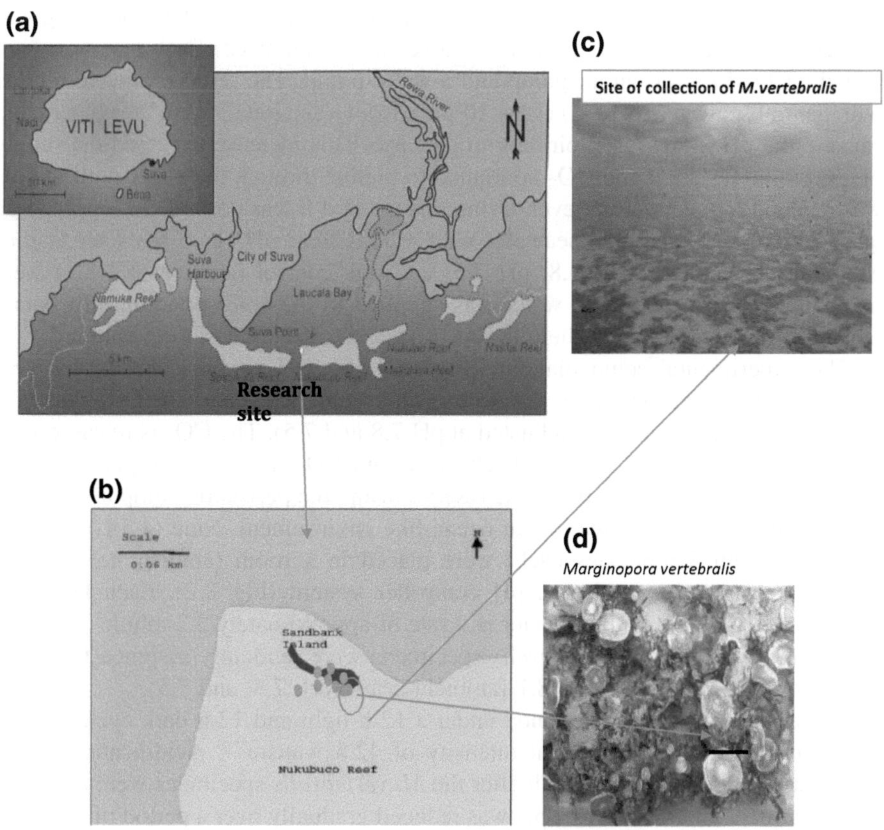

Fig. 8.1 **a** Map of Laucala Bay; **b** the mapping of the site of collection of *Marginopora vertebralis;* **c** algal flat habitat between Sandbank Island and Nukubuco Reef; and **d** *M. vertebralis* attached to algae and coral (*arrow*). *Scale bars* **a** 5 km; **b** 0.06 km (*Source*: SOPAC 2012; Sharma 2007); **d** 1 cm

(0.3 m), *M. vertebralis was* easily collected as it is large enough to be identified in situ. The foraminifers and associated sediments were collected using hand scoops and immediately placed in an aerated aquarium with filtered seawater, where they were kept overnight. The next day, *M. vertebralis* specimens had crawled upward and attached themselves to the sides of the aquarium and therefore could be easily identified and removed, based on the methods of Sharma (2007).

Natural seawater (salinity 33.0), contained in 35 L reservoir tanks, was acidified by means of a bubbling system supplying CO_2 gas (Fig. 8.2). The gas, saturated with water vapour (to limit evaporation) was injected through the water as very fine bubbles allowing the gas to rapidly dissolve. The seawater pH in the reservoirs was monitored using a flat-surface, combination pH electrode (YSI Environmental pH 100). Once the pH reached the required level, the supply of CO_2 was halted via an automated feedback relay system (Accu-Max pH controller). As the acidified water was pumped from the reservoir to supply the experimental tanks, the overflow water was returned to the reservoir. Whenever the pH increased, CO_2 bubbling was triggered to ensure that the desired pH level was maintained. Each reservoir contained a CO_2 Reactor, an air pump and a water pump. The reservoirs were refilled with natural seawater (pH 8.1) every 10 days from a separate 300 m^3 seawater tank, causing the pH in the reservoir tank to increase. This increase triggered the supply of CO_2 to be restarted and CO_2 continued to bubble through the water until the pH was reduced to the required level. Using this method it was possible to supply large quantities of CO_2 acidified seawater with a consistent pH. Two seawater acidification reservoir tanks (pH 7.8, pH 7.5) and one control (ambient pH 8.1) were maintained. From each tank, water was supplied to the corresponding three small tanks ($25 \times 10 \times 15$ cm) where the *M. vertebralis* specimens were cultured.

The experimental setup diagram (Fig. 8.2) shows how the connections were made from the reservoirs to the connecting tanks and vice versa. The CO_2 system is plugged with pH controllers (adjusted at pH 7.8 and 7.5). The CO_2 is released with the help of the bubble counter, check valve and CO_2 injector. The pH controller regulates and stabilizes the pH in the reservoirs and the peristaltic pump helps in the circulation of seawater, creating an ocean-like environment. Nine (4.5 L) tanks (3 treatments with 3 replicates each) were placed in a room (ambient temperature ~27.5 °C) and connected to a pH controller system (Fig. 8.2). Each tank was individually supplied with seawater at a rate of approximately 3.3 $mlmin^{-1}$ using a peristaltic pump. The *M. vertebralis* specimens were randomly assigned to one of the three pH treatment levels (8.1 [ambient seawater], 7.8, and 7.5).

The treatments were maintained under a 12-h light and 12-h dark cycle using a T5 Aquarium lamp with a light intensity of 42.5 $wattsm^{-2}$. Acidification of the seawater did not begin until 24 h after the *M. vertebralis* specimens were placed in their treatment tanks. Seawater pH was reduced gradually over a period of two days and the experiment started when the final water chemistry for each treatment was reached. The experiment ran for 11 weeks, during which time, the pH of the water supplied to the tanks was monitored daily via a pH controller (Accu Max I). Samples for Total Alkalinity were taken at the beginning of the experiment and then once per week throughout the duration of the experiment. The values given by the

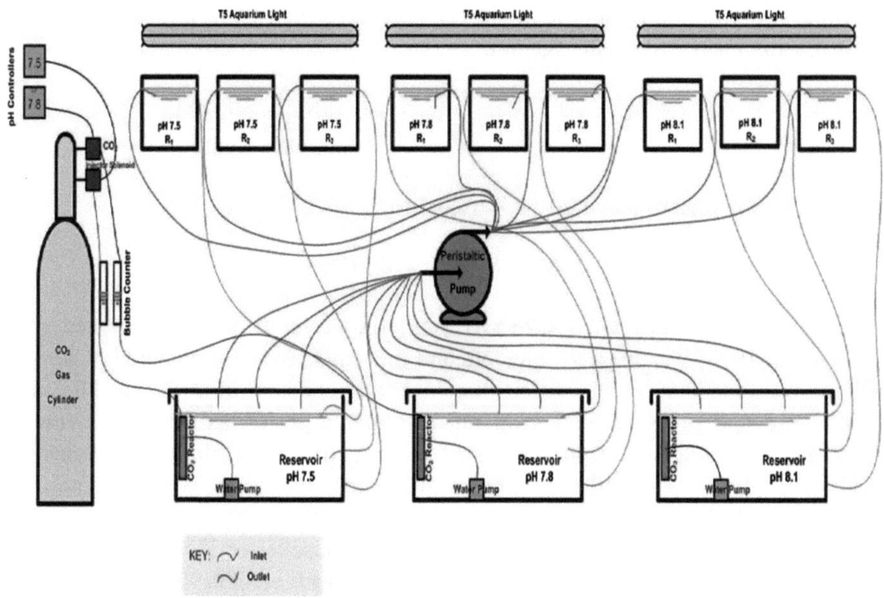

Fig. 8.2 Experimental set-up model

pH electrodes in the reservoir tanks were cross-checked every week against values measured by a regularly calibrated pH meter (InLab 413SG, Mettler–Teledo).

The carbonate chemistry in the culture media was determined using the following procedures and calculations (Pawlowski 1994). Borosilicate bottles, used to collect the water samples for alkalinity measurements, were rinsed with concentrated HCl, followed by at least five rinses with deionised water. Seawater samples were collected from the experimental aquaria. Borosilicate bottles were fully filled and tightly capped. Analysis was carried out within 6 h of sample collection. To standardize acid used in titrations, 10 ml aliquot of 0.01 M sodium tetraborate was placed into a 250 ml conical flask and a few drops of mixed indicator were added. The aliquot was titrated with 0.01 M HCl until the end point was reached (color changed from green to purple). The titre was recorded for the calculation of HCl concentration:

$$\text{Conc of HCl} = \frac{\text{Vol. of Borax} \times \text{Mass of Borax} \times 1000 \times 2}{\text{Vol. of titre} \times 381.37 \times 100}$$

To determine the total alkalinity (A_T) of a sample, 50 ml of the water sample were pipetted into a 250 ml Erlenmeyer flask and few drops of phenolphthalein indicator were added. Once the solution color turned pink, it was titrated with 0.01 M HCl until the color disappeared and the volume of added acid was recorded. Then 2–3 drops of mixed bromocrescol green/methyl red indicator were added to the same solution and titrated with standardized 0.01 M HCl to a pink color. Then A_T was calculated using this formula:

$$\text{Total Alkalinity, } A_T(\text{mg/L}) = \frac{\text{A} \times \text{M} \ \times 50{,}000}{\text{Sample volume}}$$

where A = Volume of standard HCl used (ml)

M = Molarity of HCl

The MATLAB program CO2SYS (single-input mode directly adapted from Lewis et al. 1998) was used to calculate and present seawater parameters where the salinity was constant at 35 and temperature was constant at 27.5 °C. The input temperature and pressure were from measurements performed in the laboratory. There are four measurable parameters of the aquatic carbon dioxide system: pH, pCO_2, total dissolved inorganic carbon (DIC) and total alkalinity (TA). The measured pH and alkalinity values were entered into the program. The equilibrium constants from Mehrbach et al. (1973) as refit by Lueker et al. (2000) on a total scale were used. The measurements made by Mehrbach et al. (1973) were made on real seawater.

Input variables (input conditions):

- Salinity (35), Temperature (27.5 °C) and Pressure (1 atm).
- Total Si (optional) and Total Phosphate (optional). If left empty, the total Si and total P concentrations are assumed to be zero in the calculations.
- Two (2) known CO_2 parameters (TA, pH).

Output (for both "input" and "output" conditions):

- The other CO_2 parameters (pCO_2), total dissolved inorganic carbon [DIC].
- Contributions to the alkalinity.
- Carbonate speciation.

Growth Assessments

The maximum diameter (d) of the specimens at the start of the experiments ranged from $0.4 \leq d \leq 1.4$ cm. The 25 *M. vertebralis* specimens in each treatment were weighed prior to the start of the experiment and once each week as a group using an analytical balance (Libror-Aex 200 g). The specimens were collected using hand scoops, pat dried using tissue, and wet weights of the group of specimens from each replicate of each treatment were determined.

To assess linear growth, Calcein (-bis [N, N-bis (carboxymethyl) aminomethyl]-fluorescein) was used to fluorescently label existing chambers in *M. vertebralis*. Calcein makes newly formed chambers distinguishable when the specimens are

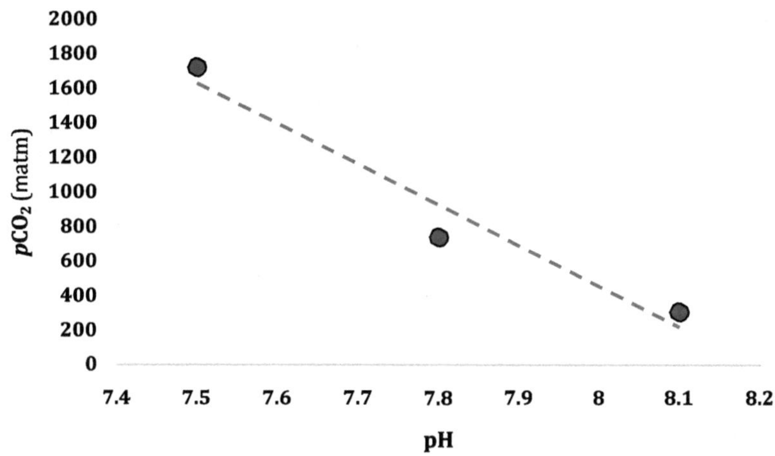

Fig. 8.3 pCO_2 in seawater of two pH treatments (7.5 and 7.8) and control (8.1)

viewed under a confocal microscope (Dissard et al. 2009). Prior to placement of the specimens into their treatment tanks, the foraminifers were placed in a solution at 40 μM of Calcein for approximately 48 h to allow the incorporation of the green fluorescent cell marker (Fig. 8.3). Once Calcein is incorporated inside the cells, Calcein AM (a nonfluorescent molecule) is hydrolyzed by endogeous esterase into the highly negatively charged green-fluorescent Calcein. The fluorescent Calcein is retained in the cytoplasm in living cells but is not incorporated into the subsequently added chambers (Papadopoulos et al. 1994). At the end of the experiment, this technique helps to distinguish between the pre-existing (stained) and the newly grown (unstained) calcite (Bernhard et al. 2004; Dissard et al. 2009).

Growth of the selected specimens was assessed using confocal imagery (Nikon ECLIPSE TE2000-U). The portion of the radius of the calcite shell added after Calcein incubation revealed growth that occurred during experimental treatments. The ruler property from EZC_1 3.91 imaging software was used to determine the increase in radius in μm. The mean increases in radius at each treatment for each replicate was recorded.

Results

Seawater Chemistry

Seawater chemistry contrasted strongly between the 2 treatments and control, with mean values ranging from 8.1 to 7.5 pH units and 309 to 1717 matm pCO_2, respectively. Mean total alkalinity was 2277 μmol kg^{-1} (n = 11, SD = 101). The alkalinity values were between 2176 and 2408 μmol kg^{-1}, a range we used together with the pH measurements from all aquaria (treatments) to calculate pCO_2 concentrations (Table 8.1).

Table 8.1 Carbonate system parameters in the experiment. Temperature (T), salinity (S), total alkalinity (A_T), and the total scale pH (pH_T) were measured directly. The treatment and control values are mean of measurements (n = 11) taken weekly over the course of the experiment. The remaining values were calculated based on methods of Lewis et al. (1998), with equilibrium constants from Mehrbach et al. (1973) as refit by Lueker et al. (2000) on total scale. Partial pressure of CO_2 in seawater (pCO_2) is in milliatmospheres (matm); $[OH^-]$ is given in mmolkg^{-1}, A_T, $[HCO_3^-]$ and $[CO_3^{2-}]$ is given in mmolkg^{-1} SW

pH treatment	T ° C	S	pH_T	A_T mmolkg^{-1} SW	pCO_2 matm	$[OH^-]$ mmolkg^{-1}	$[HCO_3^-]$ mmolkg-^{1} SW	$[CO_3^{2-}]$ mmolkg^{-1} SW
8.1	27.5	35	8.1 (0.05)	2176	309.1	9.66	1577.3	240.6
7.8	27.5	35	7.8 (0.05)	2247	738.5	4.84	1891.4	144.6
7.5	27.5	35	7.5 (0.05)	2408	1717	4.8	2203.5	84.5

The pH measurements correlated with the long term averages of pCO_2 for the two treatments and control (Fig. 8.3, $R^2 = 0.95$).

Growth in M. vertebralis

We cultured 225 *M. vertebralis* individuals in all, 25 specimens per replicate, and three replicates at three different pH values: pH 7.5, 7.8, and 8.1 (control). Specimens cultured at pH 7.5 showed a 429 μm mean increase in radius, while in those at pH 7.8, the mean increase in radius was 441 μm (Table 8.2). At pH 8.1, the mean increase in radius was 1024 μm, which was more than double the increase noted for pH 7.5 and 7.8. Similarly, the masses of the 25 *M. vertebralis* specimens from each replicate of each treatment (Table 8.2) also revealed limited growth at pH 7.5 averaging 10.2 ± 3.6 mg per specimen, with slightly more at pH 7.8, averaging 14.5 ± 3.7 mg. The specimens cultured at pH 8.1 grew an average of 30.0 ± 5.7 mg. The relationship between increase in radius and increase in mass are shown in Fig. 8.4. The relative weight gain (%) by the 25 individuals of *M. vertebralis* at each pH treatment during 11 week experiment revealed that the specimens in the pH 7.5 treatments grew on average by 3.24%, whereas individuals at pH 7.8 grew on average by 4.24%. At pH 8.1, average growth was 8.39% for the 11 weeks culture. The shell growth in *M. vertebralis* was negatively related to elevated pCO_2 (Fig. 8.5) and was positively related to increased CO_3^{2-} (Fig. 8.6).

Other Observations

At the end of the experiment, most specimens contained pale pink cytoplasm, which indicated that they were alive throughout the entire experimental period. Several specimens reproduced asexually. When an individual about to reproduce, its color changes to white with a pale purple ring in the middle. During reproduction

Table 2 The mean increase in individual test radius and test weight, both assessed after 11 weeks

Treatment	Replicate	Increase in radius (μm)	Increase in weight (mg)
7.5	R_1	471 ± 102	6.05
	R_2	489 ± 77	12.3
	R_3	326 ± 23	12.3
	Mean ± SD	429 ± 89	10.2 ± 3.6
7.8	R_1	476 ± 43	8.71
	R_2	480 ± 31	13.1
	R_3	367 ± 38	11.7
	Mean ± SD	441 ± 64.1	14.5 ± 3.7
8.1	R_1	1019 ± 346	24.3
	R_2	785 ± 107	27.3
	R_3	1267 ± 306	35.4
	Mean ± SD	1024 ± 241	30.0 ± 5.7

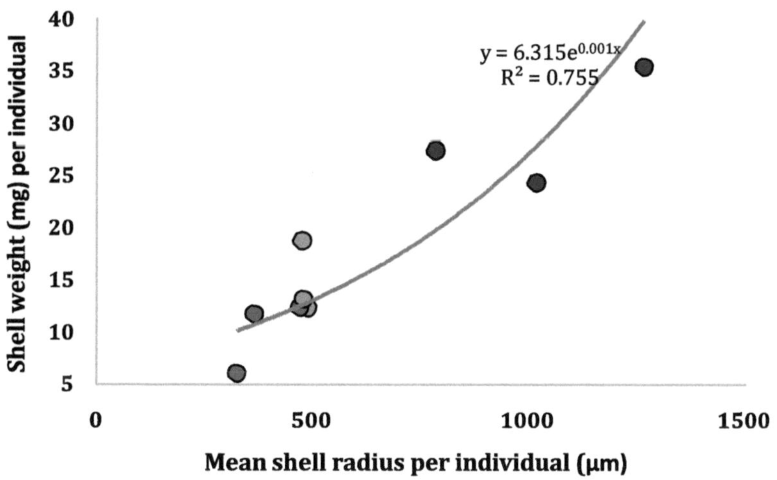

Fig. 8.4 Relationship between increase in shell radius and in average shell weight of *Marginopora vertebralis* measured after 11 weeks in culture (*red*—pH 7.5, *green*—pH 7.8, and *blue*—pH 8.1) (Color figure online)

(Fig. 8.7), the parent shell becomes white as the endoplasm containing the dinoflagellate symbionts is incorporated into the megalospheric offspring. Reproduction was only noted at the pH treatment of 8.1.

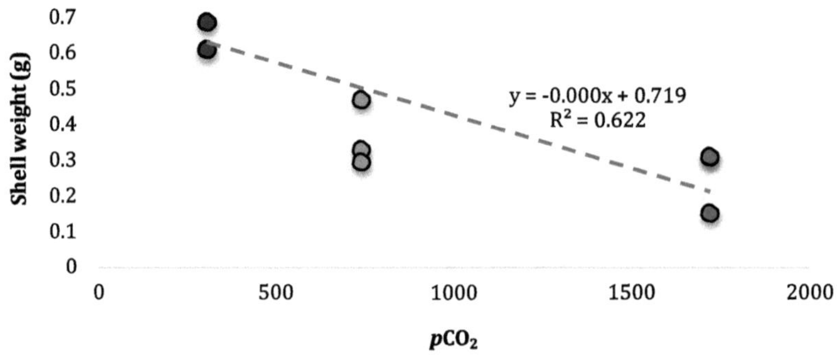

Fig. 8.5 Mean increase in shell weight of 25 *M. vertebralis*, grouped by pH level (n = 11) at three pCO_2 levels

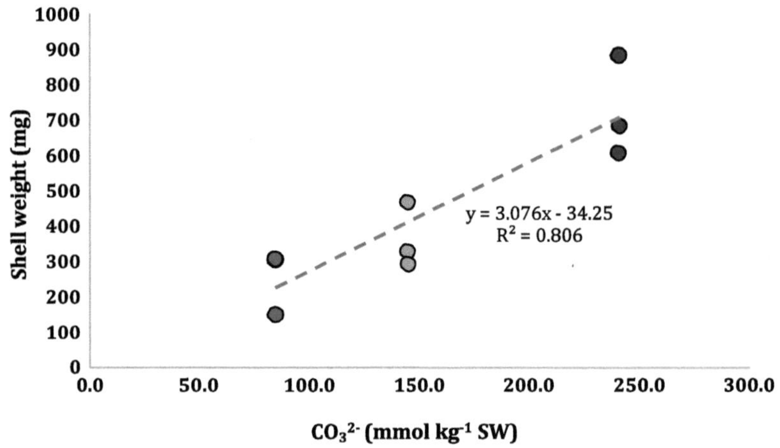

Fig. 8.6 Mean shell weight (mg) of 25 *M. vertebralis*, grouped by treatment (n = 11) against CO_3^{2-} (mmol kg^{-1} SW)

Discussion

Our experimental results revealed that growth in *Marginopora vertebralis* can be inhibited at pH of 7.5 and 7.8 compared to growth in ambient seawater at pH 8.1. This inhibition occurred with respect to growth in diameter of the experimental specimens and in addition of mass to the shells (i.e., calcification). Our results are consistent with the majority of previous studies that have shown that pH lower than 7.8 can have deleterious effects on calcareous larger benthic foraminifers (Kuroyanagi et al. 2009; Dias et al. 2010; Fujita et al. 2011; Vogel and Uthicke 2012; McIntyre–Wressnig et al. 2013; Knorr et al. 2015). These effects have included reduced calcification (Kuroyanagi et al. 2009; Moy et al. 2009; Haynert et al. 2011) and decreased growth rates (Manno et al. 2012; Reymond et al. 2012).

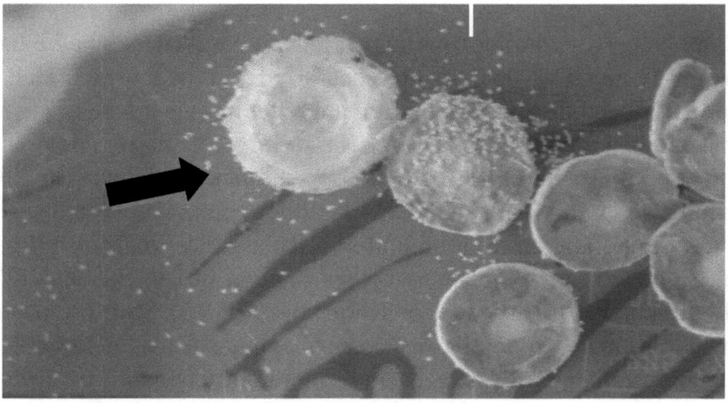

Fig. 8.7 Reproduction of *Marginopora vertebralis* at pH treatment 8.1 (*black arrow*)

However, as Doo et al. (2014) demonstrated in their review of studies examining responses of calcareous foraminifers to lower pH and elevated pCO_2, differences in responses among different taxa and even within the same species have been reported. For example, Vogel and Uthicke (2012) reported that *M. vertebralis* responded to increased pCO_2 with increased rates of calcification from laboratory experiments, while Uthicke and Fabricius (2012) reported reduced calcification in both field studies and laboratory observations.

More recent studies have added to the uncertainties and questions, as Sinutok et al. (2014) reported reduced growth under elevated pCO_2, Schmidt et al. (2014) reported somewhat mixed results, and Prazeres et al. (2015) found no influence of elevated pCO_2 on calcification. However, Schmidt et al. (2014) were largely examining response to elevated temperature, as their lowest pH was 7.9. Prazeres et al., on the other hand, were working at a relatively low temperature of 24 °C. Thus, the most logical conclusion that can be drawn from what may seem to be contradictory results are that a variety of environmental factors are influencing growth and calcification, even in well controlled experiments. The ambient seawater used in different experiments in different laboratories likely differs in salinity, nutrients, dissolved organic matter or trace element. Food available to the experimental specimens certainly differs among experiments, as do ambient light conditions.

What also remains to be determined is whether *M. vertebralis* exposed to elevated pCO_2 for an entire life cycle can acclimate or if long-term exposure will result in decreased calcification rates, as has been observed for other benthic foraminifera (Dissard et al. 2010; Fujita et al. 2011; Haynert et al. 2011; Knorr et al. 2015). Moreover, the combined effects of temperature, nutrient availability and CO_3^{2-} need to be studied to estimate the impact of oceanic environmental changes on *M. vertebralis* calcite production. The studies by Schmidt et al. (2014) and Prazeres et al. (2015) provide models in examining the interactions of pH and other parameters, and also in examining other physiological responses of the foraminifers.

Conclusions

The shells of larger foraminifers on the reef flats of coral islands in the Indo-Pacific can make up as much as 90% of the sand-sized sediments (e.g., Hallock 1981; Fujita et al. 2016). While uncertainties remain regarding the range of environmental factors that influence calcification in *M. vertebralis*, our observations that growth and calcification in *M. vertebralis* decreases as a function of decreasing $[CO_3^{2-}]$ in seawater indicate that increasing pCO_2 in this century could reduce carbonate production by important larger foraminifers such as *M. vertebralis* at least by half. While this estimate is conservative compared to those of Uthicke et al. (2013), who predicted extinction of many taxa, or Knorr et al. (2015), who estimated an 85% reduction in carbonate production by some major taxa, loss of even 50% of the carbonate sand production by larger benthic foraminifers could be devastating for low lying coral islands surrounded by reef flats. Reductions in the major source of sand- sized sediments at the same time that sea level is rising clearly compounds the threats associated with climate change for human residents of low-lying islands.

Future Research Topics

- Future studies may seek to culture other important species of foraminifera around Fiji and compare their response to increasing ocean acidity.
- Modeling and documentation of carbonate production differences in Foraminifera in a historical context, along with present and future projections will be vital to determine coastal protection and conservation strategies.

References

Bernhard, J. M., Blanks, J. K., Hintz, C. J., & Chandler, G. T. (2004). Use of the fluorescent calcite marker calcein to label foraminiferal tests. *Journal of Foraminiferal Research, 34*(2), 96–101.

Caldeira, K., & Wickett, M. (2005). Ocean model predictions of chemistry changes from carbon dioxide emissions to the atmosphere and ocean. *Journal of Geophysical Research, 110*(9), Article Number: C09S04.

Cao, L., & Caldeira, K. (2008). Atmospheric CO_2 stabilization and ocean acidification. *Geophysical Research Letters, 35*(19), Article Number: L19609.

Dias, B., Hart, M., Smart, C., & Hall-Spencer, J. (2010). Modern seawater acidification: The response of foraminifera to high-CO_2 conditions in the Mediterranean Sea. *Journal of the Geological Society, 167*(5), 843–846.

Dissard, D., Nehrke, G., Reichart, G., Nouet, J., & Bijma, J. (2009). Effect of the fluorescent indicator calcein on Mg and Sr incorporation into foraminiferal calcite. *Geochemistry, Geophysics, Geosystems, 10*(11), Article Number: Q11001.

Dissard, D., Nehrke, G., Reichart, G., & Bijma, J. (2010). Impact of seawater pCO2 on calcification and Mg/Ca and Sr/Ca ratios in benthic foraminifera calcite: Results from culturing experiments with Ammonia tepida. *Biogeosciences, 7,* 81–93.

Doo, S., Fujita, K., Byrne, M., & Uthicke, S. (2014). Fate of calcifying tropical symbiont-bearing large benthic foraminifera: Living sands in the changing ocean. *Biological Bulletin, 226,* 169–186.

Erez, J. (2003). The source of ions for biomineralization in foraminifera and their implications for paleoceanographic proxies. *Reviews in Mineralogy and Geochemistry, 54*(1), 115–149.

Feely, R., & Doney, S. (2011). Ocean acidification: The other CO_2 problem. *Limnology and Oceanography e-Lectures.*

Feely, R. A., Sabine, C. L., Lee, K., Berelson, W., Kleypas, J., Fabry, V. J., et al. (2004). Impact of anthropogenic CO_2 on the $CaCO_3$ system in the oceans. *Science, 305,* 362–366.

Fujita, K., Hikami, M., Suzuki, A., Kuroyanagi, A., Sakai, K., Kawahata, H., et al. (2011). Effects of ocean acidification on calcification of symbiont-bearing reef foraminifera. *Biogeosciences, 8*(8), 2089–2098.

Fujita, K., Otomaru, M., Lopati, P., Hosono, T., & Kayanne, H. (2016). Shell productivity of the large benthic foraminifer *Baculogypsina sphaerulata,* based on the population dynamics in a tropical reef environment. *Coral Reefs, 35*(1), 317–326.

Gattuso, J., & Lavigne, H. (2009). Technical note: Approaches and software tools to investigate the impact of ocean acidification. *Biogeosciences, 6*(10), 2121–2133.

Gattuso, J.-P., & Hansson, L. (2011). *Ocean acidification*. Oxford: Oxford University Press.

Hallock, P. (1981). Production of carbonate sediments by selected foraminifera on two Pacific coral reefs. *Journal of Sedimentary Petrology, 51,* 467–474.

Haynert, K., Schönfeld, J., Riebesell, U., & Polovodova, I. (2011). Biometry and dissolution features of the benthic foraminifer *Ammonia aomoriensis* at high pCO_2. *Marine Ecology Progress Series, 432,* 53–67.

Kleypas, J. A., Buddemeier, R. W., & Gattuso, J. P. (1999). Geochemical consequences of increased atmospheric carbon dioxide on coral reefs. *Science, 284*(5411), 118–120.

IPCC. (2013). Summary for policymakers. In T.F. Stocker, D. Qin, G.-K. Plattner, M. Tignor, S. K. Allen, J. Boschung, et al. (Eds.), *Climate change 2013: The physical science basis. contribution of working group I to the fifth assessment report of the intergovernmental panel on climate change*. Cambridge, United Kingdom and New York, NY, USA: Cambridge University Press.

Knorr, P., Robbins, L., Harries, P., Hallock, P., & Wynn, J. (2015). Response of the miliolid *Archaiasangulatus* to simulated ocean acidification. *Journal of Foraminiferal Research, 45*(2), 109–127.

Kuroyanagi, A., Kawahata, H., Suzuki, A., Fujita, K., & Irie, T. (2009). Impacts of ocean acidification on large benthic foraminifera: Results from laboratory experiments. *Marine Micropaleontology, 73*(3–4), 190–195.

Lea, D., Mashiotta, T., & Spero, H. (1999). Controls on magnesium and strontium uptake in planktonic foraminifera determined by live culturing. *Geochimica et Cosmochimica Acta, 63* (16), 2369–2379.

Lewis, E., Wallace, D., & Allison, L. (1998). *Program developed for CO_2, system calculations.* Oak Ridge, Tennessee: Carbon Dioxide Information Analysis Center, managed by Lockheed Martin Energy Research Corporation for the U.S. Department of Energy.

Lueker, T., Dickson, A., & Keeling, C. (2000). Ocean pCO_2 calculated from dissolved inorganic carbon, alkalinity, and equations for K_1 and K_2: Validation based on laboratory measurements of CO_2 in gas and seawater at equilibrium. *Marine Chemistry, 70*(1–3), 105–119.

Manno, C., Morata, N., & Bellerby, R. (2012). Effect of ocean acidification and temperature increase on the planktonic foraminifer *Neogloboquadrinapachyderma* (sinistral). *Polar Biology, 35*(9), 1311–1319.

McIntyre-Wressnig, A., Bernhard, J., McCorkle, D., & Hallock, P. (2013). Non-lethal effects of ocean acidification on the symbiont-bearing benthic foraminifer *Amphisteginagibbosa. Marine Ecology Progress Series, 472,* 45–60.

Mehrbach, C., Culberson, C., Hawley, J., & Pytkowicz, R. (1973). Measurement of the apparent dissociation constants of carbonic acid in seawater at atmospheric pressure. *Limnology and Oceanography, 18*(6), 897–907.

Moy, A., Howard, W., Bray, S., & Trull, T. (2009). Reduced calcification in modern Southern Ocean planktonic foraminifera. *Nature Geoscience, 2*(4), 276–280.

Neuer, S., Iversen, M., & Fischer, G. (2014). Ocean acidification: The other CO_2 problem. *Limnology and Oceanography e-Lectures.*

Orr, J., Fabry, V., Aumont, O., Bopp, L., Doney, S., Feely, R., et al. (2013). Benthic foraminifera show some resilience to ocean acidification in the Northern Gulf of California. *Mexico. Marine Pollution Bulletin, 73*(2), 452–462.

Papadopoulos, N., Dedoussis, G., Spanakos, G., Gritzapis, A., Baxevanis, C., & Papamichail, M. (1994). An improved fluorescence assay for thedetermination of lymphocyte-mediated cytotoxicity using flow cytometry. *Journal of Immunological Methods, 177*(1–2), 101–111.

Pawlowski, L. (1994). Standard methods for the examination of water and wastewater, 18th ed. *Science of the Total Environment, 142*(3), 227–228.

Pettit, L., Hart, M., Medina-Sánchez, A., Smart, C., Rodolfo-Metalpa, R., & Hall-Spencer, J. et al. (2013). Benthic foraminifera show some resilience to ocean acidification in the northern Gulf of California, Mexico. *Marine Pollution Bulletin, 73*(2), 452–462.

Prazeres, M., Uthicke, S., & Pandolfi J. M. (2015). Ocean acidification induces biochemical and morphological changes in the calcification process of large benthic foraminifera. *Proceedings of the royal society B: Biological sciences, 282*(1803), Article Number: 20142782.

Reymond, C., Lloyd, A., Kline, D., Dove, S., & Pandolfi, J. (2012). Decline in growth of foraminifer *Marginoporarossi* under eutrophication and ocean acidification scenarios. *Global Change Biology, 19*(1), 291–302.

Riebesell, U., Zondervan, I., Rost, B., Tortell, P., Zeebe, R., & Morel, F. (2000). Reduced calcification of marine plankton in response to increased atmospheric CO_2. *Nature, 407,* 364–367.

Sabine, C. (2004). The oceanic sink for anthropogenic CO_2. *Science, 305*(5682), 367–371.

Schmidt, C., Kucera, M., & Uthicke, S. (2014). Combined effects of warming and ocean acidification on coral reef foraminifera *Marginopora vertebralis* and *Heterostegina depressa. Coral Reefs, 33*(3), 805–818.

Sen Gupta, B. (2003). *Modern foraminifera.* Dordrecht: Kluwer Academic Publishers.

Sharma, A. (2007). *A study of the benthic foraminifera of laucala bay, with special focus on Marginopora vertebralis.* Degree of Master of Science, The University of South Pacific.

Sinutok, S., Hill, R., Doblin, M., Wuhrer, R., & Ralph, P. (2011). Warmer more acidic conditions cause decreased productivity and calcification in subtropical coral reef sediment-dwelling calcifiers. *Limnology and Oceanography, 56*(4), 1200–1212.

Sinutok, S., Hill, R., Kühl, M., Doblin, M., & Ralph, P. (2014). Ocean acidification and warming alter photosynthesis and calcification of the symbiont-bearing foraminifera *Marginopora vertebralis. Marine Biology, 161*(9), 2143–2154.

Quoy, J. R., & Gaimard, P. (1830). Mollsques, vers et Zoophytes. In H. Blainville (Ed.), *Dictionnaire des sciences naturelles.* Paris: Levrault, F.

Uthicke, S., & Fabricius, K. (2012). Productivity gains do not compensate for reduced calcification under near-future ocean acidification in thephotosynthetic benthic foraminifer species *Marginopora vertebralis. Global Change Biology, 18*(9), 2781–2791.

Uthicke, S., Momigliano, P., & Fabricius, K. (2013). High risk of extinction of benthic foraminifera in this century due to ocean acidification. *Scientific Reports*, 3, Article Number: Q11P05.

Vogel, N., & Uthicke, S. (2012). Calcification and photobiology in symbiont-bearing benthic foraminifera and responses to a high CO environment.*Journal of Experimental Marine Biology and Ecology, 424–425*, 15–24.

Chapter 9
The Role and Capacity of Government in a Climate Crisis: Cyclone Pam in Vanuatu

Johanna Nalau, J. Handmer and Malcolm Dalesa

Introduction

The Republic of Vanuatu has long been regarded as one of the most vulnerable countries in the world due to its exposure to multiple hazards such as cyclones, earthquakes and tsunamis and its social and economic vulnerabilities (World Risk Report 2015). 75% of Vanuatu's population lives in rural areas scattered across many islands with high reliance on subsistence farming for their livelihoods (Malvatumauri 2012). Such population dispersal, across islands and distant rural areas is typical of Melanesian countries, and poses challenges to service delivery and disaster recovery (Wickham et al. 2009). This has also major bearings in terms of food security, climate adaptation, and disaster risk reduction strategies given both the projected climate change scenarios and current disaster vulnerability (Government of Vanuatu 2015a).

The country is highly exposed to cyclones and other hazards (https://en.wikipedia.org/wiki/Category:Tropical_cyclones_in_Vanuatu). It is threatened by about two cyclones a year. Other regular climate hazards are heavy rain, flooding, landslides and occasionally drought. It is also prone to regular earthquakes and contains a number of active volcanoes. Vanuatu has in effect, two economies, both highly vulnerable to natural hazards. There is a market economy largely within

J. Nalau (✉)
Griffith Climate Change Response Program and Griffith Institute for Tourism, Griffith University, Gold Coast, Australia
e-mail: j.nalau@griffith.edu.au

J. Handmer
Centre for Risk and Community Safety, Mathematical and Geospatial Sciences, RMIT University, Melbourne, Australia

M. Dalesa
UNDP Pacific Risk Resilience Programme, Port Vila, Vanuatu

W. Leal Filho (ed.), *Climate Change Adaptation in Pacific Countries*, Climate Change Management, DOI 10.1007/978-3-319-50094-2_9

urban areas and a subsidence economy, which provides livelihoods for the majority of people. Within the urban economy tourism is the largest employer. The majority of people depend on their own crops for survival. If these are damaged then external assistance is needed. Even the threat of a cyclone will lead to mass cancellations disrupting tourism—and this effect is felt across the country including areas far from any direct impacts.

As a Least Developed Country (LDC), Vanuatu ranks medium-low in socioeconomic development (e.g. Human Development Index) and its economy is largely influenced by Migration, Remittances, Aid and Bureaucracy similar to other Small Island Developing States (Kuruppu and Willie 2015), with tourism, aid, and construction as the three largest economic sectors (Government of Vanuatu 2015b). In the 2015 Human Development Report Vanuatu was ranked134 out of 188 countries, which is considered to be "medium", similar to Myanmar, Ghana and Nicaragua (UNDP 2015). Yet, paradoxically Vanuatu also ranked as the world's happiest country in 2006 and 2008 based on life expectancy, ecological footprint, and experienced well-being (NEF 2016). Vanuatu's position on the Human Development Index, its very high level of exposure to natural extremes, and its vulnerability to those extremes, reinforce its situation as a small island developing state. This also highlights its vulnerability to climate change, and the potential role of climate justice in supporting the country to adapt to change.

In March 2015, category 5 Tropical Cyclone Pam (hereafter TC Pam) made landfall in Vanuatu on the 13th of March 2015 causing wide spread damage. Out of Vanuatu's population of 246 000 approximately 188 000 people were impacted by TC Pam (Shelter Cluster 2015). 75,000 people were in need of shelters, 110 000 people in need of clean drinking water (OCHA 2015), and 81% of housing had some damage (Secretariat of the Pacific Community 2015), with economic damages up to 64.1% of Vanuatu's GDP (Government of Vanuatu 2015b). The worst impacted province was the southern Tafea province as the eye of TC Pam went directly over the island of Tanna on the 14th of March stripping off all the green vegetation. Approximately 60–80% of community structures were destroyed on Tanna, with significant damage to livestock, crops, and infrastructure (Tafea Provincial Disaster Committee and CARE International 2015). Following the Cyclone, a UN General Assembly resolution (A/RES/70/78, 9 December 2015), extended Vanuatu's status as a Least Developed Country by three years (until December 2020), as a result of "the unique disruption caused to the economic and social progress of Vanuatu by Cyclone Pam".

This chapter looks at the governance responses to TC Pam and in particular at recovery coordination and management, and identifies a range of contextual issues that acted as constraints in providing effective governance of the response. In doing so, it examines three different governance models, which emerged in the post-disaster context and explain to some extent the differing views on disaster response and its perceived success. The research presented in this chapter is based on a desktop study and discussions with multiple stakeholder groups in Vanuatu who were involved in TC Pam responses and governance. The chapter is organised in the following manner: section "Recovery Coordination and Management for

Cyclone Pam" examines recovery response coordination and management and identifies a range of contextual issues, which emerged in the recovery process. Section "Vanuatu's Disaster Risk Governance" discusses the future implications of TC Pam and similar storms in the Pacific, while section "Way Forward" provides the conclusions and recommendations arising from this research.

Tropical Cyclone Pam

Despite Vanuatu's long history of coping with cyclones, TC Pam was unprecedented in many ways. First, it was unprecedented in its impacts as the first category 5 cyclone to make landfall in Vanuatu, with wind bursts 0f up to 320 km/h hitting 22 of Vanuatu's 83 islands (Government of Vanuatu 2015b), with near record breaking central pressure of 896 hPa on March 14 lowest after Zoe at 890 hPa, and lower than the 2016 category 5 TC Winston at 915 hPa. Second, it generated an unprecedented volume of international response and overwhelmed the current system in place for disaster risk reduction, recovery and management (Barber 2015). Yet, faced with a potential increase in the intensity of extreme events due to climate change (Barber 2015), such events have the potential to become the new norm in the region. The question is what would such a new norm mean for climate adaptation and disaster risk reduction and what lessons does TC Pam offers for government preparedness in dealing with mega disasters in a small island developing state context.

TC Pam had a massive impact on Vanuatu's tourism industry. Most of the large hotels in Port Vila, such as Holiday Inn and Le Lagoon, had to close down due to extensive damage, which made accommodation an issue even if tourists would have been able to travel to Vanuatu. After the cyclone, both the Australia and New Zealand governments issued official travel warnings for travel to Vanuatu, which led to a large number of cancellations across the country, even in non-impacted areas such as Santo Island (Interview 3). The cruise ship industry took Vanuatu off destination schedules, which also had a major impact in Port Vila where many small and medium enterprises (taxi and bus drivers, tour operators, restaurants), both formal and informal, found that they were starved of cash flow in a time where many needed additional finances to reconstruct their homes and livelihoods.

Recovery Coordination and Management for Cyclone Pam

Vanuatu has spent much time developing an emergency response system that has been acknowledged internationally. For example, National DRR systems and policies are in place including National Disaster Response Plan, a National Cyclone Support Plan, national-level humanitarian clusters, a National Disaster Management Organisation with Standard Operating Procedures, and the Vanuatu Humanitarian

Team (VHT) that coordinates non-government organisations' actions. These had been found to function adequately in disasters that were small in scale (Barber 2015). Vanuatu has long been a forerunner in integrating climate change adaptation and disaster risk reduction and implementing this integration agenda also in practice by, for example, establishing the National Adaptation Board to oversee all adaptation and disaster risk reduction projects, programs, and initiatives (Nalau et al. 2015). The newly developed and endorsed *Vanuatu Climate Change and Disaster Risk Reduction Policy 2016–2030* recognizes the challenges of decentralized governance and aims to provide a holistic coordination approach that recognizes also the role of traditional governance systems in parallel with official government (Government of Vanuatu 2015a).

Yet, Vanuatu did not anticipate the magnitude of the cyclone and the magnitude of the international response, leading to underutilisation and problems in using the current Disaster Risk Reduction systems (Barber 2015). There was no plan by the national government to engage with such rapid and large-scale inflow of different actors and assistance to the country. The international actors, such as the UN agencies and other external organisations, did not always have a good understanding of the national systems, reporting structures, or roles and responsibilities within the country (Barber 2015; Nalau et al. 2015). TC Pam resulted in the mobilisation of many new NGOs and organisations in the post-disaster phase that often lacked the understanding of the operational context and culture in Vanuatu and had no previous working relationships with existing organisations and groups on the ground (Interview 4). In turn, the national government and NGOs were not familiar with the UN operational systems and ways of working in disaster contexts (Barber 2015). The national coordination and decision-making system is also complex as it involved multiple government agencies e.g. Prime Minister's Office, government ministries, donors, and other actors (Secretariat of the Pacific Community 2015). All this is in a context of a small dispersed population over scattered islands with limited infrastructure resulting in significant logistical difficulties. Hence, overall effective and integrated post-disaster governance remained a challenge although some sectors, such as water, worked fairly well through the cluster approach, where all directly relevant groups work together to resolve problems and deliver the desired outcomes.

An example of actors' differing perceptions of disaster recovery priorities was food aid, its timing and distribution. The government of Vanuatu made a strategic decision not to distribute food aid for the first 2–3 weeks on the basis that most of the communities should utilise first the fallen fruits and crops. However, some NGOs identified food aid as the first priority in the recovery process and began delivering food aid one week after TC Pam, with limited communication with the government about where and what had been distributed (Interview 3). The national government was concerned that immediate aid could create dependencies and expectations that might not be able to be met. Some NGOs felt that the priority was food aid to those who appeared vulnerable. This complicated the food aid process further as some communities received aid whereas others did not, and caused tensions between the government, the non-governmental actors and communities.

Another factor contributing to challenges was the lack of up-to-date demographic statistics: the latest population census was done in 2009 (Vanuatu National Statistics Office 2009) and appears to differ substantially from the current official count (Barber 2015; Secretariat of the Pacific Community 2015). This missing information complicated adequate estimates of the number of people needing support and the extent of needs in different islands and areas. For example, the census counted 27,000 people living on Tanna island, which was hit the hardest by TC Pam, but the actual figures in Tanna during disaster relief distribution, derived from needs assessments and community reporting, suggest a much higher number.

On top of the governance issues, several factors added to the problematic of delivering the disaster response including "poor infrastructure, maritime services and telecommunication problems" (Secretariat of the Pacific Community 2015, p. 23). Vanuatu has a geographic challenge in being comprised of 83 islands, which creates significant logistical difficulties for aid and service delivery (Nalau et al. 2015; Secretariat of the Pacific Community 2015). The Australian army helped with logistics for the first month after the cyclone but in particular after it had left, there were not enough boats and trucks to transport the needed aid materials to the islands (Interviews 3 and 4). Once aid arrived on remote islands, it was difficult to get the aid delivered away from the site of arrival, as the few roads were in poor condition and transport means very few (Barber 2015).

The majority of people depend on rainwater and many of the rainwater tanks and other water sources were damaged by the cyclone. This was especially a problem for some small communities on atolls, who found that their water supplies had been contaminated by the sea, leaving them without any potable water. This was a serious situation for the communities involved and required immediate assistance. Given the twin issues of very difficult logistics and widespread need, delivering sufficient water to remote islands was especially challenging (Interview 3). Another issue possibly affecting logistics and delivery time was thought to have been a bottleneck at Brisbane airport. Vanuatu officials believe that all relief goods in transit were treated as if they were entering Australia, and therefore required clearance from customs and quarantine. There is a belief that this caused delays in the delivery of needed supplies. Also, some of the donated items were inappropriate such as food types that were not relevant to local context (e.g. different rice type than what people were used to cooking) and in some cases out-of-date food items (Interviews 7 and 10).

Telecommunication was also an issue as the storm took out most of the communications to outer islands. For example, on Tanna Island, the telecommunication tower was restored only a week after TC Pam, and there was very little communication between the provincial government, other islands within Tafea province and the national level as to the level of damage in the province (Interview 4). The internet network was down four weeks in Tafea province and although community disaster committee members in Tafea were conducting damage assessments, there were no uploading or printing facilities available to send the information to the national level (Interview 4). This issue also caused delays in getting reliable information and estimates on the losses and damages across the islands to ensure a targeted relief effort.

Nevertheless, communications across Vanuatu have improved greatly over the past few years since the introduction of commercial mobile phone networks, and satellite phones were provided to provincial governments for emergencies.

Vanuatu's Disaster Risk Governance

Governance is generally defined as the quality of public administration, legal certainty and judicial capacity (EU). It is seen as having a good process, rather than necessarily "correct" decisions (http://www.goodgovernance.org.au/about-good-governance/what-is-good-governance/). Good disaster governance is not easily achieved, with the 2011 Global Assessment Report (UNISDR 2011, p. 116) concluding that "aside from reducing disaster mortality, existing risk governance capacities and arrangements generally fail to achieve their aims." Good governance attributes for Australian local government emphasise, among other attributes, accountability, transparency, and participation especially by those affected by the decisions being taken (http://www.goodgovernance.org.au/about-good-governance/what-is-good-governance/).

The ODI—UNDP Disaster Risk Governance Index:

The Overseas Development Institute (ODI) and UNDP have developed and applied a disaster risk governance index (ODI UNDP 2014). It takes the view that human development, political stability and democracy are needed for good disaster risk governance. However, the measured outcome depends on the details of the index applied (ODI-UNDP 2014, p.12). The index rates governance measures both disaster specific actions, such as plans, regulation, and policies, and more general attributes such as accountability, transparency, and participation. The resulting index ranks many countries highly, but many achieve low scores including Vanuatu.

The index is based on three existing indicators with global coverage, that focus on "generic governance characteristics, and environmental shocks and stresses" such as those from disaster risk management and climate change adaptation. The three existing indexes are:

1. Coping and adaptive capacities as measured in the World Risk Report (2012) including perceived corruption index, good governance (failed states index), various medical facility and health outcome indicators, and a range of capacity indicators such as literacy rates and natural resource management;
2. The Readiness Score (NG-GAIN) national level scores of vulnerability and readiness to adapt to climate change, consisting of economic, governance (e.g. accountability, stability) and social indicators (e.g. education, mobile phone usage, rule of law);
3. The national monitor for the HFA—indicators from all five priority areas are included. These are: that DRR is a national priority with capacity for implementation; risks are identified, monitored and with early warning systems; a culture of safety is developed; reduce risk factors; and strengthen response capacity.

Vanuatu is ranked in the third quartile in all three measurement sectors. This is not a good result, but the third quartile contains a mix of countries and capacities, with a range of outcomes within those countries: examples include Vietnam, India, and the Philippines. A major positive element for Vanuatu is seen as the restructuring of government institutions to explicitly reflect the linkage of disaster with development (The Ministry of CC, disaster management, etc.). This is counterbalanced somewhat by the compartmentalization of donor activity on which the country depends (See also Nalau et al. 2015).

Although definitions generally include all formal and informal means of managing by government, and organisations fulfilling key social and economic roles, most of the ranking and commentary about Vanuatu in the UNDP Report draws on formal institutions only. Once we include informal support and allow for the distinctive features of Vanuatu villages, including houses made from local materials and subsistence or semi subsistence capacity for food production, and the existing traditional governance systems (including reciprocity and informal exchange economy), the picture would be expected to change. Yet, for example, after TC Pam the high reliance on subsistence farming and on traditional materials was a problem in some areas, as many of the plants for food and materials for construction were destroyed, leaving communities dependent on government handouts to some extent. What this shows is that governance is a complex and fluid, contingent on the circumstances, which are dynamic in the aftermath of major disasters.

Approaches to Disaster Risk Governance in Vanuatu for TC Pam

There appear to be three approaches or models to disaster risk governance in Vanuatu that were at work at various times when examined in the context of TC Pam. The discussion of these approaches needs to be seen in the context of the prevalent assumption of low government capacity.

The first model, 'Submerged governance', describes a model where the actors are overwhelmed by influx of aid and people and therefore are unable to ascertain the effectiveness of the pre-existing plans and mechanisms, with the results that the response is somewhat piecemeal, uncoordinated and slow. The unprecedented nature of TC Pam and its related media and humanitarian attention partly contributed for this model in the beginning of the aftermath. The second model, 'Partial strategic coordination', is also government-focused but with a focus on the government handling what it can manage and coordinating the other activities to some extent. The model is focused on encouraging the recovery process without necessarily trying to assert full control of all actors and activities.

The model, 'Government control', was favoured by those in the government in which all aid and recovery efforts are managed and controlled by the Vanuatu government. This model came into play at a later stage in the relief efforts and tried to coordinate all the hundreds of activities and different organisations who had an interest in the relief efforts. The counter argument, often noted by non-governmental organisations, is that the government does not necessarily have the capacity, or the experience, to manage the aftermath of a mega disaster. The non-governmental organisations see themselves fulfilling a role that the government struggles to do as evident in the example of food relief and the differing perceptions between actors as to when to distribute food, how and to whom. Yet, arguing that government capacity is low can become a self-fulfilling assumption as other actors step into take control over activities leading to capacity substitution rather than capacity building per se. One approach here would be to work in formal partnership.

We would argue that all three models were at work in a sequence in the aftermath of TC Pam. Yet, it would be too simplistic to claim that these are the only models at work—they are indicative only. As noted earlier, governance is a fluid process, and much of it remains hidden as it is contingent on power relationships and negotiations of values and priorities between multitude of actors and scales. While a lot was done to prepare for major events by government and NGOs, we cannot properly assess how effective this was due to the overwhelming flow of aid, which prevented deployment of local disaster plans. We do know that the system struggled to cope and one year later, where reconstruction is dependent on the government, much remains to be done. On the other hand, those communities that had established disaster committees with the help of NGOs were able to better prepare for TC Pam and respond in the aftermath; these community based structures and their subsequent operationalization could be seen in a way to fulfill disaster plan strategies.

The response and its coordination and related decision-making is now more of an internal activity for the Vanuatu government as it needs to decide which recovery efforts should be prioritized and in which sectors. Yet, much of the recovery aid has come with clear expectations and guidelines as to how the money can be accessed and what it should be spent on. This in turn is somewhat problematic for the Government-control model where it would be expected that the government would have the sole decision-making power in regards to the priority sectors, activities, and locations. In the context of the mega disaster, the slowness of constructing nation-wide recovery plans, allocating funding, and beginning the implementation therefore are not just dependent on the national government, but rather on a mixed set of actors all trying to exercise differing models of disaster risk governance.

Again, the civil society and the communities themselves have done practical projects as part of their recovery efforts at the local scale to enable livelihoods and homes to be restored. Most communities were able to re-build their homes for example in the aftermath to some extent while NGOs contributed with livestock projects, such as chicken farms, and handing out equipment and tarpaulins to quicken the recovery process. Yet the communities have been also dependent on government assistance: for the island of Tanna, for example, TC Pam represented the first cyclone event in its history when people could not recover simply just by their own means but needed external government assistance in the recovery efforts.

Way Forward

Many lessons are being learned from TC Pam that are briefly covered here. The main theme across most lessons learned is that of co-ordination of different sectors and groups, and of different levels of governance, ranging from the community level to provincial to national to international. Here we focus on three main responses that featured in the interviews and across the reports proposing recommendations after TC Pam.

Lesson 1: Integrating responses across sectors and groups

There is a need for a systematic rethink of how the food and water security nexus will be addressed in similar future disaster related conditions. More focus should also be placed on aligning humanitarian and development interventions for both food and water security. This is particularly important where during an El Nino period, impacts on food and water from extreme events such as cyclones are more profound and aggravated by the underlying slow onset effects of reduced rainfall, as has been the case following TC Pam. Therefore taking heed of climate seasonal outlook notifications during the cyclone season should form an indispensable element of both preparedness and response activities by government lead clusters and respective communities. Each cluster brings together the responsible organisations who work together to solve problems and deliver the service. The coordinating role of the Ministry of Foreign Affairs and that of government bodies should be strengthened so that there are clear lines of communications and responsibilities and roles across sectors and groups involved in pre- and post-disaster relief and response.

Lesson 2: Integrating risk into formal and informal disaster risk governance

The demonstration of the significant impact made by the cluster approach to managing specific areas of governance during TC Pam's response phase needs to be translated into effective mechanisms for long term recovery or development planning and implementation. Importantly the clear allocation of roles and responsibilities between cluster specific or traditional development focused agencies will be vital in all aspects of the disaster risk management cycle. This essentially needs to entail emphasis on risk integration' into cluster activities especially cluster agencies or sectors outside the cyclone season as well. Whilst the cluster system was more prevalent at the national level, strengthening cluster approach and humanitarian development synergies at the provincial level is similarly required. Investing in capacity at the provincial level is crucial given that this is often the level of formal governance closest to the people. By the same token, at the community level there is need to make best use of and strengthen existing traditional governance configurations where disaster governance is concerned. Clear allocation of roles is essential for effective disaster risk governance: if similar event was to occur in the future, knowing who to contact for each sector and activity is likely to increase the speed of coordination and implementation of recovery efforts.

Lesson 3: Integrating multiple knowledges for effective DRR

Additionally the established role of traditional knowledge relative to cyclone preparedness and response cannot be disputed (Government of Vanuatu 2015b). However, the utility of appropriate traditional knowledge requires strengthening and advocacy. This is particularly the case for food preservation before and after a cyclone, and the use of traditional cyclone resistant housing styles. The onset of El Nino has moreover pushed a number of communities to rely on traditional coping mechanisms for both water management and use. These practices need to be further studied and where necessary encouraged to be incorporated within broader cluster led preparedness messages in future.

Conclusion

While TC Pam demonstrated many issues regarding the nexus of global-national-provincial disaster risk governance, it is unclear whether the current systems in place would have worked well without the massive aid and involvement of so many actors. The government representatives interviewed in this study were more prone to advocate for the 'Government-control' model whereas other stakeholders seemed to prefer either partial strategic coordination or a situation where everyone could pitch in and decide the priorities. One important unresolved question concerns which model best represents the interests of those most affected by the cyclone or other disasters?

TC Pam, as the first category 5 cyclone, also raises questions about climate justice and limits to adaptation. Essentially this argues that as those who contribute the least to climate change are bearing the largest and most immediate burdens, they are entitled to support from those who contribute the most to climate change. If such megastorms are linked to anthropogenic climate change, then the losses and damages accruing from such events might become windows for discussing compensation in the form of support for adaptation. In the context of climate justice, small island developing states and extreme events, capacity building and support form an important opportunity to enable countries to have their own discussion in regards to adaptation limits.

Acknowledgements We thank those who assisted with this study including all stakeholders who contributed their time, and in particular for the assistance of Paul Nalau, Government of Vanuatu, and Hannah Iveson of the Centre for Risk & Community Safety, Mathematical and Geospatial Sciences, at RMIT University. We acknowledge with gratitude the funding provided by the Centre for Risk & Community Safety.

References

Barber, R. (2015). One size doesn't fit all: Tailoring the international response to the national need following Vanuatu's Cyclone Pam. A contribution to the Pacific Regional Consultation for the World Humanitarian Summit Save the Children Australia, CARE International, Oxfam Australia, and World Vision.

Government of Vanuatu (2015a). *Vanuatu Climate Change and Disaster Risk Reduction Policy 2016-2030*. Port Vila, Republic of Vanuatu

Government of Vanuatu (2015b). Post Disaster Needs Assessment. Vanuatu's Prime Minister's Office, Port Vila, Vanuatu

Kuruppu, N., & Willie, R. (2015). Barriers to reducing climate enhanced disaster risks in least developed country-small islands through anticipatory adaptation. *Weather and Climate Extremes, 7*, 72–83.

Malvatumauri National Council of Chiefs (2012). *Alternative Indicators for Well-being for Melanesia,* Vanuatu Pilot Study Report.

Nalau, J., Handmer, J., Dalesa, M., Foster, H., Edwards, J., Kauhiona, H., et al. (2015). The practice of integrating adaptation and disaster risk reduction in the south-west Pacific. *Climate and Development, 8*, 1–11.

NEF. (2016). Happy Planet Index. New Economics Foundation (NEF). http://www.happyplanetindex.org.

OCHA. (2015). *Vanuatu: Tropical Cyclone Pam Situation Report No. 12 (as of 26 March 2015).* United Nations Office for the Coordination of Humanitarian Affairs (OCHA). http://reliefweb.int/sites/reliefweb.int/files/resources/OCHA_VUT_TCPam_Sitrep12_20150326.pdf

ODI UNDP. (2014). Disaster Risk Governance: Unlocking progress and reducing risk. http://www.undp.org/content/undp/en/home/librarypage/crisis-prevention-and-recovery/disaster-risk-governance-unlocking-progress-and-reducing-risk.html

Secretariat of the Pacific Community. (2015). *Tropical Cyclone Pam: Lessons learned workshop report June 2015* (66 pp.). Suva, Fiji. http://reliefweb.int/sites/reliefweb.int/files/resources/tc_pam_lessons_learned_report_final_170316.pdf.

Shelter Cluster. (2015). Shelter and settlements vulnerability assessment—Final report: Cyclone Pam Response. Retrieved October 8, 2015 from https://www.sheltercluster.org/sites/default/files/docs/shelter_cluster_report_shelter_and_settlements_vulnerability_assessment_after_cyclone_pam_may2015.pdf.

Tafea Provincial Disaster Committee and CARE International. (2015). *Lessons Learnt from Tropical Cyclone Pam for Tafea Provincial Disaster Committee*. Tafea Provincial Disaster Committee and CARE International, Republic of Vanuatu

UNDP. (2015). Country Notes Vanuatu. http://hdr.undp.org/sites/all/themes/hdr_theme/country-notes/VUT.pdf

UNISDR. (2011). Global Assessment Report. https://www.unisdr.org/we/inform/publications/19846

Vanuatu National Statistics Office. (2009). Vanuatu 2009 National Population and Housing Census. Ministry of Finance and Economic Management, Republic of Vanuatu, Port Vila

Wickham, F., Kinch, J., & Padma, L. (2009). *Institutional capacity within Melanesian countries to effectively respond to climate change impacts, with a focus on Vanuatu and the Solomon Islands*. Secretariat of the Pacific Regional Environment Programme.

Chapter 10
Response and Adaptation to Climate Change in the South China Sea and Coral Sea

Rongshuo Cai, Haixia Guo, Di Fu, Xiuhua Yan and Hongjian Tan

Introduction

The South China Sea (SCS) and the Coral Sea (CS) are respectively located in the northwest and southwest tropical Pacific Ocean, but share similar geographies and habitat characteristics. The SCS contains areas with significant commercial fisheries and includes coral reefs, mangroves, and seagrass beds, covering about 350.0×10^4 km^2 (Feng et al. 1999). The CS lies to the east of Australia's Great Barrier Reef and covers a total area of 479.3×10^4 km^2 (Ceccarelli 2012; Welch et al. 2015), containing rich coral-dominated reef ecosystems, and related benthic and fish communities. Both of these areas are marginal seas of the western tropical Pacific Ocean and critically important for coastal communities and industries because of their biological diversity and high productivity. The Fifth Assessment Report (AR5) of the Intergovernmental Panel on Climate Change (IPCC) pointed out that, in addition to anthropogenic disturbances, the western Pacific Ocean is greatly impacted by climate change and has already experienced significant change over the past decades, putting ecosystems and fisheries at risk (Hoegh-Guldberg et al. 2014). Like other tropical areas, changes in physical and chemical ocean conditions, e.g., increasing water temperatures, rising sea level, and ocean acidification, have caused the recent northward shift of warm-water species from tropical to higher latitudes and caused coral bleaching and mortality (Cai et al. 2015; Hoegh-Guldberg et al. 2014).

Even though IPCC AR5 and other previous studies have included the SCS and CS (Cai et al. 2009, 2010, 2011, 2015; Hoegh-Guldberg et al. 2014; Welch et al. 2015), information about the responses and adaptations of these tropical marginal seas to climate change is still limited and needs further analysis. Particularly,

R. Cai (✉) · H. Guo · D. Fu · X. Yan · H. Tan
Key Laboratory of Global Change and Marine-Atmospheric Chemistry, Third Institute of Oceanography, State Oceanic Administration of China, Xiamen 361005, China
e-mail: rscai@163.com; cairongshuo@tio.org.cn

W. Leal Filho (ed.), *Climate Change Adaptation in Pacific Countries*,
Climate Change Management, DOI 10.1007/978-3-319-50094-2_10

although both the SCS and CS are located in similar tropical ocean environments and are part of the Asian–Australia monsoon region, climate conditions in these areas are dominated by different climate patterns (Hoegh-Guldberg et al. 2014). The spatial patterns of sea surface temperature (SST), sea surface salinity (SSS), and chlorophyll-a concentration as a proxy for marine productivity, and their relationships on longer time scales, including their response to global warming and the hiatus epochs before and after 1999 are not fully examined and understood. Hence, the intent of this study is to investigate the physical, chemical, and biological changes caused by climate changes in the SCS and CS, to compare their similarities and differences, and to summarize related adaptation measures in response to climate change. In this article, the spatial and temporal patterns of sea surface temperature (SST), sea surface salinity (SSS), chlorophyll-a, sea surface level (SRL), and ocean acidification will be analyzed, discussed, and compared for both seas.

Additionally, climate change and ocean acidification increase the stress on these seas, in addition to local human activities such as overfishing and pollution. Therefore, it is imperative to better understand the environmental changes in these marginal seas and to provide recommendations to better mitigate future climate-related problems.

Data and Methods

The Hadley Centre Global Sea Ice and Sea Surface Temperature (HadISST) dataset with a horizontal resolution of 1° × 1° (Rayner et al. 2003) and the Simple Ocean Data Assimilation (SODA) dataset with a horizontal resolution of 0.5° × 0.5° (Carton and Giese 2008) were used to analyze SST and SSS changes in the SCS and CS for 1958–2014 and 1958–2008, respectively (Fig. 10.1). The satellite remote sensing datasets of SeaWiFs (NASA) with a horizontal resolution of 9 km × 9 km were used to investigate the annual averaged concentration of chlorophyll-a in the SCS and CS for 1998–2014 and their relationship to the time series of SST and SSS. The OAFlux3 dataset with a horizontal resolution of 1° × 1° (Yu et al. 2008) and the GPCP dataset with a horizontal resolution of 2.5° × 2.5° (Adler et al. 2003) were used to investigate the climatological surface water flux [evaporation and precipitation (E-P)] for 1979–2014, which is closely linked to the water cycle and ocean salinity.

First, the spatial patterns of SST and SSS changes were analyzed using a linear fitting method. For a variable Y, its linear trend k can be obtained by applying a linear fit to the time series of Y_i, $Y_i = Y_0 + kX_i + \varepsilon_i$, where k is the linear variation trend of the variable Y, X_i is the time corresponding to the Y_i value in yearly records for the 1958–2014 period, and ε_i is the error term introduced to estimate the uncertainty of the fit. Then, the linear change ΔY with time can be obtained using $\Delta Y = k(X_i - X_0)$. Second, the empirical orthogonal function (EOF) analysis method was applied to analyze the changes of SST and SSS in the SCS and CS. The EOF method has been commonly used to enable analysis of data with spatial and temporal structures, which are decomposed into representative modes. The first few

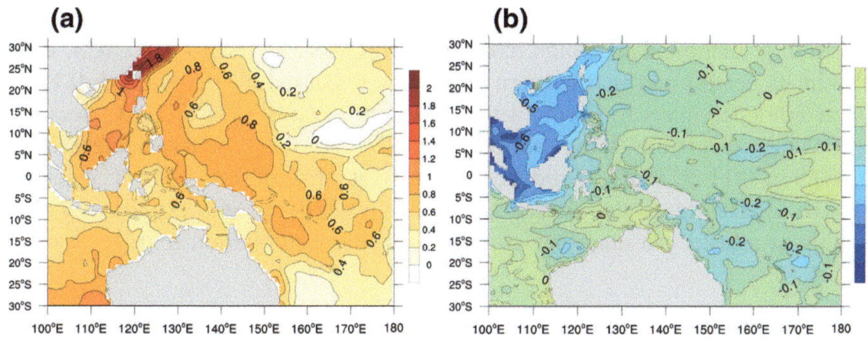

Fig. 10.1 Maps of the South China Sea (SCS) and Coral Sea (CS) in the tropical western Pacific; **a** spatial patterns of linear SST changes for 1958–2014 and **b** SSS for 1958–2008 in the SCS and CS; contour intervals for SST and SSS are 0.2 °C and 0.1, respectively. The uncertainty ranges of linear changes for SST and SSS in the colored areas are ±0.11 °C and ±0.087 g kg^{-1}, respectively. Data for SST and SSS are from the HadISST and SODA reanalysis datasets, respectively

modes of EOF can contain a significant portion of the total variance and can be ordered according to the percentage of the total variance. The modes are statistically uncorrelated with one another, and hence each isolated mode includes phenomena with differing spatial and temporal scales. We performed an EOF analysis on SST and SSS anomalies of the SCS and CS to identify modes of the variability pattern for the 1958–2014 and 1958–2008 periods, respectively. Next, we used the linear fitting method to analyze the spatial pattern of linear averaged chlorophyll-a concentration changes in the SCS and CS for 1998–2014 and compared these changes with the first EOF mode temporal coefficient of SST and SSS anomalies for the period of 1998–2014.

Due to the limited availability of data, we focused our analyses on SST, SSS, and chlorophyll-a changes in the tropical western Pacific marginal seas, particularly in the SCS and CS. In addition, we also reviewed published data on sea level rise and ocean acidification and related impacts on both study areas with the aim to propose adaptation strategies to mitigate the effect of climate change in the SCS and CS.

Results and Discussion

Responses of the Marine Environment in the SCS and CS to Climate Change

Sea Surface Temperature (SST)

The IPCC AR5 concluded that SST in the Pacific Ocean has increased by 0.31 °C for 1950–2009 due to global climate change (Hoegh-Guldberg et al. 2014). In this study, based on the annual mean HadISST and SODA analysis, long-term linear changes of SST in the SCS and the CS for 1958–2014 were analyzed using the linear fitting method. The results indicate that the SST in most of the SCS, especially in its central and northeastern areas, has increased by 0.8 – 1.8 ± 0.11 °C (95% confidence interval) during 1958–2014, at a rate of 0.14–0.31 °C decade^{-1} (Fig. 10.1a). Over the same period, the SST in the CS has increased by 0.4 – 0.8 ± 0.11 °C (95% confidence interval), at a rate of 0.07–0.14 °C decade^{-1} (Fig. 10.1a). The figure also shows that surface water in the SCS experienced more warming at a faster rate than that in the CS for the study period (Fig. 10.1a).

Sea Surface Salinity (SSS)

Changes in sea surface salinity (SSS) can influence both the chemical characteristics of a water mass and the marine ecosystem. The long-term linear changes of SSS in the SCS and the CS were also investigated using the linear fitting method described above. The results indicate that the SSS in most of the SCS has decreased by −0.5 to −0.6 ± 0.087 g kg^{-1} at a rate of −0.09 to −0.11 g kg^{-1} decade^{-1} during 1958–2008 (color shading indicates 95% confidence interval; Fig. 10.1a). In most of the CS, SSS has decreased by −0.2 g kg^{-1} at a rate of −0.04 g kg^{-1} decade^{-1} (grid shading indicates 95% confidence interval; Fig. 10.1b). Similarly, it is easy to see that the surface water in the SCS shows greater freshening at a faster rate than that in the CS (Fig. 10.1b).

Chlorophyll-a

As a proxy signal for marine productivity, the long-term linear changes of annual averaged concentrations of chlorophyll-a in the SCS and CS for 1998–2014 were investigated using the linear fitting method. Chlorophyll, which is an easily

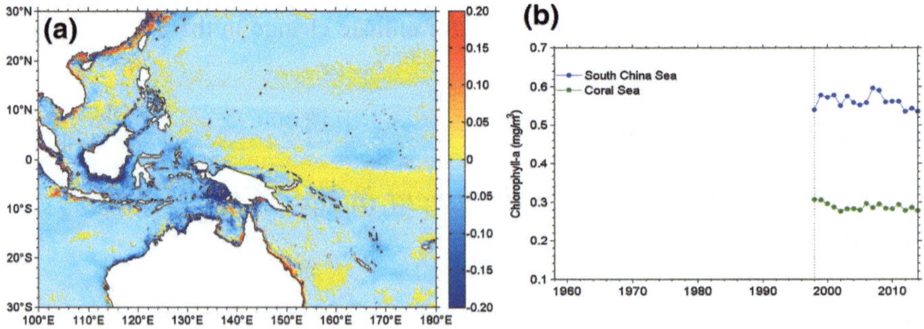

Fig. 10.2 **a** Spatial pattern of linear chlorophyll-a concentrations changes in the SCS and CS during the period of 1958–2014; **b** time series of annual averaged chlorophyll-a concentrations in the SCS and CS. Uncertainty ranges of linear changes in the SCS and CS are ±0.0008 and ±0.0004 mg m^{-3}, respectively. Data are from the SeaWiFs (NASA) datasets

measured pigment specific to photosynthesis, has become the most widely measured indicator of phytoplankton abundance, even though its relation to phytoplankton carbon changes significantly with light intensity, nutrient availability, and species composition (Geider 1987). The results indicate that the annual mean concentrations of chlorophyll-a in the SCS and CS experienced uneven linear spatial changes during the period of 1998–2014, as shown in Fig. 10.2a. The statistical analysis shows that the annual mean concentrations of chlorophyll-a in the SCS and CS were 0.562 and 0.288 mg m^{-3}, respectively, for 1998–2014 (Fig. 10.2b). The linear changes of mean concentrations of chlorophyll-a in the SCS and CS during this period are -0.0182 ± 0.0008 and -0.0076 ± 0.0004 mg m^{-3}, respectively (Fig. 10.2a). A time series of chlorophyll-a concentrations shows slightly decreasing trends from 0.541 mg m^{-3} to 0.536 mg m^{-3} in the SCS and from 0.308 mg m^{-3} to 0.278 mg m^{-3} in the CS during the study period (Fig. 10.2b). In addition, the results show that chlorophyll-a concentrations in the coastal areas have increased much more obviously than those in the offshore areas (Fig. 10.2a).

Relationship of Annual Chlorophyll-a Concentrations to SST and SSS

In order to further understand the relationship of the annual chlorophyll-a concentrations to the changes in SST and SSS, the spatial–temporal characteristics of SST and SSS changes in the SCS and CS were investigated with an empirical orthogonal function (EOF) analysis (Figs. 10.3, 10.4, 10.5 and 10.6). The first EOF modes (EOF1) of SST anomalies explain about 78.9 and 78.8% of the SST variability in the SCS and the CS, respectively (Figs. 10.3 and 10.4). The EOF1 indicates that SSTs in the SCS and the CS had shown a clear warming trend since the mid-1980 or mid-1990, with the most significant warming in the central and northern regions (Figs. 10.3 and 10.4). The SSTs in both regions show a plateau after 2000, reflecting the global warming trend before 1999 and the hiatus after 2000 (Kerr 2009; Met Office 2013).

The EOF results indicate that the EOF1 of SSS anomalies explains about 53.6 and 51.9% of the SSS variability in the SCS and the CS, respectively, during 1958–2008. The EOF1 of the SSS anomaly (SSSA) shows that the SSS concentrations had clearly decreased since the 1980s in both the SCS and CS (Figs. 10.5 and 10.6).

It is noted that the annual averaged concentrations of chlorophyll-a in the SCS and CS experienced obvious changes on an inter-annual time scale in the past decade (Figs. 10.2b, 10.3, 10.4, 10.5, 10.6, and 10.7). The correlation coefficients between the annual mean chlorophyll-a concentration and the time coefficient of the SSTA EOF1 mode (PC1) in the SCS and CS are −0.28 and 0.43, respectively, for the corresponding periods (Figs. 10.3b, 10.4b, and 10.7a). The correlation coefficient between PC1 of SSSA and the annual chlorophyll-a concentration in the SCS and the CS are −0.52 and −0.40, respectively, for the corresponding periods (Figs. 10.5b, 10.6b, and 10.7b). It appears that the annual averaged concentrations of chlorophyll-a vary with changes of SST in the CS, but with changes of SSS in the SCS (Figs. 10.7a and b). This suggests that the change of chlorophyll-a is

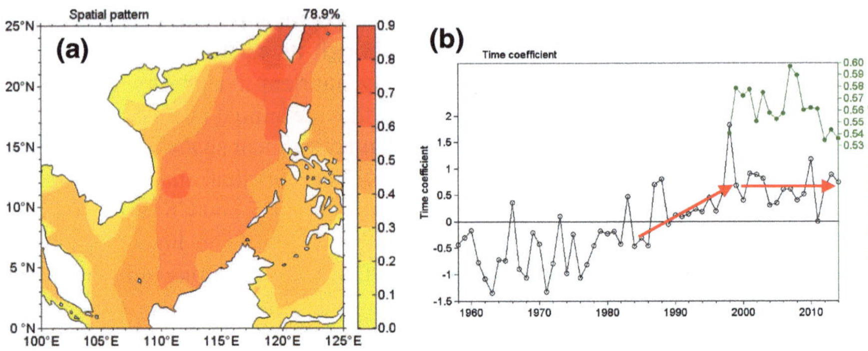

Fig. 10.3 Spatial pattern (**a**) and time coefficient (**b**) of the empirical orthogonal function (EOF) mode of the annual mean sea surface temperature anomaly (SSTA) in the SCS that explains 78.9% of the variance during 1958–2014; data are from the HadISST dataset. The *green line* in (**b**) indicates the time series of the annual mean chlorophyll-a concentration in the SCS; data are from SeaWiFs (NASA) datasets. The correlation coefficient between the chlorophyll-a concentration and time coefficient of SSTA is only −0.28

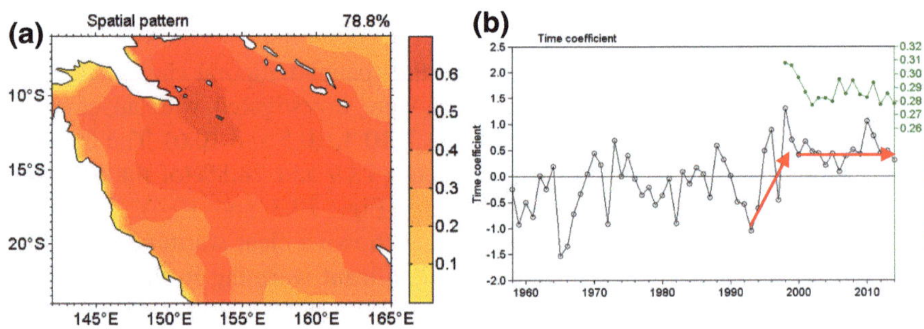

Fig. 10.4 Spatial pattern (**a**) and time coefficient (**b**) of the empirical orthogonal function (EOF) mode of the annual mean sea surface temperature anomaly (SSTA) in the CS that explains 78.8% of the variance during 1958–2014; data are from the HadISST dataset. The *green line* in (**b**) indicates the time series of the annual mean chlorophyll-a concentration in the Coral Sea; data are from SeaWiFs (NASA) datasets. The correlation coefficient between the chlorophyll-a concentration and time coefficient of SSTA EOF1 is 0.43

linked to the SST in the CS, but to the SSS in the SCS. Accordingly, SSS plays a greater role than SST in the changes of chlorophyll-a in the SCS, while this relationship is reversed in the CS.

Secondly, as mentioned before, the time series of the chlorophyll-a concentrations in the SCS and CS (Fig. 10.2) show tendencies to slightly drop but then level off after 2000 at concentrations of 0.58–0.54 mg m^{-3} in the SCS and 0.31–0.28 mg m^{-3} in the CS, which compares with the PC1 s of SSTA and SSSA shown in Figs. 10.3, 10.4, 10.5, 10.6, and 10.7. It can be concluded that the time series of chlorophyll-a in the SCS and CS seemingly varied in phase with the ocean warming

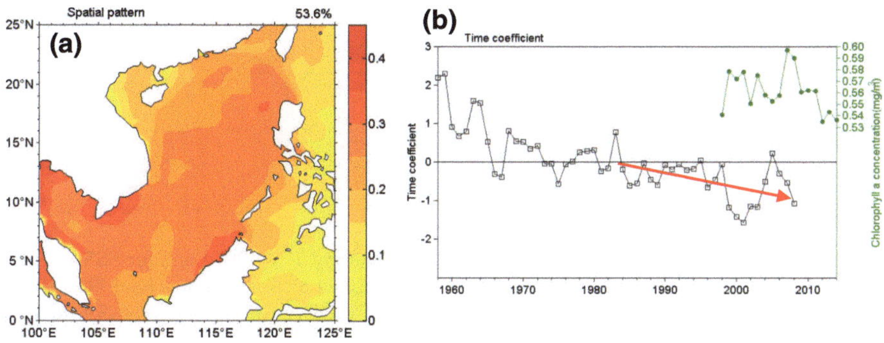

Fig. 10.5 Spatial pattern (**a**) and time coefficient (**b**) of the empirical orthogonal function (EOF) mode of the annual mean sea surface salinity anomalies (SSSA) in the SCS that explains 53.6% of the variance during 1958–2008; data are from the SODA dataset. The *green line* in (**b**) indicates the time series of the annual mean chlorophyll-a concentration in the SCS; data are from SeaWiFs (NASA) datasets. The correlation coefficient between the chlorophyll-a concentration and time coefficient of SSSA is −0.52

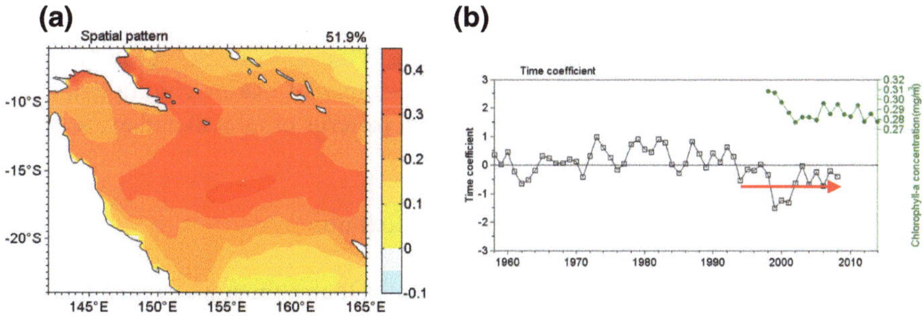

Fig. 10.6 Spatial pattern (**a**) and time coefficient (**b**) of the empirical orthogonal function (EOF) mode of the annual mean sea surface salinity anomaly (SSSA) in the CS that explains 51.9% of the variance during 1958–2008; data are from the SODA dataset. The *green line* in (**b**) indicates the time series of the annual mean chlorophyll-a concentration in the Coral Sea; data are from SeaWiFs (NASA) datasets. The correlation coefficient between the chlorophyll-a concentration and time coefficient of SSSA is −0.40

before 1999 and the hiatus after 2000, despite the observation that the role of SST in changes of the chlorophyll-a concentrations is weaker than that of SSS in the SCS.

Previous studies indicated that rainfall not only increased with rising SSTs that exceed the mean surface warming in the tropics (Xie et al. 2010; Johnson and Xie 2010), but also increased in presently rainy regions (Neelin et al. 2003; Held and Soden 2006; Chou et al. 2009), which is often referred to as "warmer-get-wetter and wet-get-wetter" (Huang et al. 2013). Our results indicate that the sea surface in the SCS experienced a linear change toward warmer and fresher water at a faster rate (0.18–0.31 °C decade^{-1} and 0.088–0.11 g kg^{-1} decade^{-1}) than that of the CS (0.14 °C

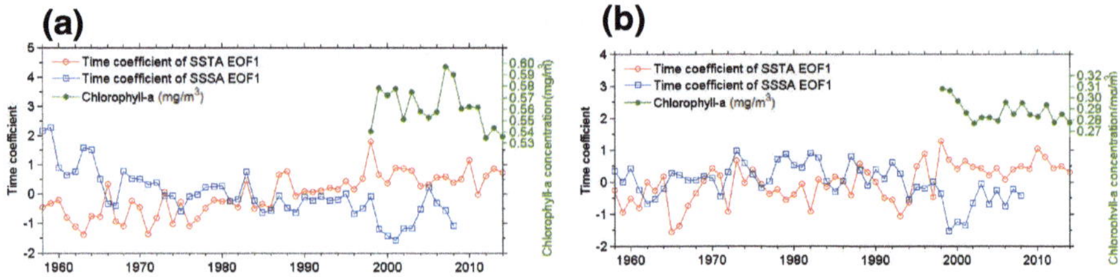

Fig. 10.7 Time coefficients of PC1s for SSTAs EOF1 during 1958–2014 and for SSSAs EOF1 during 1958–2008 and time series of annual mean chlorophyll-a concentrations for 1998–2014 in the SCS (**a**) and in the CS (**b**); data are from the HadISST, SODA, and SeaWiFs (NASA) datasets

decade^{-1} and 0.04 g kg^{-1} decade^{-1}) for the period of 1958–2014 (Fig. 10.1). It can be inferred that the SCS would be located in a rainfall-dominated region, while the CS would be located in an evaporation-dominated region (Fig. 10.8a), which actually has also been demonstrated by the IPCC AR5 WGII (Fig. 10.8b) (Hoegh-Guldberg et al. 2014). Therefore, changes in rainfall and evaporation should be the main factors affecting chlorophyll-a in the SCS and the CS, respectively, although tropical rainfall changes are positively correlated with spatial deviations of SST warming from the tropical mean (Xie et al. 2010).

In brief, climate change, particularly changes in the atmosphere–ocean system in the western tropical Pacific seas, could affect changes of marine chlorophyll-a and primary production via the spatial and temporal patterns of rainfall and evaporation.

Review of Sea Level Rise and Ocean Acidification

The report of the IPCC-AR5 WGII (2014) indicated that, although current rates of sea level rise vary geographically, those rates are up to three times higher than the global mean in the western tropical Pacific for the period of 1993–2012 as measured by satellite altimetry (Fig. 9) (IPCC-AR5 WGII, Fig. 30.5; WGI 13.6.5). It is noted that sea level rise (SLR) in the western tropical Pacific Ocean presents a distinct spatial pattern. Particularly, SLR to the east of the SCS and the north of the CS appears much more pronounced than in other areas (Fig. 9). Cai et al. (2015) showed that the sea surface height (SSH) along the coastal China Sea was increasing, especially in the East China Sea (ECS) with a long-term rise of more than 3.2 mm yr^{-1}. Their data demonstrate that small islands and the coastal region in the western tropical Pacific Ocean, e.g., in the SCS, face more severe threats than other areas.

Numerous studies indicate that ocean acidification in the SCS and the Great Barrier Reef of the CS accelerates with increasing atmospheric CO_2 (Liu et al. 2014; Pelejero et al. 2005; Wei et al. 2009), and varies in phase with the inter-decadal weakening of the Asian Winter monsoon, which could exacerbate acidification in the SCS and CS (Liu et al. 2014). Pelejero et al. (2005) pointed out that the extent to which corals and other calcifying reef organisms can adapt to such

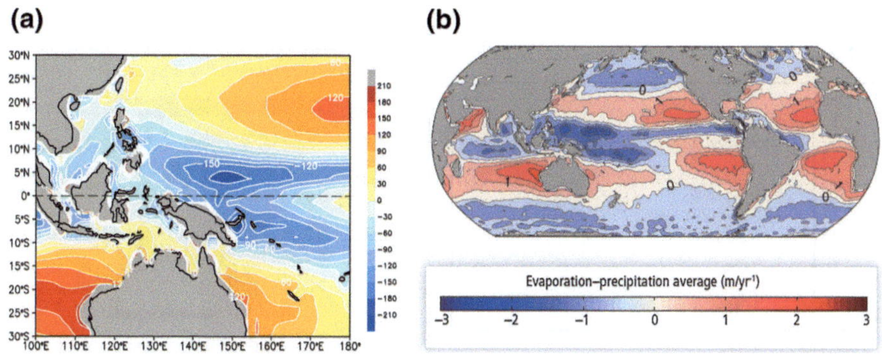

Fig. 10.8 **a** Annual mean evaporation–precipitation (m yr^{-1}) for 1979–2014; data are from the Objectively Analyzed air-sea Fluxes (OAFlux3) and the Global Precipitation Climatology Project (GPCP); **b** Annual mean evaporation–precipitation averaged for the 1950–2000 period and color contoured at 0.5 m yr^{-1} intervals (*black lines*); data are from the National Centers for Environmental Prediction (NCEP) (IPCC-AR5 WGII, Fig. 30.6) (Figure 8b is cited from Hoegh-Guldberg et al. 2014)

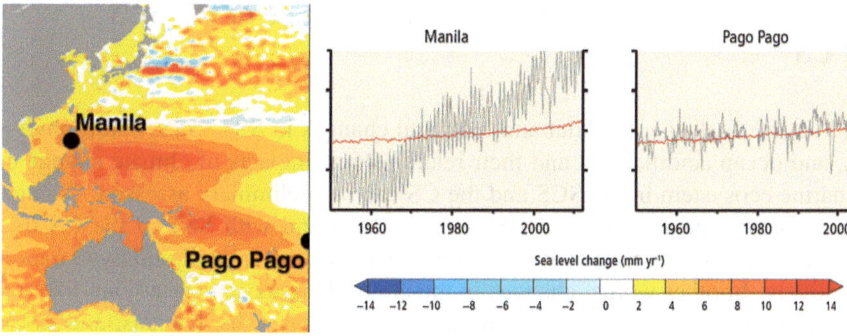

Fig. 10.9 Rate of change in sea surface height (geocentric sea level) for the period of 1993–2012. Relative sea level fluctuations (*gray lines*) are shown for selected tide gauge stations for the 1950–2012 period. An estimate of global mean sea level change is shown (*red lines*) for each tide gauge time series. Note the relatively large short-term oscillations in local sea level (*gray lines*) shown for Pago Pago and Manila. Figure 9 was originally presented in the report of IPCC-AR5 WGI (FAQ 13.1, Fig. 1) (cited from IPCC-AR5 WGII, Hoegh-Guldberg et al. 2014)

rapid decreases in pH is largely unknown. The surface pH in the tropical ocean (20°S–20°S), especially in the CS, is predicted to decrease in the future (Fig. 10.10). Thus, tropical marine habitats such as coral-dominated reef ecosystem will experience dangerous conditions and marine goods and services important for coastal communities will be threatened.

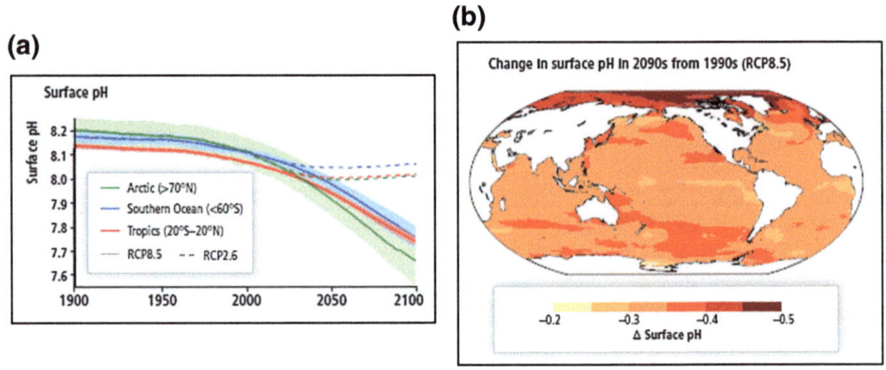

Fig. 10.10 Projected model results of ocean acidification under RCP8.5: **a** Time series of surface pH shown as the mean (*solid line*) and range of models (*shaded area*), given as area-weighted averages over the tropical oceans (*red*), and the Southern Ocean (*blue*). **b** Maps of the median model's change in surface pH in 2090s from 1990s (RCP8.5). Panel (a) also includes mean model results from RCP2.6 (*dashed lines*). Figure was originally presented in IPCC WGI AR5 as Fig. 6.28 (cited from IPCC-AR5 WGII, Hoegh-Guldberg et al. 2014)

Impacts of and Adaptation to Climate Change in the SCS and CS

Based on the analysis of spatial and temporal changes in SST, SSS, chlorophyll-a, SRL, and ocean acidification and their relationships, impacts of climate change on the marine ecosystem in the SCS and the CS can be preliminary assessed. Changes in chlorophyll-a have varied with SSS in the SCS and with SST in the CS over recent decades, which appear to be mainly controlled by rainfall and evaporation, respectively. In other words, changes in the atmosphere–ocean system are likely to affect chlorophyll-a concentrations and, hence, the food webs, habitats, and stocks important for fisheries and aquaculture in these regions (Bell et al. 2013). Chlorophyll-a is not only a proxy for marine primary production, but also indicates that marine and coastal ecosystems are sensitive to changes in ocean temperature, salinity, rainfall and evaporation patterns, and acidification. In addition, because of sea level rise, acidification, and ocean warming, coastal systems will continue to experience significant negative consequences such as coastal flooding, coastal submergence, coastal erosion, and a decreased ability of many calcifying organism to construct their shells, which makes coral reefs the most vulnerable marine ecosystem (Wong et al. 2014). Like other tropical areas in the ocean, coral reefs in the SCS and CS have experienced elevated water temperatures and acidification since the early 1980s (China-SNAP 2011; Hoegh-Guldberg et al. 2009).

Low-lying coastal areas, particularly on small islands and atolls in the SCS and CS will face increasing threats from faster rates of sea level rise, ocean warming, and marine habitat destruction, in addition to non-climate-related factors such as overfishing, pollution, and ocean acidification. Changes in marine primary production and associated ecosystem shifts, especially for commercial fish species such as tuna have occurred since the mid-20th century (Bell et al. 2013; Hoegh-Guldberg et al. 2014; Cai et al. 2015). The changes severely impact food chains and commercially harvested fish, with consequences for global human societies and industries. Therefore, it is imperative to propose marine adaptation policies and measures addressing climate change, although the complex interactions and indirect pathways within the climate system complicate our understanding of climate change impacts and challenge our ability to plan adaption measures (Allison and Bassett 2015).

In view of the responses of the SCS and CS to climate change over recent decades, proposed adaptation policies and measures at various scales can be summarized as follows: first, in rainfall-dominated regions such as the SCS, an increase in freshwater aquaculture would benefit from regional climate change, i.e., greater rainfall will allow local communities to construct fish ponds in more locations, including some islands as suggested by Bell et al. (2013). Second, considering chlorophyll-a variability and marine species associations, key marine protection areas for vulnerable species and sustainable harvesting strategies should be identified. Third, as adaptation strategies to sea level rise and coastal flooding, protective zones such as marches, mangrove forests, and oyster reefs should be ecological engineered, which would also benefit local communities through increased opportunities for aquaculture and employment (Cheong et al. 2013). Next, adaptation strategies and actions should be formulated by individuals, communities, governments, and international bodies such as regional and global frameworks for decision-making, e.g., the Coral Triangle Initiative on Coral Reefs, Fisheries, and Food Security (CTI) for reversing the decline in coastal ecosystems and the Global Partnership for Oceans for adaptation measures to climate change (Hoegh-Guldberg et al. 2014). Finally, key data must be acquired to fill gaps in our understanding of past and future changes in the atmosphere–ocean system, which will allow us to find targeted solutions to mitigate the impacts of climate change in different regions of the world.

Probably the greatest challenges for adapting to climate change are those faced by people who either live in low-lying coastal areas or close enough to the coast to be directly affected by a combination of sea level rise, ocean warming, coastal flooding and submergence, coastal erosion, and other related events, or whose livelihoods are closely linked to changes in the marine ecosystem, or who are nutritionally dependent on access to the marine food web and other marine resources.

Conclusion

Over the past decades, both the SCS and CS show clear responses to climate change, including surface warming, decreases in sea surface salinity, changes to evaporation–rainfall patterns, sea level rise, and ocean acidification. The main findings of this study can be summarized as follows:

(a) The SST in the SCS has increased by $0.8 - 1.8 \pm 0.11$ °C during 1958–2014, at a rate of $0.14 - 0.31$ °C decade^{-1}. Over the same period, the SST in the CS has increased by $0.4 - 0.8 \pm 0.11$ °C at a rate of $0.07 - 0.14$ °C decade^{-1}.
(b) The SSS in the SCS has decreased by −0.5 to -0.6 ± 0.087 g kg^{-1} at a rate of −0.09 to −0.11 g kg^{-1} decade^{-1} during 1958–2008. In the CS, SSS has decreased by −0.2 g kg^{-1} at a rate of −0.04 g kg^{-1} decade^{-1}.
(c) The linear changes of mean concentrations of chlorophyll-a in the SCS and CS during this period are -0.0182 ± 0.0008 and -0.0076 ± 0.0004 mg m^{-3}, respectively.
(d) It is suggested that the change of chlorophyll-a concentration is linked to the SST in the CS, but to the SSS in the SCS.
(e) Climate change, particularly changes in the atmosphere–ocean system in the western tropical Pacific seas, could affect changes of marine chlorophyll-a and primary production via the spatial and temporal patterns of rainfall and evaporation.
(f) The current rates of sea level rise in the SCS and CS are higher than the global mean for the period of 1993–2012, which indicates that small islands and the coastal region in these areas face more severe threats than other areas.
(g) Ocean acidification in the SCS and the Great Barrier Reef of the CS has accelerated with increasing atmospheric CO_2 over the past decades and is predicted to increase in the future. Marine habitats such as coral-dominated reef ecosystems will experience dangerous conditions and marine goods and services important for coastal communities will be threatened.

In brief, the SCS and CS show similar, but also somewhat different responses to climate change. The impacts of climate change on the SCS and CS are not only closely linked to individuals, communities, nations, or regions, but are also reflected in sectors such as fisheries, aquaculture and tourism and hence greatly affect the livelihood of people living in low-lying coastal regions and small islands or atolls. How to select effective adaptations to climate change in different regions depends on those regions' responses to climate-related drivers. This has profound implications for the sustainable development of low-lying coastal areas and small islands. Developing strategies how to respond to present and future climate change impacts in the SCS and CS is imperative, although changes in the atmosphere–ocean system and the effectiveness of present adaptation strategies are not fully understood.

For this reason, further research on climate change and its impacts in the SCS and CS are needed in the following areas:

(a) Higher resolution datasets of spatial and temporal changes in SST, SSS, and chlorophyll-a over the past decades.
(b) Changes in the atmosphere–ocean system such as spatial and temporal pattern of SST, SSS, chlorophyll-a, rainfall and evaporation, sea level, and pH under different climate change scenarios such as Representative Concentration Pathways (RCP 2.6, 4.5, 6.5, 8.5).
(c) Relationships between SST, SSS, chlorophyll-a, rainfall, and evaporation.

With the availability of these data and based on projected climate change impacts, adaptation measure in response to future climate change in the SCS and CS can be proposed.

Acknowledgements This work was supported by the Grant China Clean Development Mechanism (CDM) Fund Project (2014112), the Scientific Research Foundation Project of Third Institute of Oceanography, SOA (TIO2015030) and Chinese Public Science and Technology Research Funds Project of Ocean (201005019).

References

Adler, R. F., Huffman, G. J., Chang, A., Ferraro, R., Xie, P. P., Janowiak, J., et al. (2003). The version-2 global precipitation climatology project (GPCP) monthly precipitation analysis (1979–present). *Journal of Hydrometeorology, 4*(6), 1147–1167.

Allison, E. H., & Bassett, H. R. (2015). Climate change in the oceans: Human impacts and responses. *Science, 350*(6262), 778–782. doi:10.1126/science.aac872.

Bell, J. D., Ganachaud, A., Gehrke, P. C., Griffiths, S. P., Hobday, A. J., & Hoegh-Guldberg, O. (2013). Mixed responses of tropical pacific fisheries and aquaculture to climate change. *Nature Climate Change, 3*(6), 591–599.

Cai, R. S., Chen, J. L., & Tan, H. J. (2011). Variations of the sea surface temperature in the offshore area of China and their relationship with the East Asian monsoon under the global warming. *Climatic and Environmental Research, 16*(1), 94–104. (in Chinese, with English abstract).

Cai, R. S., & Tan, H. J. (2010). Influence of interdecadal climate variation over East Asia on offshore ecological system of China. *Journal Oceanography in Taiwan Strait, 29*(2), 173–183. (in Chinese, with English abstract).

Cai, R. S., Tan, H. J., & Qi, Q. H. (2015). Impacts of and adaptation to inter-decadal marine climate change in coastal China seas. *International Journal of Climatology*. doi:10.1002/JOC.4591.

Cai, R. S., Zhang, Q. L., & Qing-Hua, Q. I. (2009). Spatial and temporal oscillation and long-term variation in sea surface temperature field of the South China Sea. *Journal of Oceanography in Taiwan Strait, 28*(4), 559–568. (in Chinese, with English abstract).

Carton, J. A., & Giese, B. S. (2008). A reanalysis of ocean climate using Simple Ocean Data Assimilation (SODA). *Monthly Weather Review, 136*(8), 2999–3017.

Ceccarelli, D. M. (2012). Australia's Coral Sea—how much do we know? In *Proceedings of the 12th International Coral Reef Symposium, Cairns, Australia, 9–13 July 2012 (18E The Management of the Coral Sea Reefs and Sea Mounts).*

Cheong, S. M., Silliman, B., Wong, P. P., Wesenbeeck, B. V., Kim, C. K., & Guannel, G. (2013). Coastal adaptation with ecological engineering. *Nature Climate Change, 3*(9), 787–791.

China-SNAP. (2011). *The second national Chinese assessment report on climate change Chinese government* (pp. 245–256). Beijing: Science Press. (in Chinese, with English abstract).

Chou, C., Neelin, J., Chen, C., & Tu, J. (2009). Evaluating the 'rich-get-richer' mechanism in tropical precipitation change under global warming. *Journal of Climate, 22*, 1982–2005.

Feng, S. Z., Li, F. Q., & Li, S. J. (1999). *An introduction to marine science* (pp. 434–436). Beijing: Higher Education Press. (in Chinese).

Geider, R. J. (1987). Light and temperature dependence of the carbon to chlorophyll a ratio in microalgae and cyanobacteria: Implications for physiology and growth of phytoplankton. *New Phytologist, 106*(1), 1–34.

Held, I. M., & Soden, B. J. (2006). Robust responses of the hydrological cycle to global warming. *Journal of Climate, 19*, 5686–5699.

Hoegh-Guldberg, O., Hoegh-Guldberg, H., Veron, J. E. N., Green, A., Gomez, E. D., Lough, J. M., et al. (2009). *The coral triangle and climate change: Ecosystems, people and societies at risk* (276 pp). World Wide Fund For Nature (WWF) Australia, Brisbane, QLD, Australia.

Hoegh-Guldberg, O., Cai, R. S., Poloczanska, E. S., Brewer, P. G., Fabry, V. J., Hilmi, K., et al. (2014). The ocean. In: C. B. Field, V. R. Barrows, D. J. Dokken, K. J. Mach, M. D. Mastrandrea, T. E. Bilir, et al. (Eds.), *Climate change 2014: Impacts, adaptation, and vulnerability. Part B: regional aspects. Contribution of working group II to the fifth assessment report of the intergovernmental panel on climate change* (pp. 1655–1731). UK and New York, NY, USA: Cambridge University Press.

Huang, P., Xie, S. P., Hu, K., Huang, G., & Huang, R. H. (2013). Patterns of the seasonal response of tropical rainfall to global warming. *Nature Geoscience, 6*(5), 357–361.

Johnson, N. C., & Xie, S. P. (2010). Changes in the sea surface temperature threshold for tropical convection. *Nature Geoscience, 3*(12), 842–845.

Kerr, R. A. (2009). What happened to global warming? Scientists say just wait a bit. *Science, 326* (5949), 28–29.

Liu, Y., Peng, Z., Zhou, R., Song, S., Liu, W., You, C. F., et al. (2014). Acceleration of modern acidification in the South China Sea driven by anthropogenic CO_2. *Scientific Reports, 4*(6), 1158–1159.

Met Office. (2013). *The recent pause in global warming (1): What do observations of the climate system tell us?* Met Office: Exeter, Devon, UK, 1–2.

Neelin, J., Chou, C., & Su, H. (2003). Tropical drought regions in global warming and El Nino teleconnections. *Geophysical Research Letters, 30*, 2275.

Pelejero, C., Eva, C., Mcculloch, T. M., Marshall, J. F., Gagan, M. K., Lough, J. M., et al. (2005). Preindustrial to modern interdecadal variability in coral reef pH. *Science, 309*, 2204–2207.

Rayner, N. A., Parker, D. E., Horton, E. B., Folland, C. K., Alexander, L. V., Rowell, D. P., et al. (2003). Global analyses of sea surface temperature, sea ice, and night marine air temperature since the late nineteenth century. *Journal of Geophysical Research: Atmospheres, 108*(14), 1063–1082. doi:10.1029/2002JD002670.

Wei, G., McCulloch, M. T., Mortimer, G., Deng, W., & Xie, L. (2009). Evidence for ocean acidification in the Great Barrier Reef of Australia. *Geochimica et Cosmochimica Acta, 73*, 2332–2346.

Welch, H., Pressey, R. L., Heron, S. F., Ceccarelli, D. M., & Hobday, A. J. (2015). Regimes of chlorophyll-a in the Coral Sea: Implications for evaluating adequacy of marine protected areas. *Ecography, 38*, 001–016. doi:10.1111/ecog.01450.

Wong, P. P., Losada, I. J., Gattuso, J. P., Hinkel, J., Khattabi, A., McInnes, K. L., et al. (2014). Coastal systems and low-lying areas. In: C. B. Field, V. R. Barrows, D. J. Dokken, K. J. Mach, M. D. Mastrandrea, T. E. Bilir, et al. (Eds), *Climate change 2014: Impacts, adaptation, and vulnerability. Part a: global and sectoral aspects. Contribution of working group II to the fifth assessment report of the intergovernmental panel on climate change* (pp. 361–409). Cambridge, UK and New York, NY: Cambridge University Press.

Xie, S. P., Deser, C., Vecchi, G. A., Ma, J., Teng, H. Y., & Wittenberg, A. T. (2010). Global warming pattern formation: Sea surface temperature and rainfall. *Journal of Climate, 23*(4), 966–986.

Yu, L., Jin, X., & Weller, R. A. (2008). *Multidecade global flux datasets from the Objectively Analyzed air-sea Fluxes (OAFlux) Project: Latent and sensible heat fluxes, ocean evaporation, and related surface meteorological variables*. OAFlux Project Technical Report, Woods Hole Oceanographic Institution, Massachusetts, US.

Chapter 11
Drought Modelling Based on Artificial Intelligence and Neural Network Algorithms: A Case Study in Queensland, Australia

Kavina Dayal, Ravinesh Deo and Armando A. Apan

List of Acronyms

ANN	Artificial Neural Networks
AWAP	Australian Water Availability Project
BFGS quasi-Newton	Broyden-Fletcher-Goldfarb-Shanno quasi-Newton
d	Willmott's Index of Agreement
E	Nash-Sutcliffe Coefficient of Efficiency
LM	Levenberg-Marquardt training algorithm
Logsig	Logarithmic sigmoid
MAE	Mean Absolute Error
NRM	Natural Resource Management
PE	Prediction Error
PET	Potential Evapotranspiration
Q1	First quartile (25th percentile)
Q2	Second quartile (50th percentile or median)
Q3	Third quartile (75th percentile)
R^2	Coefficient of Determination
RMSE	Root Mean Squared Error
SPEI	Standardized Precipitation-Evapotranspiration Index
Tansig	Tangent sigmoid
Trainbfg	Training BFGS quasi-Newton
Trainlm	Training Levenberg-Marquardt

K. Dayal (✉) · R. Deo
School of Agricultural Computational and Environmental Sciences,
International Centre of Applied Climate Sciences (ICACS),
University of Southern Queensland, Springfield, Australia
e-mail: KavinaShaanu.Dayal@usq.edu.au

A.A. Apan
School of Civil Engineering and Surveying, University of Southern Queensland,
Springfield, Australia

W. Leal Filho (ed.), *Climate Change Adaptation in Pacific Countries*,
Climate Change Management, DOI 10.1007/978-3-319-50094-2_11

Introduction

A major challenge societies in the Pacific Island nations are facing in the current era is the threat of climate change. It has become an international priority to improve our understanding of the climate system, which is described as a complex phenomenon that is imperfectly observed and imperfectly simulated even more (Monteleoni et al. 2011). A fundamental tool that climate scientists use for understanding and predicting climate is climate models, which are mathematical models for producing computer simulations. Each climate model is a highly complex system that climate scientists and meteorologists use to model processes such as sea-ice melt, ocean circulation, and so on. The climate models have evolved over time since 1970s when the global effort on climate modelling began, becoming ever more complex. Given the complexity and number of assumptions required for input parameters, the variance in the prediction among existing climate models remains high (Reichler and Kim 2008; Reifen and Toumi 2009). Scientists are developing and exploring relatively simple tools and ideas to analyse climate systems, such as using artificial intelligence and machine learning approach. One such area of interest in climate studies is drought.

A drought is a natural climatic feature, described as rainfall amounts below average for more than three months in a given area (Mpelasoka et al. 2008). The shortage of water is also affected by aridity, which is a permanent climatic feature with low average annual rainfall and soil moisture content, expressed in terms of precipitation and evapotranspiration ratios (Arora 2002). Both drought and aridity are driven by higher surface temperature and evapotranspiration. Thus, a standardized metric with a combination of changes in precipitation and evapotranspiration is a potential indicator for studying aridity and dry conditions, especially in an area of ecological and agricultural significance. One particular drought indicator that incorporates precipitation and potential evapotranspiration (PET) in its formulation is the Standardized Precipitation—Evapotranspiration Index (SPEI) (Vicente-Serrano et al. 2010a, b). The SPEI is a statistically based index that only requires climatological information (rainfall and temperature), thus making its application to climate studies a very useful tool. Unlike other drought indices, SPEI has the strength to capture increased impact of temperature on water resources. Merits of SPEI has been discussed in a number of studies as it combines multi-timescale aspects of the Standardized Precipitation Index (SPI) with additional information about evapotranspiration that makes SPEI more useful for climate change studies (Vicente-Serrano et al. 2010a).

In literature, there are two kinds of models commonly used for forecasting key parameters of drought such as rainfall and evaporation: (1) physical models that are based on interactive behaviour of ocean, atmosphere, land surface and sea ice, and (2) data-driven models that learn from the trends and behaviour of the observed climatic parameters such as rainfall, temperature and evaporation. The major drawback of physical models is that they require information on physical properties of parameters and in drought forecasting it does not provide reliable information about rainfall, which is the crucial determinant of drought (Hudson et al. 2011; Kuligowski and Barros 1998). In Australia, the two models used for predicting seasonal climates are Predictive Ocean Atmosphere Model for Australia (POAMA) (Zhao and Hendon 2009) and Seasonal Pacific Ocean Temperature Analysis (SPOTA-1) (Day et al. 2010). Studies that have compared precipitation forecast using the POAMA and data-driven models have found the latter with significant improvement in the prediction of the variables tested (Abbot and Marohasy 2012, 2014; Hudson et al. 2011; Inquiry and Holmes 2012). Given the challenges with physical model accuracy and reliability, researchers are testing the viability of data-driven models as alternative tools for forecasting atmospheric, climatic and hydrological variables.

A data-driven model is basically composed of and utilizes computational capacity of artificial intelligence algorithms that are not based on physical properties of atmospheric and oceanic data as with the case of physical models, but instead uses historical patterns and data attributes to learn the linear or non-linear relationship between predictors and the predictand (Deo and Şahin 2015a, b; Şahin et al. 2013). Thus compared to physical models, data-driven models have the ability to explain future trends in climate variables with less complexity in comprehending outputs, requires less information about data, has inexpensive computational cost, is efficient in training and testing datasets, and has simpler computation and evaluation procedures (Abbot and Marohasy 2012, 2014). Data-driven models have shown significant improvement in predicting temperature and precipitation patterns (Abbot and Marohasy 2012, 2014) and drought related atmospheric variables (Deo and Şahin 2015a, b; Deo et al. 2015) in Australia.

A popular, and well-established data-driven model is the Artificial Neural Network (ANN). It has been used is many fields ranging from financial, medical, industrial, data-mining and operational analysis. The ANN mimics biological structure of the brain (McCulloch and Pitts 1943, 1990). It operates like a black box where the user may consider it either an advantage or a disadvantage of ANN, mainly because the internal processing of the networks is not revealed to the user. However, ANN has the ability to learn from the relationship between input (predictor) and output (predictand) variables under supervised or unsupervised learning environment. Moreover, ANN has the capability to work with large, non-linear and non-stationary datasets (Şahin et al. 2013). In eastern Australia, an ANN algorithm has been applied to prediction of SPEI using hydro meteorological and climatic parameters where ANN shown to have exhibited a very good prediction skill for various candidate stations tested (Deo and Şahin 2015b). Likewise, Mekanik et al. (2013) used ANN and Multi-Regression models to predict rainfall in Victoria using lagged relationship of El-Nino Southern Oscillation (ENSO) and Indian Ocean Dipole (IOD) as predictors and found ANN superior to Multiple Linear Regression model in terms smaller prediction errors. Therefore, the applicability of ANN model for prediction of meteorological or climatic parameters is not a new concept, but an enlightening everlasting research endeavour.

In light of the increasing demand for practical applications of ANNs for forecasting of drought events, this study has tested the feasibility of an ANN model for predicting drought using historical climate data. In the present investigation, there are three main objectives: (1) To develop and apply ANN model for predicting SPEI using four meteorological parameters (precipitation, maximum and minimum temperature, potential evapotranspiration) for 1915–2013 period for two locations in Queensland: grassland and temperate. (2) To apply key statistical metrics to evaluate and determine optimum ANN model. (3) To assess the model's performance for drought periods only and compare with predictions in the overall test period used for evaluating the model. The rest of the paper is structured as follows. The theoretical background of the SPEI, ANN and statistical metrics used for model evaluation is discussed in the next section followed by a description of the data and the study regions. Next, the discussion of results and finally the summary and conclusions are provided in the closing section.

Theoretical Framework

Standardized Precipitation-Evapotranspiration Index

In this study, the SPEI was the main output parameter in ANN model. It is calculated using the *'spei'* library, developed by Vicente-Serrano et al. (2010a), freely available in *R* program (https://cran.rproject.org/web/packages/SPEI/index.html).

This section describes the basic SPEI computational approach, but for a detailed theory the readers are referred to works of SPEI developer experts (Vicente-Serrano et al. 2010a, b, 2011, 2012, 2013). As per the index title, the SPEI computation comprises precipitation and PET data. While PET data is not available at hand, it has to be estimated using either Thornthwaite (1948), Hargreaves (1994) or Penman (1948) method where each method has its own pros and cons in terms of data requirements. The PET estimates the amount of evaporation and transpiration that could occur if sufficient water was available, representing the significance of evapotranspiration in the hydrologic cycle (Hanson 1988).

Of the three PET estimation methods, *Hargreaves* is the most data-expansive approach as it requires data information such as cloud cover, incoming solar radiation, saturation water pressure, dew point temperature, humidity, and atmospheric surface pressure for calculating the psychometric constants for PET. Similarly, the *Penman* method requires mean incoming solar radiation, which is not readily available. Thus, this study adopts *Thornthwaite* method that only requires mean temperature and latitude of the site. The PET is calculated as:

$$PET = 16K\left(\frac{10T}{I}\right)^{m} \quad (1)$$

where T is the monthly mean temperature (°C); m deduced empirically can be stated as $m = 6.75 \times 10^{-5}, I^3 + 7.75 \times 10^{-7},\ I^2 + 1.79 \times 10^{-2}, I + 0.492$; I is the heat index that is calculated as the sum of 12-monthly index values i, where i is derived from monthly mean temperature as:

$$i = 16K\left(\frac{T}{5}\right)^{1.514} \quad (2)$$

Note that K is the correction coefficient derived as a function of latitude and month:

$$K = \left(\frac{N}{12}\right)\left(\frac{NDM}{30}\right) \quad (3)$$

The NDM is the number of days in the particular month and N is the maximum number of sun hours, calculated as:

$$N = \left(\frac{24}{\pi}\right)\omega_s \quad (4)$$

and ω_s is the hourly angle of the sun determined as $\omega_s = \arccos(-\tan\phi \tan\delta)$, where ϕ is the latitude in radians; and δ is the sun declination angle in radians.

The SPEI represents the climatological water balance between the precipitation (PCN) and PET, i.e., $D_i = PCN_i - PET_i$, where i is the month. The D_i can be calculated on different timescales, k; a procedure similar to Standardized Precipitation Index (SPI).

The SPEI utilizes probability density function of a three parameter log-logistic distributed variable, expressed as:

$$f(x) = \frac{\beta}{\alpha}\left(\frac{x-\gamma}{\alpha}\right)^{\beta-1}\left(1+\left(\frac{x-\gamma}{\alpha}\right)^{\beta}\right)^{-2} \quad (5)$$

where α, β and γ are scale, shape and origin parameters, respectively, for D values in the range ($\gamma > \mathrm{D} > \infty$).

The parameters (α, β, γ) of the log-logistic distribution can be obtained using different procedures. Among them, the probability-weighted moments (PWMs) procedure is a robust and easy approach (Vicente-Serrano et al. 2010a) where the parameters of the log-logistic distribution can be estimated as:

$$\beta = \frac{2w_i - w_o}{6w_i - w_o - 6w_2} \quad (6)$$

$$\alpha = \frac{(w_o - 2w_i)\beta}{\Gamma\left(1+\frac{1}{\beta}\right)\Gamma\left(1-\frac{1}{\beta}\right)} \quad (7)$$

$$\gamma = w_o - \alpha\Gamma\left(1 + \frac{1}{\beta}\right)\Gamma\left(1 - \frac{1}{\beta}\right) \tag{8}$$

where $\Gamma(B)$ is the gamma function of B.

The probability distribution function of the D series according to the log-logistic distribution is given by:

$$F(x) = \left[1 + \left(\frac{\alpha}{x - \gamma}\right)^{\beta}\right]^{-1} \tag{9}$$

With $F(x)$, the SPEI can easily be obtained as the standardized values of $F(x)$:

$$SPEI = W - \frac{C_o + C_1 W + C_2 W^2}{1 + d_1 W + d_2 W^2 + d_3 W^3} \tag{10}$$

where

$$W = \sqrt{-2\ln(P)} \tag{11}$$

for $P \leq 0.5$, and P is the probability of exceeding a determined D value, $P = 1-F(x)$. If $P > 0.5$, P is replaced by $1-P$ and the sign of the resultant SPEI is reversed. The constants are: $C_o = 2.515517$, $C_1 = 0.802853$, $C_2 = 0.010328$, $d_1 = 1.432788$, $d_2 = 0.189269$, $d_3 = 0.001308$. The average value of SPEI is 0, and the standard deviation is 1. The SPEI is a standardized variable, and it can therefore be compared with other SPEI values over time and space. SPEI of 0 indicates a value corresponding to 50% of the cumulative probability of D, according to a log-logistic distribution (Vicente-Serrano et al. 2010a, b).

Artificial Neural Network

An ANN model is a computational algorithm that has interconnections between the various neurons in different layers of the ANN system. This study uses a 3-layer neural network system. Figure 11.1 illustrates the basic 3-layer ANN topology. The first layer consists of input neurons that send data through synapses to the second layer of hidden neurons, from which a single output is generated through a non-linear activation function in the third layer of output neurons.

The ANN model has the fundamental processing unit, a neuron k, described by:

$$u_k = \sum_{j=1}^{m} w_k x_j \tag{12}$$

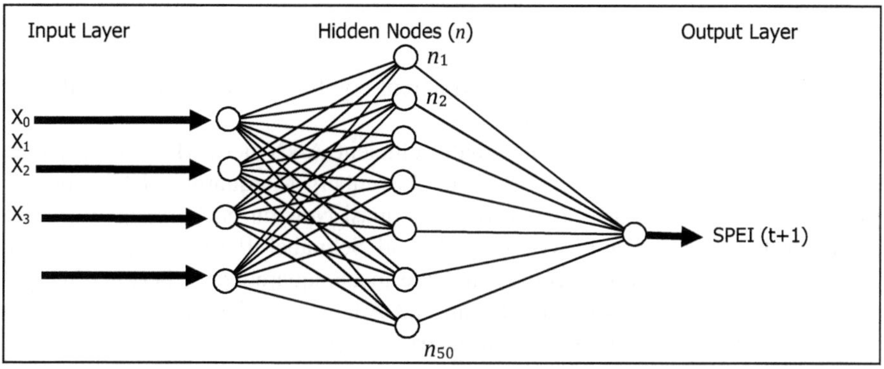

Fig. 11.1 The topological structure of the 3-layer Artificial Neural Network used for prediction of the Standardized Precipitation and Evaporation Index (*SPEI*). The first layer consists of 4 inputs, "*X*" represent precipitation (X_0), maximum temperature (X_1), minimum temperature (X_2) and potential evapotranspiration (X_3). The second layer consists of 50 hidden neurons (n_1:n_{50}). The third layer is the output layer, the predicted SPEI (t+1)

$$y_k = \Phi(u_k + b_k) \tag{13}$$

where $x_1, x_2, \ldots, x_m$ are the input signals; $w_{k1}, w_{k2}, \ldots, w_{km}$ are the synaptic weights of neuron k; u_k is the linear combiner output due to input signals; b_k is the bias; $\Phi(.)$ is the activation function; and y_k is the output signal of the neuron.

The bias b_k has the effect of transforming the output u_k of the linear combiner by shifting the activation functions to the right or to the left, hence it is very useful in successful learning of the model, shown by:

$$v_k = u_k + b_k \tag{14}$$

Thus, the combination of Eqs. (12) and (13) may be formulated as:

$$v_k = \sum_{j=0}^{m} w_{kj} x_j \tag{15}$$

$$y_k = \Phi(v_k) \tag{16}$$

The three transfer functions: tangent sigmoid, $\phi(x)$ logarithmic sigmoid, $\psi(x)$ and linear, $\chi(x)$ are described as follows (Vogl et al. 1988):

$$\phi(x) = \frac{2}{1 + e^{-2x}} - 1 \tag{17}$$

$$\psi(x) = \frac{1}{1 + e^{-x}} \tag{18}$$

$$\chi(x) = linear(x) = x \tag{19}$$

where (17)—(19) may be trailed with different combinations to determine the best predictive model (Şahin et al. 2013).

In order to develop a computationally efficient ANN network, a second-order training algorithms, known as the Levenberg-Marquardt (LM) or the Broyden-Fletcher-Goldfarb-Shanno (BFGS) quasi-Newton backpropagation learning algorithms can be employed (Dennis and Schnabel 1983; Marquardt 1963). These training algorithms help minimize the mean squared error between the predicted and the observed variable (Tiwari and Adamowski 2013). The LM method uses an approximation to the Hessian matrix, given as:

$$x_{k+1} = x_k - \left[J^T J + \mu I\right]^{-1} J^T e \tag{20}$$

where J is the Jacobian matrix that is calculated using standard backpropagation techniques and contains first derivatives of network errors with respect to the weights and biases (Hagan and Menhaj 1994). The computation of Jacobian matrix is simpler than the Hessian matrix (Marquardt 1963). The term e is a vector of errors.

An alternative to the conjugate gradient methods for fast optimization is the BFGS quasi-Newton. It uses the following equation:

$$x_{k+1} = x_k - A_k^{-1} g_k \tag{21}$$

where A_k^{-1} is the Hessian matrix (second derivatives) of the performance index at the current values of the weights and biases (Dennis and Schnabel 1983).

Materials and Methods

Study Regions and Climate Data

In order to model the SPEI using an ANN algorithm, data from two climatic regions in Queensland, Australia has been used in this study. The 1° × 1° longitude/latitude grid from two regions namely: (i) grassland (27.75°S–28.75°S, 148°E–149°E) and (ii) temperate (27.75°S–28.75°S, 151.75°E–152.75°E) regions are used as study regions. Figure 11.2 shows location of the study regions on a geographical map. Both regions, grassland and temperate, are located on the west side of Great Diving Range. These regions are mainly dominated by landforms such as tablelands, slopes and plains. The regional Natural Resource Management organization has identified a range of climate change impacts and adaptation challenges in this region. Some of these challenges include understanding the likely changes to agricultural production including changes to the growing conditions and yields for key crops; management of invasive species, soil erosion and land degradation; and improving the resilience of riparian, aquatic and terrestrial ecosystems (NRM 2015). The grassland region is

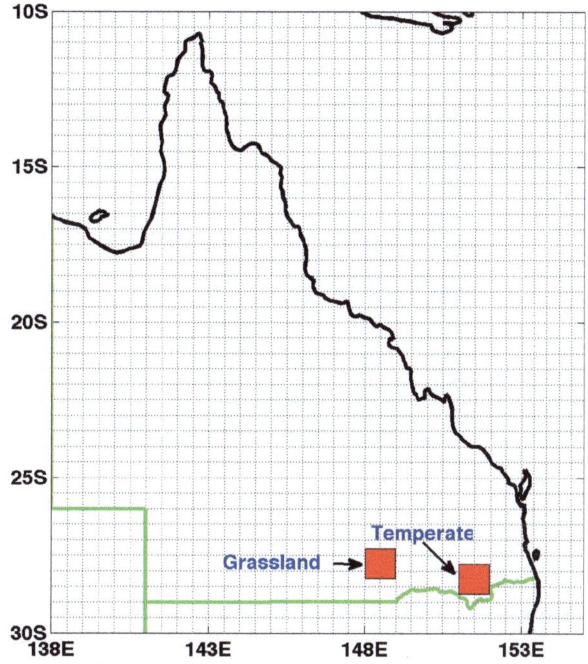

Fig. 11.2 Location of the two study regions: grassland (27.25°S–28.25°S, 148°E–149°E) and temperate (27.75°S–28.75°S, 151.75°E–152.75°E)

noted for its diverse agriculture, tourism and energy sectors. The temperate region is known as the major food basket of the state. The landscape of temperate region comprises pastures, cattle, and vegetable farming.

The climatic characteristics between grassland and temperate regions in terms of seasonal rainfall amount differ to some extent. Figure 11.3 shows a bar graph of seasonal precipitation climatology. In general, the summer season is the wettest and winter season is the driest at both locations, but the amount of rainfall received by the two locations differs significantly. For instance, grassland receives 61.39 mm rainfall on average in summer, whereas temperate region receives 45.76% more with 113.18 mm. Similarly, during winter season grassland receives 28.17 mm of rainfall on average, whereas temperate region receives 33.92% more with an amount of 42.63 mm. Likewise, autumn and spring season in temperate region is wetter compared to the grassland region. The autumn (spring) season receives 36.56 mm (30.66 mm) and 64.93 mm (57.78 mm) on average in grassland and temperate region, respectively.

For the prediction of the SPEI, a total of four meteorological input (predictor) variables were used: monthly precipitation (in millimeters), maximum temperature (in °C), minimum temperature (in °C) and potential evapotranspiration (PET). While PET data was not readily available, it was calculated with mean temperature and mean latitude of the two study locations using *Thornthwaite* approach. The gridded at 0.05° by 0.05° monthly precipitation, maximum and minimum temperature were obtained from Australian Water Availability Project (AWAP)

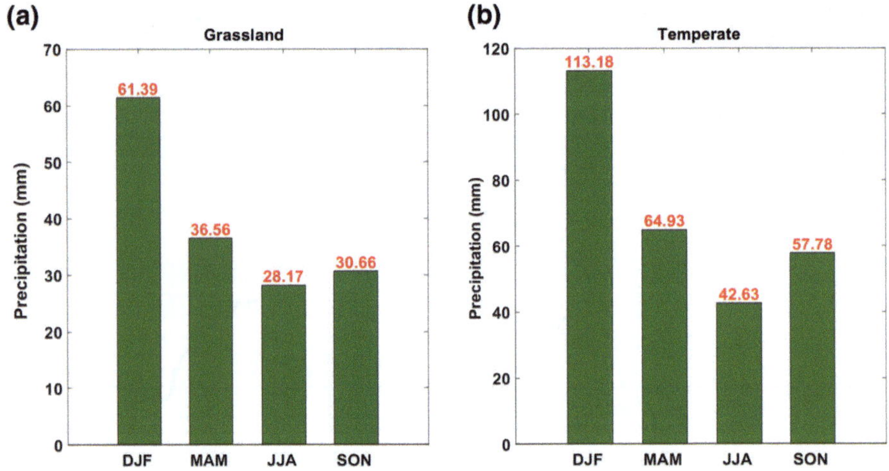

Fig. 11.3 Histogram of seasonal precipitation (mm) patterns averaged over 1915–2013. Acronyms are: DJF (summer); MAM (autumn); JJA (winter); SON (spring)

(Raupach et al. 2008) for the period 1915 to 2013. On a 1° × 1° longitude/latitude grid, there were 441 points of data, which were averaged to obtain a single time-series of precipitation, maximum and minimum temperature. The model output (predictand) was the monthly SPEI, calculated using precipitation and PET for the 1915–2013 period.

ANN Model Development

For this study, the ANN model was developed using MATLAB. As shown in Fig. 11.1, the 3-layer neuron in the ANN model's topology consisted of an input layer, hidden nodes, and an output layer. In the input layer, the four pre-selected input variables (*X*: precipitation, maximum temperature, minimum temperature and PET) were used. Since there was no mathematical formulae for determining the neuronal structure in hidden nodes layer, the optimal number of hidden nodes were decided by trial and error method (Şahin 2012) where a maximum of 50 neurons were tested for development of the network architecture.

It is important to note that this study only used a standalone ANN model, therefore comparison of its performance with other artificial intelligence models (e.g. support vector machines or multiple linear regressions) were not made in this investigation. In addition, one practical limitation could have been the amount of time it requires to train the network, i.e., the training time would increase as the number and size of input increases. In this study however, the efficiency of the model in terms of running time was not a factor for determining the optimum model, instead minimum prediction error was used as a determinant of the best neural network configurations. For the model simulations, the size and number of inputs remained unchanged while only the number of hidden neurons were ranged from 1 to 50 with alternating training and transfer functions. The size of the input variables was 1188 months (from 1915 to 2013), from which 80% (950 months) was used for training the network, 10% (119 months) was used to validate the model, and last 10% (119 months) was used for testing where predictions were made for the test datasets and compared with the actual observed values. The input and output data had to be standardized first before using it for the training purpose. Standardizing the input and output variables would help reach global minima faster and reduce the chances of the networks getting stuck in the local minima. Since only four input variables were used, there were four columns of input data (precipitation, maximum temperature, minimum temperature and PET) and one column of output data (SPEI). Moreover, in developing the ANN model for the current investigation, it was not possible to predetermine the optimal design of the neural networks for the four inputs and an output variable. To overcome this, trial and error approach was employed where number of neurons was tested from 1 to 50 while alternating the training algorithms (LM and BFGS quasi-Newton) and hidden transfer and output functions (tangent sigmoid, logarithmic sigmoid and linear) to evaluate performance of the model for different neural network configurations. The *'newff'* command in MATLAB was used to create a feed-forward backpropagation network.

Table 11.1 lists the ANN model architecture of the 12 model runs for each study region. The training functions were alternated between the LM *(trainlm)* and BFGS quasi-Newton *(trainbfg)*, hidden transfer functions were alternated between tangent sigmoid *(tansig)* and logarithmic sigmoid *(logsig)*, and the output transfer functions were alternated between tangent sigmoid *(tansig)*, logarithmic sigmoid *(logsig)* and linear *(purelin)*. In each trial, the number of hidden neurons ranged from 1 to 50. Therefore, in total a 12 $\times$ 50 model runs were performed for each study region in order to determine the best architecture of the optimum ANN model in this study. To determine an optimum architecture of the ANN model, a total of 12 model runs (M1-M12) for each study region (grassland: Table 11.1a and temperate: Table 11.1b) were conducted one by one by using unique combinations of training (LM *(trainlm)* and BFGS quasi-Newton *(trainbfg)*, and transfer functions (tangent sigmoid, *tansig*; logarithmic sigmoid, *logsig*; and linear, *purelin*). From the training (learning) phase, a final weight matrix was obtained that was then applied to the individual inputs independently in the testing phase. The optimum ANN model for each study region was obtained after evaluating the model outcomes using a set of statistical metrics, discussed in the next section.

Statistical Evaluation of Model Performance

The performance of ANN models over the testing period were evaluated with five statistical methods: (i) Root-Mean Square Error (*RMSE*), (ii) Mean Absolute Error (*MAE*), (iii) Coefficient of Determination (R^2) (Paulescu et al. 2011; Ulgen and

Table 11.1 Quantitative measures of the ANN model performance based on overall prediction score metrics over the test period 2004–2013. The boldfaced rows are combinations for the best architecture (input-hidden neurons-output) of ANN models: grassland (4-28-1) and temperate (4-27-1)

Model number	Training function	Output transfer function	Hidden transfer function	Network architecture	R^2	d	E	RMSE	MAE
(a)									
M1	Trainlm	Purelin	Logsig	4-33-1	0.9740	0.9845	0.9737	0.1704	0.1148
M2	Trainlm	Purelin	Tansig	4-12-1	0.9748	0.9853	0.9746	0.1674	0.1183
M3	Trainbfg	Purelin	Tansig	4-40-1	0.9547	0.9723	0.9534	0.2267	0.1789
M4	Trainbfg	Purelin	Logsig	4-23-1	0.9565	0.9737	0.9555	0.2216	0.1775
M5	Trainlm	Logsig	Logsig	4-29-1	0.9795	0.9884	0.9796	0.1502	0.1078
M6	Trainlm	Tansig	Logsig	4-17-1	0.9767	0.9867	0.9766	0.1608	0.1186
M7	Trainlm	Logsig	Tansig	4-20-1	0.9750	0.9856	0.9745	0.1678	0.1246
M8	**Trainlm**	**Tansig**	**Tansig**	**4-28-1**	**0.9839**	**0.9909**	**0.9838**	**0.1338**	**0.0882**
M9	Trainbfg	Logsig	Logsig	4-39-1	0.9590	0.9751	0.9584	0.2144	0.169
M10	Trainbfg	Tansig	Logsig	4-47-1	0.9575	0.9738	0.9566	0.219	0.1723
M11	Trainbfg	Logsig	Tansig	4-29-1	0.9524	0.9716	0.9518	0.2306	0.175
M12	Trainbfg	Tansig	Tansig	4-42-1	0.9624	0.9763	0.9606	0.2085	0.168
(b)									
M1	Trainlm	Purelin	Logsig	4-11-1	0.9787	0.9879	0.9768	0.1629	0.1197
M2	Trainlm	Purelin	Tansig	4-14-1	0.9823	0.9899	0.9805	0.1492	0.1135
M3	Trainbfg	Purelin	Tansig	4-36-1	0.9530	0.9731	0.951	0.2367	0.185
M4	Trainbfg	Purelin	Logsig	4-48-1	0.9575	0.9762	0.9556	0.2252	0.1631
M5	**Trainlm**	**Logsig**	**Logsig**	**4-27-1**	**0.9886**	**0.9935**	**0.9874**	**0.1198**	**0.0814**
M6	Trainlm	Tansig	Logsig	4-14-1	0.9872	0.9926	0.9856	0.1282	0.0852
M7	Trainlm	Logsig	Tansig	4-35-1	0.9797	0.989	0.9788	0.1557	0.1087
M8	Trainlm	Tansig	Tansig	4-34-1	0.9868	0.9926	0.9859	0.1271	0.0828
M9	Trainbfg	Logsig	Logsig	4-31-1	0.9516	0.9732	0.9504	0.238	0.1804
M10	Trainbfg	Tansig	Logsig	4-48-1	0.9569	0.9752	0.9532	0.2312	0.1616
M11	Trainbfg	Logsig	Tansig	4-34-1	0.9512	0.9737	0.9504	0.238	0.1685
M12	Trainbfg	Tansig	Tansig	4-44-1	0.9559	0.9753	0.9531	0.2316	0.1773

Hepbasli 2002), (iv) Willmott's Index of Agreement (Acharya et al. 2013; Willmott 1981, 1982) and (v) Nash-Sutcliffe Coefficient of Efficiency (*E*) (Krause et al. 2005; Nash and Sutcliffe 1970). The mathematical formulas are:

i. Root Mean Squared Error *(RMSE)*:

$$RMSE = \sqrt{\frac{1}{N}\sum_{i=1}^{n}\left(SPEI_{p_i} - SPEI_{o_i}\right)_t^2} \quad (22)$$

ii. Mean Absolute Error *(MAE)*:

$$MAE = \frac{1}{N}\sum_{i=1}^{n}\left|\left(SPEI_{p_i} - SPEI_{o_i}\right)_t\right| \tag{23}$$

iii. Coefficient of Determination *(R^2)*:

$$R^2 = \left(\frac{\sum_{i=1}^{n}\left(SPEI_{o,i} - \overline{SPEI_{o,i}}\right)\left(SPEI_{p,i} - \overline{SPEI_{p,i}}\right)}{\sqrt{\sum_{i=1}^{n}\left(SPEI_{o,i} - \overline{SPEI_{o,i}}\right)^2}\sqrt{\sum_{i=1}^{n}\left(SPEI_{p,i} - \overline{SPEI_{p,i}}\right)^2}}\right)^2 \tag{24}$$

iv. Wilmott's Index *(d)*:

$$d = 1 - \left[\frac{\sum_{i=1}^{N}\left(SPEI_{o,i} - SPEI_{p,i}\right)^2}{\sum_{i=1}^{N}\left(\left|SPEI'_{p,i} - \overline{SPEI_0}\right| - \left|SPEI'_{o,i} - \overline{SPEI_o}\right|\right)^2}\right],\ 0 \le d \le 1 \tag{25}$$

v. Nash–Sutcliffe Coefficient *(E)*:

$$E = 1 - \left[\frac{\sum_{i=1}^{N}\left(SPEI_{o,i} - SPEI_{p,i}\right)^2}{\sum_{i=1}^{N}\left(SPEI_{o,i} - \overline{SPEI_o}\right)^2}\right],\ 0 \le E \le 1 \tag{26}$$

In Eqs. (22)–(26), the $SPEI_{o,i}$ is the monthly observed SPEI and $SPEI_{p,i}$ is the monthly predicted SPEI. The '*i*' refers to month of the testing period dataset, '*o*' refers to observed and '*p*' refers to predicted.

Out of 12 × 50 model runs, the best combinations are summarized in Table 11.1 for each study region. For grassland region (Table 11.1a), the best ANN model (denoted M8) was trained using Levenberg-Marquardt (LM; *trainlm*) method and transfer function for both hidden and output layer was tangent sigmoid (*tansig*) Eq. (17). The M8 had input-hidden-output layer architecture of 4-28-1, i.e., 4 input variables (precipitation, maximum temperature, minimum temperature and PET), 28 hidden neurons, and 1 output (SPEI). The M8 was selected based on maximum R^2, *d* and *E* and minimum *RMSE* and *MAE*. The R^2, *d, E, RMSE* and *MAE* were 0.9839, 0.9909, 0.9838, 0.1338 and 0.0882, respectively. For temperate region (Table 11.1b), the best ANN model (denoted M5) was also trained using Levenberg-Marquardt (LM; *trainlm*) method while the transfer function for both hidden and output layer was logarithmic-sigmoid (*logsig*) Eq. (18). The M5 had input-hidden-output layer architecture of 4-27-1, i.e., 4 input variables (precipitation, maximum temperature, minimum temperature and PET), 27 hidden neurons, and 1 output (SPEI). The R^2, *d, E, RMSE* and *MAE* for M5 were 0.9886, 0.9935, 0.9874, 0.1198 and 0.0814, respectively.

Results and Discussion

The figures were generated based on the optimum ANN model M8 (model architecture: 4-28-1) and M5 (model architecture: 4-27-1) for grassland and temperate regions, respectively. Figure 11.4 shows a time series of the monthly observed and predicted SPEI with overlapping prediction error (PE; PE = $SPEI_p - SPEI_o$) bars for the test period (2004–2013) for each tested month. The positive $SPEI_{o,p}$ correspond to wet period and negative $SPEI_{o,p}$ correspond to dry period. The absolute PE (|PE|) ranged from 0.0017 to 0.8224 for grassland (Fig. 11.4a) and from 0.0039 to 0.6999 temperate (Fig. 11.4b). The negative PE on the graph refers to under-prediction while positive PE refers to over-prediction of the SPEI values. In general, a good visual agreement between the observed and predicted SPEI is noted for both study regions. However, the PE yield for each month is discernible,

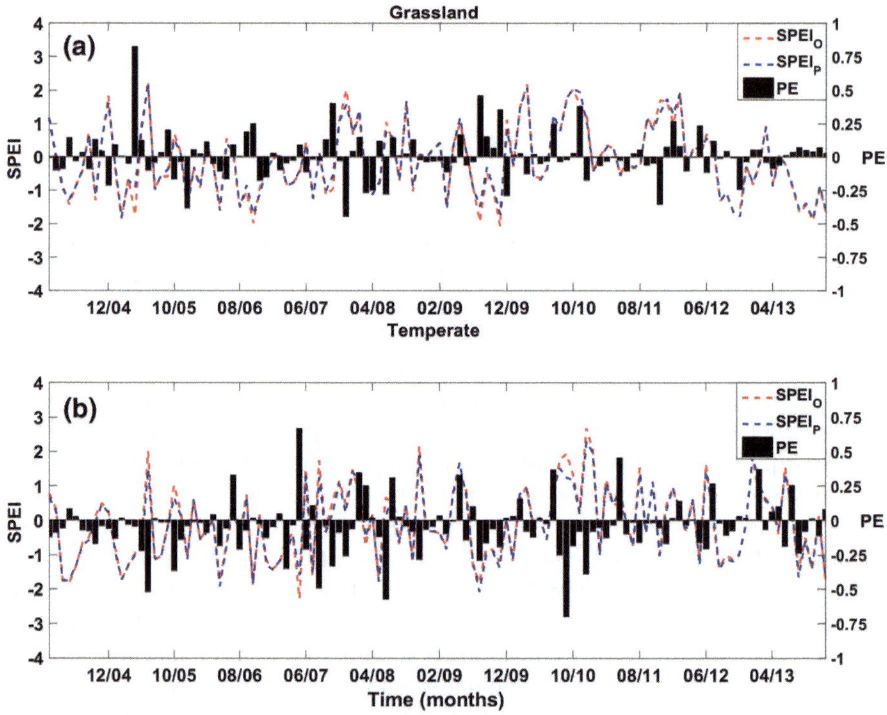

Fig. 11.4 The observed and the predicted *SPEI* plotted with the corresponding prediction error (*PE*) in the test period 2004–2013. **a** Grassland and **b** temperate

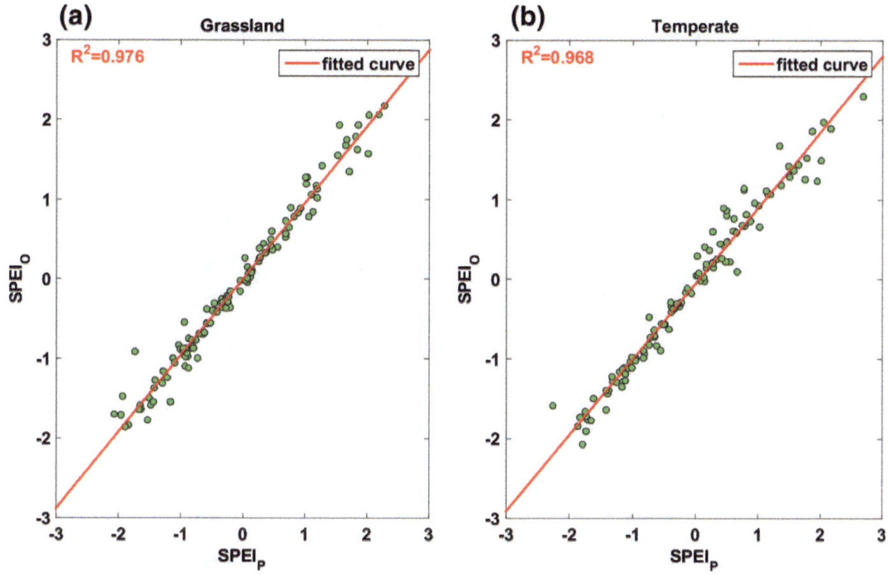

Fig. 11.5 Scatter plot of observed and the predicted *SPEI* over the test period 2004–2013. **a** Grassland and **b** temperate

especially it can be noted that the large errors in the prediction exist for extreme dry (negative SPEI) or wet (positive SPEI) months, whereas small PE exist for months with small dry or wet events. A possible reason for large errors could be due to neural networks not being able to capture extreme (dry or wet) events during the learning phase that may have affected the test dataset with errors being large in the prediction. As mentioned earlier, neural networks learn from the relationship between input and output variables, and if the learning phase had not captured large peaks in the dataset time series, then it would repeat the same in the testing/prediction phase. Nonetheless, the prediction errors were still reasonably small in general. By comparing the prediction error range, the ANN model performed slightly better for temperate region with largest error of magnitude 0.6999 whereas for grassland it was 0.8224. This was also confirmed with largest R^2, d and E and smallest *RMSE* and *MAE* values for temperate region, as shown in Table 11.1. Overall, a good prediction skill of the ANN model was obtained for both study regions in the present investigation.

Another way to demonstrate a visual representation of the observed and predicted SPEI is a scatterplot. Figure 11.5 shows scatterplots with a linear fit of the observed and predicted SPEI. The linear fit is close to perfect at both locations. For a closer look at the PE yield over the test period, a frequency distribution map of PE was produced for each study region, as shown in Fig. 11.6. A close to Gaussian distribution of PE is observed for grassland region (Fig. 11.6a). It can also be confirmed that the PE value of 0.8224 is essentially an outlier. If this outlier is disregarded, the new range for PE for grassland region is 0.0017 to 0.4569, which

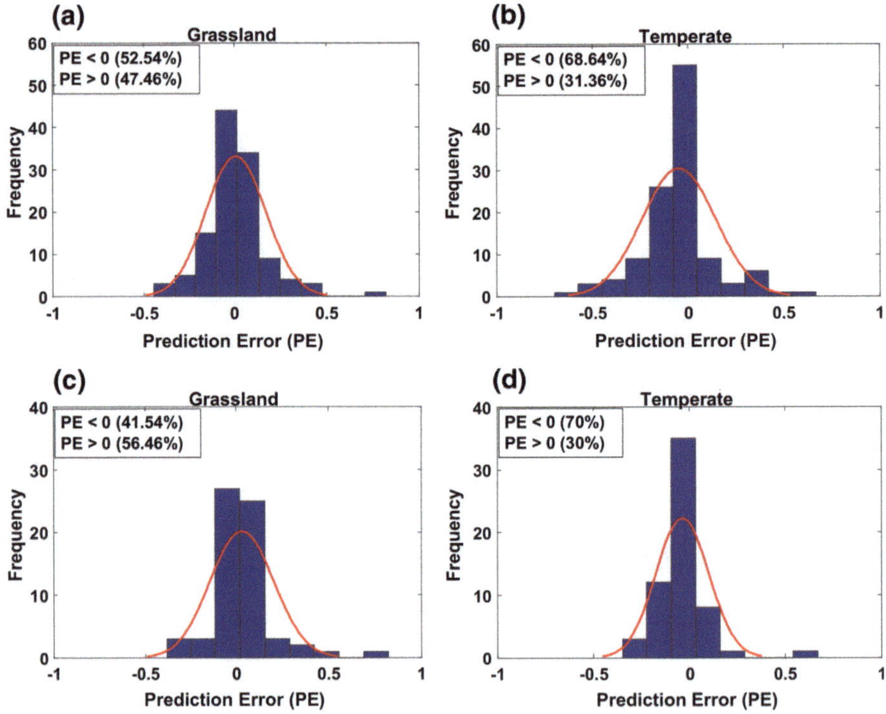

Fig. 11.6 Histogram of the frequency distribution of the Prediction Error (*PE*) calculated for the test period **a** grassland, and **b** temperate, and for the drought periods only **c** grassland and **d** temperate. The cumulative frequency deduced for the respective bin represent under-predictions ($PE < 0$) and over-predictions ($PE > 0$) by the ANN models and comparisons with the Gaussian distribution

then makes the ANN model performance for grassland region better than that of temperate region. The PE spread for temperate region does not show presence of any outliers (Fig. 11.6b). In terms of over- and under-prediction, ANN models appeared to have under-predicted SPEI values at both locations, with 52.54 and 68.64% in grassland and temperate regions, respectively, where 68.64% in temperate region shows a larger discrepancy in the model performance. In terms of over-prediction, 47.46 and 31.36% of the time the ANN model predicted smaller SPEI values than the observed values over the test period in grassland and temperate regions, respectively.

For a better understanding of how the ANN model performed only for drought events over the testing period that consisted 118 months, the prediction error for the corresponding $SPEI_o$ for drought events (i.e., $SPEI_o < 0$) were separated from the wet events ($SPEI_o > 0$). This yielded 65 and 60 months with drought events for grassland and temperate regions, respectively. Figure 11.6c, d show the frequency distribution of PE for drought months only over the test period. In grassland region (Fig. 11.6c), ANN model over predicted 58.46% of the SPEI values and under-predicted 41.54% of SPEI values. However, a large discrepancy is noticed for the temperate region (Fig. 11.6d). The ANN model has under-predicted 70% of SPEI values for drought months and over-predicted 30% of the SPEI values. Comparing the frequency of PE between overall test period and only for the drought periods, the under- and over-prediction of SPEI values in temperate region remain nearly consistent (under-prediction: 68.64 and 70%, over-prediction: 31.36 and 30%) whereas for grassland region the ANN model under-predicted slightly fewer SPEI values with 41.54% when compared with overall test period (52.54%).

In studying droughts using standardized drought indicators (e.g., SPEI), previous studies have classified drought events into different categories, such as extreme droughts, severe droughts, moderate droughts and normal conditions e.g., (Dogan et al. 2012; Morid et al. 2006; Pandey et al. 2008). The categories of drought events are identified as extreme when SPEI ≤ -2, severe when SPEI is between -1.99 to -1.50, moderate when SPEI is between -1.49 to -1.00 and a normal condition when SPEI is between -0.99 to 0.99. Considering the PE values obtained from the ANN model outputs and the SPEI range for each drought category, the PE values are small enough to maintain the $SPEI_p$ in the same category as $SPEI_o$. In other words, the small magnitudes of prediction errors have demonstrated very good prediction skill of the proposed ANN model in this study.

To illustrate the spread of the observed and predicted SPEI, a boxplot for grassland and temperate regions is produced, shown in Fig. 11.7. A boxplot represents the degree of spread in the observed and predicted SPEI using respective

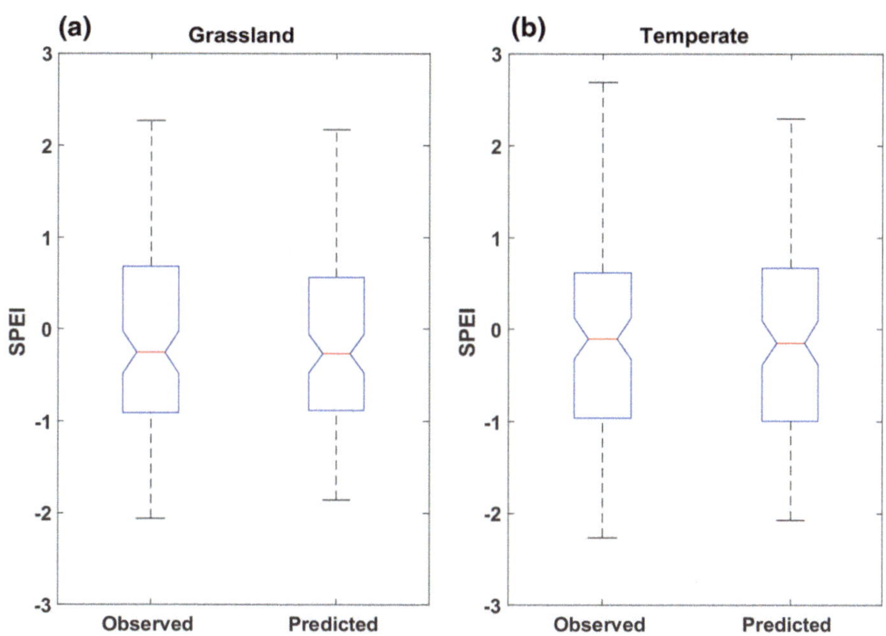

Fig. 11.7 Boxplot of the Standardized Precipitation and Evaporation Index (*SPEI*) from the observed and the predicted dataset using the artificial neural network (ANN) models for the testing period. **a** Grassland and **b** temperate

Table 11.2 The quartile Q1 (25th percentile), Q2 (50th percentile; median), and Q3 (75th percentile) values for observed ($SPEI_o$) and predicted ($SPEI_p$) values over the test period 2004–2013

Grassland			Temperate	
Quartiles	$SPEI_o$	$SPEI_p$	$SPEI_o$	$SPEI_p$
Q1	−0.9097	−0.8822	−0.9632	−0.9959
Q2	−0.2515	−0.2628	−0.0988	−0.1446
Q3	0.6858	0.5677	0.6197	0.6693

quartiles, where first quartile (Q1) represent 25th percentile, second quartile (Q2) represent 50th percentile or the median value, and the third quartile (Q3) represent the 75th percentile. The two whiskers extend from Q1 to the smallest non-outlier value and from Q3 to the largest non-outlier value in the dataset. It can be noticed that the median of the observed and predicted SPEI are nearly identical at both study regions whereas the whiskers extend to slightly different values. Table 11.2 lists the actual quartile values of the observed and predicted SPEI. A very subtle difference is noted for each quartile between observed and predicted SPEI at both locations. This result confirmed that the ANN model developed for the present investigation has exhibited very good skill in predicting a drought indicator, SPEI. Thus the ANN models can also be used for predicting other meteorological variables associated with drought events in climate change adaptation studies that can help prepare for the upcoming drought conditions.

Conclusion

The modelling of drought events using a drought indicator has shown a significant step forward into climate forecasting. In this study, we have applied the state-of-the-art artificial neural networks to predict Standardized Precipitation-Evapotranspiration Index (SPEI) using four meteorological variables (monthly precipitation, maximum and minimum temperature and potential evapotranspiration) for two study locations: grassland and temperate regions. The artificial intelligence technique was the artificial neural networks (ANN) where the model was trained using input parameters (precipitation, maximum and minimum temperature and PET) as predictors, and SPEI as the output parameter or predictand. In developing the ANN model, 80% (1915–1993) of the monthly input data were used to train the ANN network where the analysis of data pattern and adjustment of model weights and biases were undertaken, 10% (1994–2003) of the input data were used to validate the network to minimize overfitting, and the last 10% (2004–2013) of the input data were used to test the model output/prediction for comparison with observed SPEI.

For the optimum ANN architecture of 2-28-1 and 4-27-1, the prediction metrics for grassland and temperate regions in order of R^2, *d, R, RMSE* and *MAE* were 0.9839, 0.9909, 0.9838, 0.1338, 0.0882 and 0.9886, 0.9935, 0.9874, 0.1198, 0.0814, respectively. Based on quantitative statistics, the prediction error calculated from the predicted SPEI compared with the observed values in the test period ranged from 0.0025 to 0.8224 (grassland region) and from 0.0113 to 0.6667 (temperate region). The magnitudes of prediction errors are small enough to maintain the category of drought event as that of observed SPEI at both study locations. The over- and under-prediction of SPEI in the ANN model over the test period varied between study regions, where temperate region recorded higher percentage of under-predicted values (68.64%) and grassland region had near normal distribution of over-prediction (52.54%). For the drought cases only, the percentage of under- and over-prediction in temperate region recorded a very subtle change with under-prediction by 70% and over-prediction by 30%. On the other hand, the under-prediction for drought cases only in grassland region decreased to 41.54% and over-prediction increased to 58.46%. In further examination of the spread of observed and predicted SPEI over the test period, the quartile values were compared between $SPEI_o$ and $SPEI_p$. The analysis showed the quartile values to be nearly equal between observed and predicted SPEI.

The developed ANN models demonstrated good ability for predicting the monthly Standardized Precipitation-Evapotranspiration Index for grassland and temperate regions considered in this study. The success of using ANN models that yielded reasonably small prediction errors is a promising result for forecasting drought events on a real-time (monthly) basis. The ANN models are therefore good tools for scientists and engineers in different areas of climate predictions. This study has some limitations. For example, the optimum model was selected based on best combinations of training algorithms and transfer functions but a range of data training and testing periods and their impacts on model's accuracy was not investigated. In addition, this study used only four input variables, however drought is also associated with a number of other climatic variables that must be tested to establish a set of versatile predictive models. Therefore, a rigorous study that is able to test the sensitivity of various inter-related predictor variables to develop drought forecasting models can be useful in climate risk management, climate change adaptation and mitigation.

Acknowledgements The data were acquired from the Australian Water Availability Project (AWAP). Kavina Dayal was supported by the University of Southern Queensland Postgraduate Research Scholarship (USQ-PRS) and School of Computational, Agricultural and Environmental Sciences. We thank the reviewers and editor for their comments and feedback.

References

Abbot, J., & Marohasy, J. (2012). Application of artificial neural networks to rainfall forecasting in Queensland, Australia. *Advances in Atmospheric Sciences, 29*(4), 717–730.

Abbot, J., & Marohasy, J. (2014). Input selection and optimisation for monthly rainfall forecasting in Queensland, Australia, using artificial neural networks. *Atmospheric Research, 138,* 166–178.

Acharya, N., Shrivastava, N. A., Panigrahi, B., & Mohanty, U. (2013). Development of an artificial neural network based multi-model ensemble to estimate the northeast monsoon rainfall over south peninsular India: An application of extreme learning machine. *Climate Dynamics, 43*(5), 1303–1310.

Arora, V. K. (2002). The use of the aridity index to assess climate change effect on annual runoff. *Journal of Hydrology, 265*(1), 164–177.

Day, K., Ahrens, D., & Peacock, A. (2010), Seasonal Pacific Ocean Temperature Analysis-1 (SPOTA-1) as at November 1, 2010 (2 pp). *Report issued by the Queensland Climate Change Centre of Excellence, Queensland Government, Brisbane, Australia.* Retrieved from https://www.longpaddock.qld.gov.au/spota1-getpassword.html.

Dennis, J. E., & Schnabel, R. B. (1983). *Numerical methods for unconstrained minimization.* Englewood Cliffs: Prentice-Hall.

Deo, R. C., & Şahin, M. (2015a). Application of the extreme learning machine algorithm for the prediction of monthly Effective Drought Index in eastern Australia. *Atmospheric Research, 153*, 512–525. item: WOS:000347264600041.

Deo, R. C., & Şahin, M. (2015b). Application of the artificial neural network model for prediction of monthly standardized precipitation and evapotranspiration index using hydrometeorological parameters and climate indices in eastern Australia. *Atmospheric Research, 161*, 65–81.

Deo, R. C., Samui, P., & Kim, D. (2015). Estimation of monthly evaporative loss using relevance vector machine, extreme learning machine and multivariate adaptive regression spline models. *Stochastic Environmental Research and Risk Assessment, 30,* 1–16.

Dogan, S., Berktay, A., & Singh, V. P. (2012). Comparison of multi-monthly rainfall-based drought severity indices, with application to semi-arid Konya closed basin, Turkey. *Journal of Hydrology, 470–471,* 255–268.

Hagan, M. T., & Menhaj, M. B. (1994). Training feedforward networks with the Marquardt algorithm. *IEEE Transactions on Neural Networks, 5*(6), 989–993.

Hanson, R. L. (1988). Evapotranspiration and droughts (pp. 99–104). *Paulson, RW, Chase, EB, Roberts, RS, and Moody, DW, Compilers, National Water Summary.*

Hargreaves, G. H. (1994). Defining and using reference evapotranspiration. *Journal of Irrigation and Drainage Engineering, 120*(6), 1132–1139.

Hudson, D., Alves, O., Hendon, H. H., & Marshall, A. G. (2011). Bridging the gap between weather and seasonal forecasting: intraseasonal forecasting for Australia. *Quarterly Journal of the Royal Meteorological Society, 137*(656), 673–689.

Inquiry, QFCo, & Holmes, CE. (2012). *Queensland Floods Commission of Inquiry: Final Report.* Queensland Floods Commission of Inquiry.

Krause, P., Boyle, D., & Bäse, F. (2005). Comparison of different efficiency criteria for hydrological model assessment. *Advances in Geosciences, 5*(5), 89–97.

Kuligowski, R. J., & Barros, A. P. (1998). Experiments in short-term precipitation forecasting using artificial neural networks. *Monthly Weather Review, 126*(2), 470–482.

Marquardt, D. W. (1963). An algorithm for least-squares estimation of nonlinear parameters. *Journal of the Society for Industrial and Applied Mathematics, 11*(2), 431–441.

McCulloch, W. S., & Pitts, W. (1943). A logical calculus of the ideas immanent in nervous activity. *The Bulletin of Mathematical Biophysics, 5*(4), 115–133.

McCulloch, W. S., & Pitts, W. (1990). A logical calculus of the ideas immanent in nervous activity. *Bulletin of Mathematical Biology, 52*(1–2), 99–115.

Mekanik, F., Imteaz, M., Gato-Trinidad, S., & Elmahdi, A. (2013). Multiple regression and Artificial Neural Network for long-term rainfall forecasting using large scale climate modes. *Journal of Hydrology, 503,* 11–21.

Monteleoni, C., Schmidt, G. A., Saroha, S., & Asplund, E. (2011). Tracking climate models. *Statistical Analysis and Data Mining, 4*(4), 372–392.

Morid, S., Smakhtin, V., & Moghaddasi, M. (2006). Comparison of seven meteorological indices for drought monitoring in Iran. *International Journal of Climatology, 26*(7), 971–985.

Mpelasoka, F., Hennessy, K., Jones, R., & Bates, B. (2008). Comparison of suitable drought indices for climate change impacts assessment over Australia towards resource management. *International Journal of Climatology, 28*(10), 1283–1292.

Nash, J., & Sutcliffe, J. (1970). River flow forecasting through conceptual models part I—A discussion of principles. *Journal of Hydrology, 10*(3), 282–290.

NRM. (2015). *Climate change in Australia, impacts & adaptation information for Australia's NRM regions*. Commonwealth Scientific and Industrial Research Organisation. Accessed April 01, from http://www.climatechangeinaustralia.gov.au/en/impacts-and-adaptation/central-slopes/.

Pandey, R. P., Dash, B. B., Mishra, S. K., & Singh, R. (2008). Study of indices for drought characterization in KBK districts in Orissa (India). *Hydrological Processes, 22*(12), 1895–1907.

Paulescu, M., Tulcan-Paulescu, E., & Stefu, N. (2011). A temperature-based model for global solar irradiance and its application to estimate daily irradiation values. *International Journal of Energy Research, 35*(6), 520–529.

Penman, H. L. (1948). Natural evaporation from open water, bare soil and grass. In *Proceedings of the Royal Society of London A: Mathematical, Physical and Engineering Sciences* (pp. 120–45). The Royal Society.

Raupach, M., Briggs, P., Haverd, V., King, E., Paget, M., & Trudinger, C. (2008). Australian water availability project (AWAP) (Vol. 2, p. 38). *CSIRO Marine and Atmospheric Research Component: Final report for phase.*

Reichler, T., & Kim, J. (2008). How well do coupled models simulate today's climate? *Bulletin of the American Meteorological Society, 89*(3), 303–311.

Reifen, C., & Toumi, R. (2009). Climate projections: Past performance no guarantee of future skill? *Geophysical Research Letters, 36*(13), L13704.

Şahin, M. (2012). Modelling of air temperature using remote sensing and artificial neural network in Turkey. *Advances in Space Research, 50*(7), 973–985.

Şahin, M., Kaya, Y., & Uyar, M. (2013). Comparison of ANN and MLR models for estimating solar radiation in Turkey using NOAA/AVHRR data. *Advances in Space Research, 51*(5), 891–904.

Thornthwaite, C. W. (1948). An approach toward a rational classification of climate. *Geographical Review, 38,* 55–94.

Tiwari, M. K., & Adamowski, J. (2013). Urban water demand forecasting and uncertainty assessment using ensemble wavelet-bootstrap-neural network models. *Water Resources Research, 49*(10), 6486–6507.

Ulgen, K., & Hepbasli, A. (2002). 'Comparison of solar radiation correlations for Izmir, Turkey. *International Journal of Energy Research, 26*(5), 413–430.

Vicente-Serrano, S. M., Beguería, S., & López-Moreno, J. I. (2010a). A multiscalar drought index sensitive to global warming: the standardized precipitation evapotranspiration index. *Journal of Climate, 23*(7), 1696–1718.

Vicente-Serrano, S. M., Beguería, S., López-Moreno, J. I., Angulo, M., & El Kenawy, A. (2010b). A new global 0.5 gridded dataset (1901–2006) of a multiscalar drought index: Comparison with current drought index datasets based on the Palmer Drought Severity Index. *Journal of Hydrometeorology, 11*(4), 1033–1043.

Vicente-Serrano, S. M., Beguería, S., Lorenzo-Lacruz, J., Camarero, J. J., López-Moreno, J. I., Azorin-Molina, C., et al. (2012). Performance of drought indices for ecological, agricultural, and hydrological applications. *Earth Interactions, 16*(10), 1–27.

Vicente-Serrano, S. M., Gouveia, C., Camarero, J. J., Beguería, S., Trigo, R., López-Moreno, J. I., et al. (2013). Response of vegetation to drought time-scales across global land biomes. *Proceedings of the National Academy of Sciences, 110*(1), 52–57.

Vicente-Serrano, S. M., Beguería, S., & López-Moreno, J.I. (2011). Comment on "Characteristics and trends in various forms of the Palmer Drought Severity Index (PDSI) during 1900–2008" by Aiguo Dai. *Journal of Geophysical Research: Atmospheres, 116*, no. D19.

Vogl, T. P., Mangis, J., Rigler, A., Zink, W., & Alkon, D. (1988). Accelerating the convergence of the back-propagation method. *Biological Cybernetics, 59*(4–5), 257–263.

Willmott, C. J. (1981). On the validation of models. *Physical Geography, 2*(2), 184–194.

Willmott, C. J. (1982). Some comments on the evaluation of model performance. *Bulletin of the American Meteorological Society, 63*(11), 1309–1313.

Zhao, M., & Hendon, H. H. (2009). Representation and prediction of the Indian Ocean dipole in the POAMA seasonal forecast model. *Quarterly Journal of the Royal Meteorological Society, 135*(639), 337–352.

Chapter 12
Adapting Climate Change Projections to Pacific Maritime Supply Chains

Jack Dyer

Introduction and Contribution to Literature

According to UNCTAD (2014), over 90% of the world's trade is seaborne; based on the global connections of seaports, vessels, maritime economic hinterlands and their interconnecting supply chains. Global climate change is considered to provide the most significant threat to maritime supply chains from an increasing majority of stakeholders, who favour climate change adaptation including the International Association of Ports and Harbours (IAPH) (2013), the Secretariat of the Pacific Community (SPC) (2015), Australian Bureau of Meteorology (2015) and the Intergovernmental Panel on Climate Change (IPCC 2015). One related theoretical gap identified in this paper as a research point of departure from assessing academic research such as Regmi and Hanaoka (2009), Knapp et al. (2011), Becker et al. (2011) and Kong et al. (2013), is how comparatively little is known about the potential impact of climate change on maritime supply chains (Bojinski et al. 2014; Thomas et al. 2013). This paper identifies a significant constraint to maritime supply chain adaptation for dependent stakeholders in existing research literature, includes the substantial risk of uncertainty. Types of uncertainty includes diverse methodologies, risks identified, impact costs, underlying assumptions and response strategy solutions based on various climate change risks, whether modelling climate change for river basin flooding planning (Hansen 2013), food security in the Philippines, (Carandang et al. (2015)) or harvest disruption risks in the Australian sugar industry (Sexton et al. 2015) as current literature gaps.

Pacific Climate Change Symposium Revised Paper Submission 02.

J. Dyer (✉)
University of Tasmania, Maritime Way, Launceston, TAS 7250, Australia
e-mail: Jack.Dyer@utas.edu.au

W. Leal Filho (ed.), *Climate Change Adaptation in Pacific Countries*,
Climate Change Management, DOI 10.1007/978-3-319-50094-2_12

A further gap in contemporary literature includes asymmetrical information over the exact effects of climate change complicates proposed adaptation solutions for vulnerable stakeholders experiencing the pressures and constraints of small developing nation states including tropical Pacific Ocean, sovereign islands and dependent territories. Currently stakeholders require significant time, financial and other resources to investigate potential specific climate change data scenario assumptions/impact costs, risks and consequences that directly affect them. These stakeholders wary of maladaptation costs from asymmetrical information over issues of timing, intensity and actual consequences and perceptions of scientific uncertainty, are often risk averse in pursuing a response to the ultimate survival threat of climate change. Yet this paper's response to these stakeholder challenges is to propose a Pacific Climate Futures tool with climate change projections, along with screening criteria to improve the reliability of projections for climate change impact studies.

This paper contrasts from previous climate change impact studies such as Marra (2014), which argue that a significant constraint to supply chain stakeholder adaptation consists of insufficient or inconsistent information. Therefore, models and methodologies continue to require even more research before any adaptation action can be undertaken. Previous studies e.g. Simpson et al. (2007) and Walsh et al. (2013) have indicated this as a recurrent stakeholder concern, arguing that scientific climate change projections, underlying data and research serves no purpose if it conflicts between different authorities and is overly technically complex. In contrast, this paper advocates the IPCC (2015) report contains the globally most reliable, consistent, accessible, cost effective approach to forecasting climate change with the most robust estimates and accepted by a majority of nations with internationally more reputable and accurate scientific resources. This paper aims to simplify technical, IPCC climate change projections so that they can be downscaled to maritime supply chains. Each projection and underlying assumption has been ascertained for consistency and reliability with leading established scientific institutions, individuals, research sources, meteorological agencies and empirical observations from the Pacific, accepted and utilised by SPREP (2014), Pacific Island Forum Secretariat (2012), and SPC (2015), the most frequently cited and significantly active organisations concentrating on Pacific climate change.

This tool could assist Pacific supply chain stakeholders to minimise maladaptation costs, identify and prioritise risks and implement adaptation solutions. It aims to reduce factors of uncertainty that prompt moral hazard and inertia by potentially affected stakeholders, frequently cited by existing literature. Therefore this paper's conceptual contribution to the gap in existing literature will be to propose an online tool and screening criteria (section "Implications for Maritime Supply Chains") that academics and Pacific maritime supply chain stakeholders can utilise to effectively evaluate diverse Pacific climate change scenarios, methodologies and adaptation solutions at minimal opportunity cost to improve upon existing climate change projection methods i.e. the simulation modelling proposed in Lutz et al. (2013) for southern Africa. This paper's method simulates Pacific climate change scenarios with underlying assumptions and screening criteria that can be applied generically to climate change impact studies. Its data collection instrument relies on the IPCC (2015), Australia Bureau of Meteorology and CSIRO (2014) with consistent observations but aims to be sufficiently applicable to any generic tool or method meant to ascertain the reliability of climate change projections. It aims to do so through providing certain global and Pacific climate change projections (section "Pacific Climate Change Projections and Scenario Assumptions") and evaluating implications for maritime supply chain stakeholders (section "Climate Change Implications for Pacific Maritime Supply Chains"). This paper's conceptual contribution to established literature could potentially extend to identifying potential vulnerability-risk assessments in climate change impact studies such as Metternicht et al. (2014) or to improve data information available for stakeholder's perceptions and preparations for climate change as in Nagy et al. (2014) for coastal Uruguay.

Pacific Climate Change Projections and Scenario Assumptions

In the absence of complete information and perfect stochastic forecasting capabilities towards the projected impact of climate change in the Pacific, research such as Gero et al. (2013) and Goodrich et al. (2015), has traditionally focussed on modelling projected risks, associated impact costs and consequences with underlying assumptions. These sources aim to provide supply chain stakeholders with specific guidance over which adaptation actions to prioritise pre and post-disruption event and where to efficaciously allocate resources. The assumptions below and scenarios are based on the above sources, and an ever increasing, substantial number of research sources (World Bank 2012; International Climate Change Adaptation Initiative 2013; Australia Bureau of Meteorology and CSIRO 2014; World Meteorological Organisation 2015). These assumptions possess a significant number of research advantages to satisfy maritime supply chain stakeholder requirements of sufficient information, flexibility, available and consistently updated. This paper's conceptual contribution to existing literature establishes the following assumptions accepted to present direct sudden and gradual climate change disruption threats towards the future survival of maritime supply chains, particularly in the Pacific. The potential threat of climate change for Pacific maritime supply chain stakeholders, further justifies the need for tools, screening criteria and other solutions, to enable climate change adaptation at minimal opportunity cost across divergent climate change methods (Fig. 12.1).

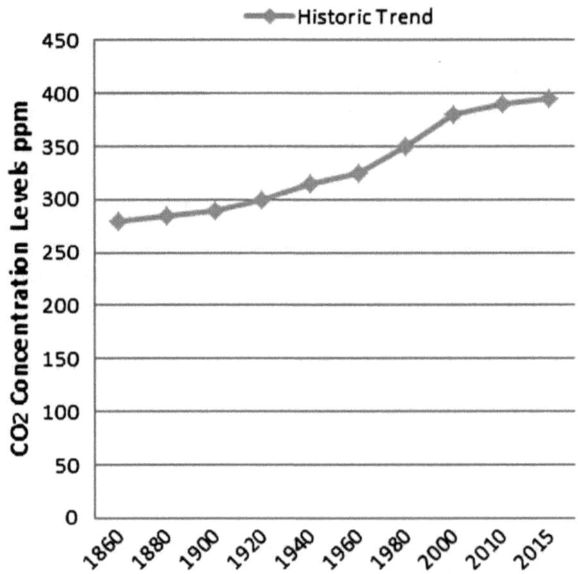

Fig. 12.1 Historic CO_2 emissions scenario growth

These produce the following gradual climate change disruption risk consequences

- An increase in global average land surface, atmosphere and sea level temperature levels, of 1.5–2 °C (B1 scenario), even if emissions were to cease based on historic inventory levels.
- An increase of 2.5–4 °C (IPCC 2015 A1B scenario) if emissions are stabilised at the current medium growth rate by 2100.
- Increases of 4–7 °C (IPCC 2015 A2 scenario) if emissions are not reduced.
- A 0.5 m global average sea level rise is projected for a low risk, current growth, scenario where emissions are highly reduced, 0.8 m rise—medium risk if emissions are stabilised and up to 1.1 m high for a high risk, continued emissions increase scenario by 2100, in pursuing current global GDP growth rates of 3–5% annually.
- Other anticipated gradual global, Pacific regional and individual increases in sea level, temperature, humidity, precipitation and wind speed along with potential variations in wind direction, current, ocean swell, wave energy and sedimentation, as gradual climate change disruption risks.
- Greenhouse gas CO_2 emissions would have to stabilise around 450 (ppm) (430–480) at present; no higher than 550 ppm (530–580) by 2100, to ensure this.
- A projected increase in the frequency, duration and intensity of sudden climate-change related natural disaster disruption risks including storms, flooding, superstorms, tsunamis, hurricanes, typhoons, heatwaves and landslides, with higher precipitation and wind speeds.

This section refers to the three most common scenarios that Pacific and other maritime supply chain stakeholders are likely to encounter when accessing divergent research sources, methodologies and adaptation solutions: B1, A1B and A2 scenarios over 3 time horizons. B1 is used by the IPCC (2015) and international climate change policy makers (Pacific Islands Forum Secretariat 2012; Pacific Island Climate Change Cooperative 2015; SPREP and UNDP 2015) to refer to a low greenhouse gas emissions growth scenario, if humanity were to become substantially more environmentally sustainable, convert from an industrial to a services based economy less resource and greenhouse gas emissions intensive and restricting population growth to reduce emissions. A1B refers to a medium emissions growth scenario or "business as usual" if population and economic activity were to continue at current growth levels. A2 refers to a projected high emissions growth scenario if developing countries don't stabilise population, dramatically reduce emissions and pursue globalisation economic activity levels of developed nations. The three projected time horizons (2030, 2055 and 2090) are advised by IPCC (2015) to consider as short, medium and long term periods for maritime supply chain stakeholders to adapt.

Based on IPCC (2015) data estimates and Fig. 12.2, global mean surface temperature rises are projected to increase from a baseline of 0 °C in 2000 to 0.85° by 2030 under all 3 scenarios. However by 2055, emissions are projected to diverge, around 1.2° under a B1, 1.59° for an A1B and 1.86° for an A2 scenario, increasing to an average of 2, 3 and 4.5° respectively by a 2100 long term projection. Figure 12.3 provides an alternative visual representation of how specific world regions will be affected and vulnerable under an A2 scenario. Increased global mean temperature implications for maritime supply chains are indicated throughout this paper possessing significant climate change disruption risks, direct and indirect impact/adaptation costs. Higher temperatures contribute towards an increased frequency of droughts, greater maximum and minimum temperatures; reduced water; higher evaporation and evapotranspiration rates, affecting future climates, natural resources and productivity, (Simpson et al. 2007; Collins et al. 2010; Matear 2014).

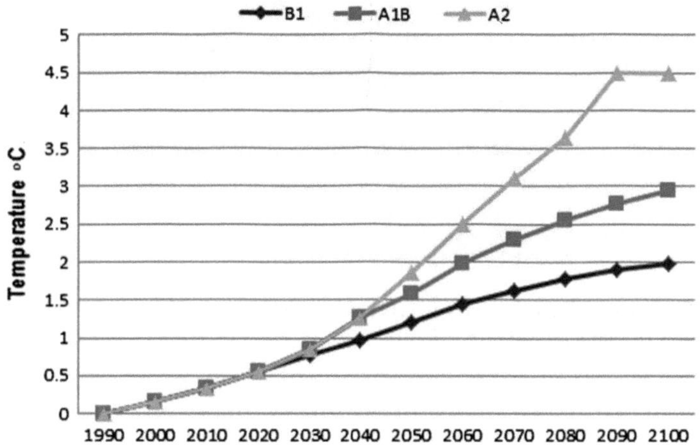

Fig. 12.2 Global mean surface temperature, climate change projections

Climate change projections may include slower ocean currents/thermohaline circulation, complicating navigation. These projections illustrate how vital it is for these stakeholders globally to adapt, enhancing vessel and infrastructure resilience to higher temperatures and increased salinity.

IPCC (2015) data estimates in Fig. 12.3, estimate historic global average sea level rise rose from a 0 m baseline in 1900 to 0.030 m by 2000. The rate of increase has substantially accelerated from an average of 1–1.5 mm per year (1900–1980) (Australia Bureau of Meteorology 2015) to 3–3.5 mm per year (1980–2014) and is projected to reach up to 8–10 mm per year by 2100, if global climate change trends are not stabilised. From IPCC (2015) data and Figs. 12.4 and 12.5, global mean sea level rises are projected to increase from a 0 baseline in 2000 under all 3 scenarios. However by 2030, scenarios are projected to diverge around 0.12 m under a B1, 0.16 for an A1B and 0.33 m for an A2 scenario, increasing to an average of 0.23, 0.3 and 0.62 m respectively by 2055 By 2100, maritime supply chain stakeholders are anticipated to experience a projected mean sea level rise of 0.6 (B1), 0.8 (A1B) and 1.5 m.

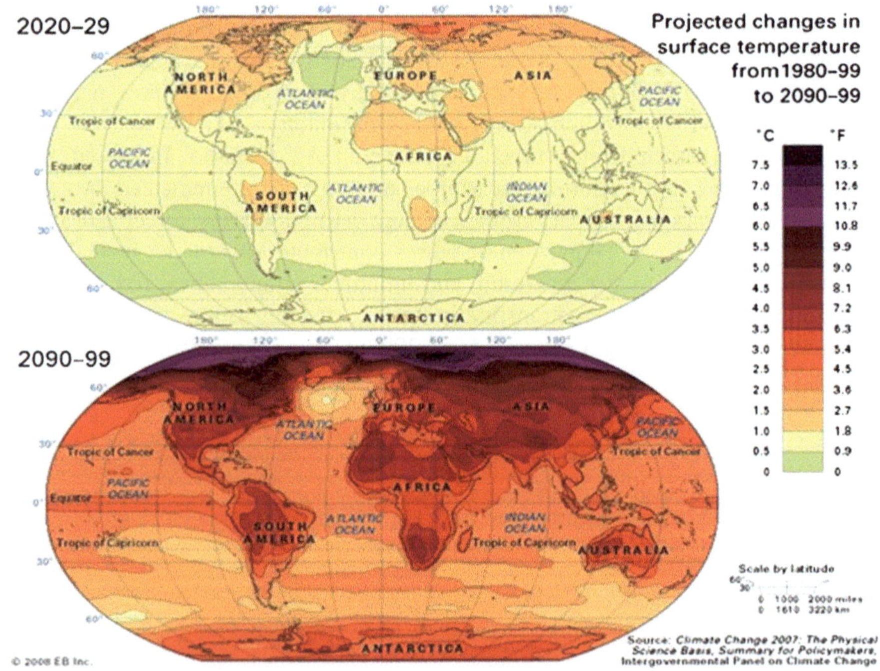

Fig. 12.3 Projected climate change, surface temperature changes 1999–2090. *Source* Encyclopaedia Britannica (2008), p. 72

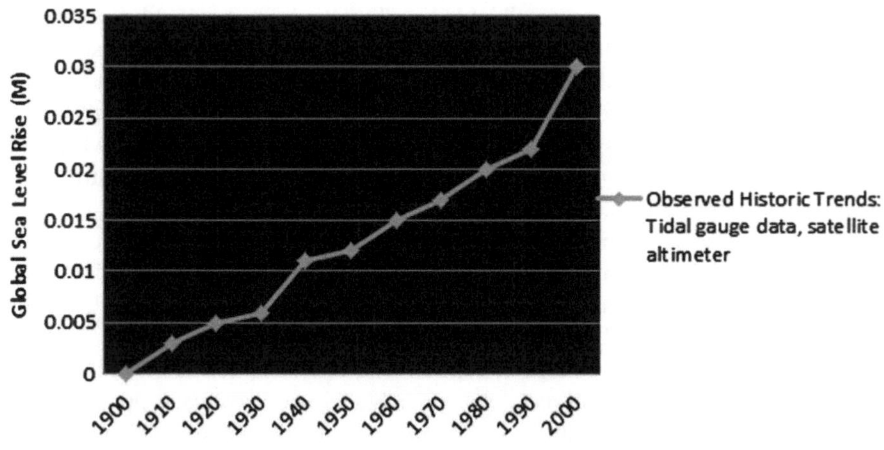

Fig. 12.4 Historic average global sea level rise 1900–2000

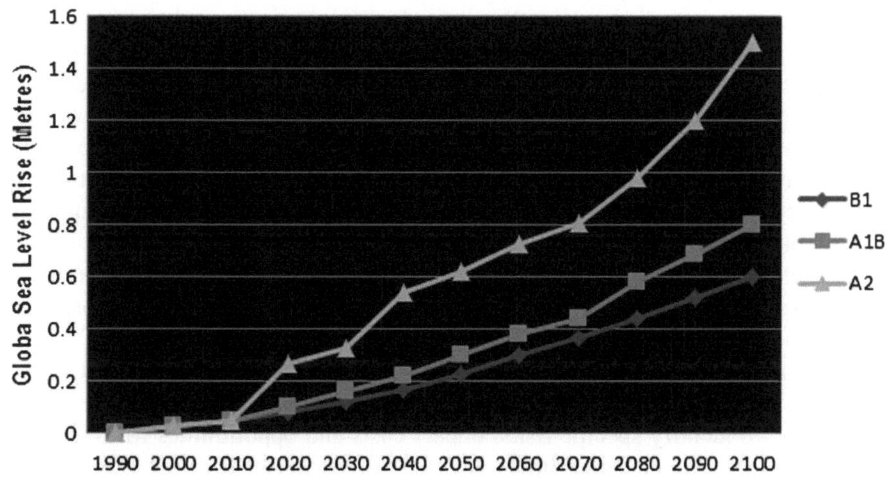

Fig. 12.5 Global sea level rise, climate change risk projections

Climate Change Implications for Pacific Maritime Supply Chains

Apart from specific Pacific nations whose physical, economic and environmental survival prospects are threatened under all three projected scenarios, global and regional maritime supply chain stakeholders may have to adapt to a world where low altitude countries, ports, populations, infrastructure, resources and coastlines experience substantial gradual and sudden, climate change related disruption risk shocks. Thorpe and Fennel (2012) conclude sufficient consensus exists over climate change, for these stakeholders to adapt even where various climate change models provide a range of confidence intervals, projected risk, impact costs and solutions. This paper's method includes climate change projections to assist adaptation strategies and in response to the following stakeholder concerns expressed as existing literature weaknesses or possible directions for future research, which can be applied to various climate change impact studies' data gathering, instruments, methods and observations.

This paper's proposed method of hypothetical scenarios and screening criteria as a theoretical study, aims to assist maritime supply chain stakeholders in identifying potential climate change risks in response to the gaps identified by existing literature. Savonis and Potter (2012) argue for transparent data that allows the potential uncertainty of both climate change disruption risks and increased information on the likelihood/extent of extreme climate change, related disaster events as possible risks. Kinrade and Justus (2006) argue for higher existing data model resolutions to enhance accuracy and that a flaw of most global circulation models is that they fail to consider localised/sub-regional projected impacts of climate change. Inoue (2012) in reviewing existing IAPH preparations for climate change; considers a lack of climate change projection studies exists globally, which concentrate on localised coastal areas, ports and maritime supply chains. Koshy (2008), Becker et al. (2011), Krämer et al. (2013) and Asian Development Bank (2013) further indicate the dearth of localised climate change projections and models in existing climate change and supply studies chains that consider localised interdependent environment-ocean-land-atmosphere climate change factor; to ascertain an inventory of exposed coastal assets/supply chain vulnerabilities. This paper therefore suggests that supply chain stakeholders would benefit from more representative studies utilising climate change projections for specific regions, islands and maritime supply chain case studies, to identify and minimise potential risks, impact costs and adaptation solutions more accurately.

Beerman (2010), CSR Asia (2011) and BSR (2014) present an emergent supply chain stakeholder requirement for more specific/localised climate change, practical projections, to identify specific risks, impact costs and opportunities for individual maritime supply chain stages including businesses (SPREP and CSIRO 2011; Wallingford 2014). A lack of global studies that specifically focus on private sector climate change adaptation rather than for governments and local communities is further criticised by Aggarwal et al. (2011) in reviewing private sector, climate change, supply chain adaptation for Organisation for Economic Cooperation and Development countries. From reviewing Pacific Catastrophe Risk Assessment and Financing Initiative (2013), this paper identifies accurate projections might further incentivise private sector funding of enhanced supply chain resilience and other potential adaptation solutions, especially for Pacific nations with limited government revenue funding, through situational awareness and accurate information. Simpson et al. (2007) points out stakeholder moral hazard as a reluctance to invest in supply chain adaptation without more certain information, in considering specific sea level rise at local level. As a method, accurate, detailed projections can further aid potential and subsequent impact damage cost estimation and the replacement value of various Pacific ocean/island ecosystems. Providing specific global, regional and local climate change projections can also assist stakeholders to determine the relationships between key climate change variables and maritime supply chain stages to minimise opportunity, disruption, delay, externality and maladaptation costs for anticipated climate change, impact events. This paper however, identifies a significant constraint to implementing supply chain stakeholder adaptation solutions, is that most supply chain business planning horizons are short term: perhaps 1, 5 even 10 years, whereas current change reviews such as Garnaut (2008), Australian Bureau of Meteorology and CSIRO (2014) and IPCC (2015), envision 100 years for projected climate change.

Although past research studies including Australia Government Bureau of Meteorology and CSIRO (2014), SPREP (2014) and IPCC (2015), have provided global climate change projections; an increasing number of sources have realised the inadequacy of global general circulation models whose baseline data and satellite observations range in cells 70–200 km^2 wide. This complicates identifying specific climate change risks, plus associated impact costs for maritime supply chain stakeholders seeking information at the Pacific regional and individual island scale. This section aims to improve its significance and value to maritime supply chain stakeholders for the Pacific region and globally, through projecting and downscaling potential climate change effects to a Pacific regional model. To satisfy the screening criteria summarised in 1.2, this section utilises the projections and field observations of the main Pacific political, economic, academic and environmentalist/community organisations such as SPC, USP, SPREP, AOSIS, SOPAC, individual governments, the UNDP, Pacific Climate Change Adaptation Initiatives and World Bank. Downscaling becomes more practical with improved data quality/accuracy over a longer time period using the 2014 completed Pacific Climate Change Futures projection tool to model potential climate change scenarios for the main gradual, climate change disruption risks for individual Pacific nations and interdependent maritime supply chain stakeholders. This identifies projected specific Pacific Ocean and climate hazards that might potentially be affected by climate change along with further potential implications for maritime supply chains to reduce potential stakeholder uncertainty, moral hazard, risk aversion, asymmetrical information and maladaptation costs (Fig. 12.6).

Based on the sources cited above and references; Pacific regional climate change is projected to include an increase in heatwaves. The recurrence intervals of temperature maximum days are expected to increase. PACCAP (2014) observed since 1951, the number of temperature days exceeding 35 °C has increased from an average of 20 to 45–80 across the Pacific region, increasing the probability and associated disruption supply chain costs of heatwaves/lower productivity. However, regional climate change circulation models differ from global projections primarily in emphasising the particular physical vulnerability of the Pacific to global climate change. Precipitation and other variables change over regions and specific Pacific islands based on local geophysical, climate, environment and human conditions. scale, timing, format and language. This is partially emphasised through Fig. 12.7, where projected Pacific sea level rise is likely to increase at a higher rate than the current projected global average

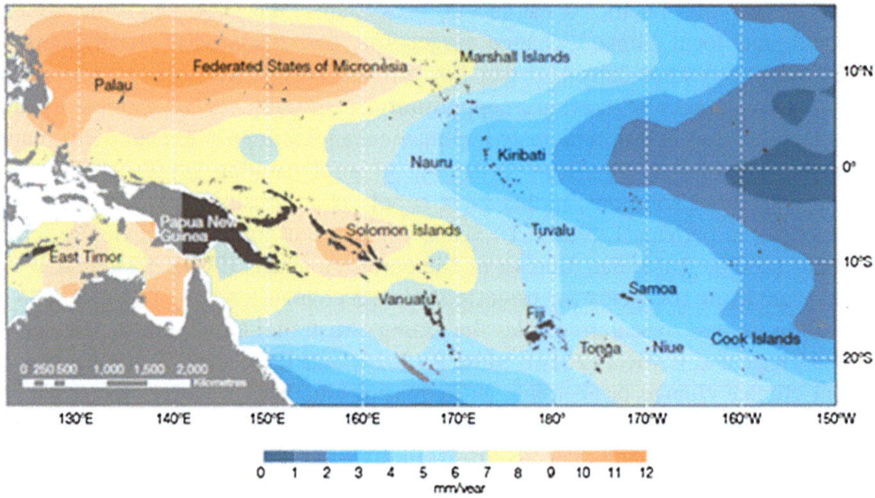

Fig. 12.6 Pacific regional projected mean sea level rise. *Source* Australia Government Bureau of Meteorology and CSIRO (2014), p. 10

of 3.2 mm per year from 3–5 mm per year in the Cook Islands to an average 4–7 mm for most Pacific nations up to 9–12 mm per year for the Solomon Islands, Palau and the Federated States of Micronesia. Regional sea level rise is influenced by Pacific ocean dynamics including the regional mass distribution of Earth's crust but also currents, localised surface winds, changes in salinity, bottom pressure, and sea surface temperature, which could alter through climate change (Fletcher and Richmond 2010; Hemer et al. 2011, Mori et al. 2013). From 1900 to 2000 average ocean surface temperatures rose 0.7 °C in the tropical Pacific Ocean from global climate change. The Pacific has increased its potential capacity to forecast regional sea level rise and temperature through aid agencies prioritising information through 12 stations of the Australian funded South Pacific Sea Level and Climate Monitoring Project (Australia Government Bureau of Meteorology 2015). Higher projected rates of sea level rise increase the predicted probability of flooding, increased wave energy, sedimentation, eroding existing coastal and engineering protection and other gradual climate change disruption risks. These further indicate the need to prioritise adaptation solutions to minimise potential risks to the future of Pacific maritime supply chains, wherever practically possible.

Projected Pacific regional climate change may further affect maritime supply chains and stakeholders through Pacific regional climate change influences (Fig. 12.8), including the South Pacific, West Pacific Monsoon and Intertropical Convergence Zone. Subtropical high pressure zones are indicated by H, yellow arrows indicate surface winds. Moser et al. (2012), Australia Government Bureau of Meteorology (2015) and IPCC (2015) project minimal variations to these influences, regulating climate variability over decades rather than climate change, apart from a slight increase in average wet season/reduction in low season precipitation.

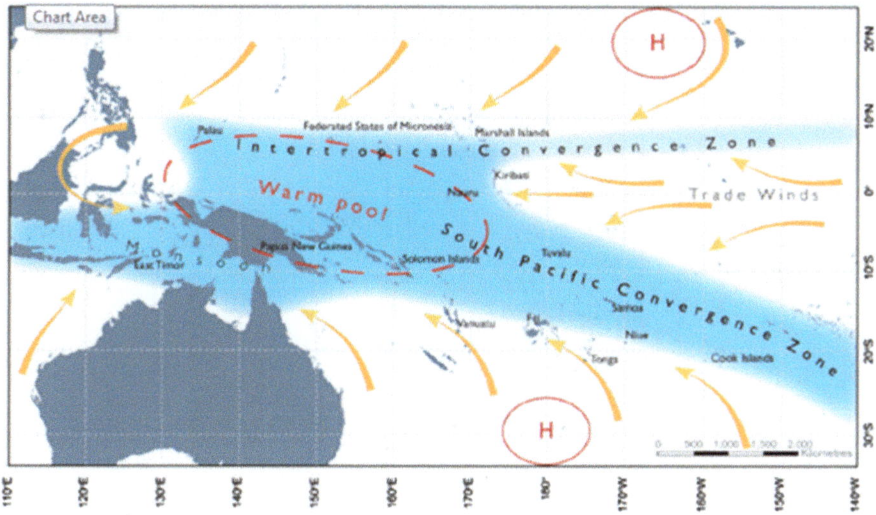

Fig. 12.7 Pacific regional climate change influences. *Source* Australia Government Bureau of Meteorology and CSIRO (2014), p. 4

Using CMIP5 models, these sources indicate a high probability of a reduced risk frequency of Pacific tropical zone cyclones north of 20° south latitude but an increased frequency, duration and intensity in potential disruption impact costs below this interval. Possible changes in wind may influence slightly El Nino Southern Oscillation (ENSO), Interdecadal Pacific and Pacific Decadal Oscillations, currents and cyclone formation but projection estimates remain inconsistent from observed sources (Collins 2010; Pacific Climate Change Science Programme 2013; Jia et al. 2015).

Implications for Maritime Supply Chains

The advantages of these particular Pacific Climate Future models and scenarios is that they consider divergences in the Pacific's regional climate and ocean ecosystem. This paper's method considers accurate climate change projections as essential to understand specific implications of climate change for Pacific maritime supply chains, to provide sufficient information to enable all stakeholders to determine their vulnerability to local climate change. It understands that each risk impact cost and adaptation response is not homogenous. Each supply chain stage, nation, stakeholder and commodity may differ in its potential effect necessitating statistical and dynamic downscaling from global scenarios and local datasets of a bottom up approach, to identify regional and individual nation effects. This represents this paper's conceptual departure from the majority of established sources of climate

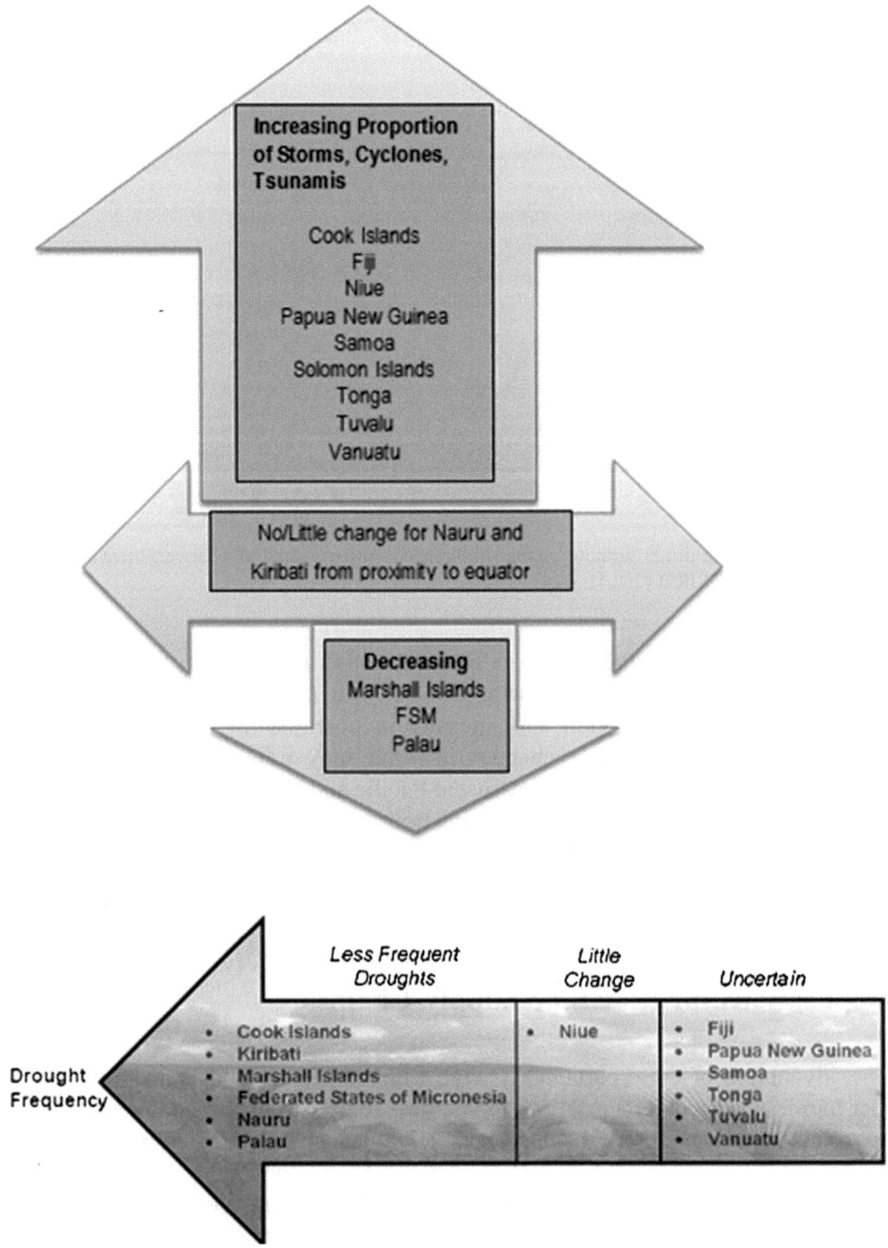

Fig. 12.8 Pacific Island climate change changes projection frequency. Based on Australia Government Bureau of Meteorology and CSIRO (2014)

change impacts on potential ports, transport and supply chains that generalise climate change effects globally, without considering specific consequences may vary. This increases the probability of maladaptation costs and risk underestimation, which this paper proposes developing countries, particularly Pacific Islands with significant time, fiscal, labour and other identified constraints cannot afford.

Further Pacific regional climate change implications for maritime supply chains from Fig. 12.7, include minor projected increased seasonal/interdecadal and other climate change variability potentially influencing shipping movements and corresponding future trade patterns of demand/supply (Matear 2014). Projected increase in annual mean wave significant heights throughout the Southern Pacific Ocean from increases in wind speed (Pacific Islands Climate Change Cooperative 2015) may increase physical risks to vessels and the probability of flooding/storm surges to coastal economies, infrastructure and ecosystems (Matear 2014; WMO 2015; SPREP and UNDP (2015), anticipate only minor variations in Pacific regional wind direction, speed and currents, which this paper considers might primarily influence maritime supply chains through altering navigation/trade routes and distribution/location of maritime resources. Melanesia is anticipated to experience greater rates of ocean acidification from higher temperatures/greater coral reef access, than Micronesia and Polynesia (Australian Bureau of Meteorology 2015). The projected increase in Pacific ocean temperatures and regional PH balance and salinity from a 0.1 increase (190–2000) to reach 0.3–0.5 by 2100, increasing ocean acidification, further threatens the natural coastal protection and maritime resource functions of coral reefs and other maritime ecosystems, along with increasing vessels corrosion rates and coastal infrastructure. This subsequently increases potential maintenance/repair costs for these supply chain stakeholders.

Specific implications of sudden, regional Pacific climate change risks, projected to lower in frequency but increase in duration and intensity, are anticipated to possess associated higher proportional impact costs over a smaller land surface and lower altitude, more physically exposed than non-Pacific, non-Caribbean regions. A projected increase in sudden climate change, related disruption shocks may further increase port closure/disrupt the functions of port, interconnected economic hinterlands and maritime supply chain functions; significantly damage exposed coastal infrastructure, assets and ecosystems, providing further justification for supply chain stakeholders to prioritise supply chain adaptation strategies. To adapt further, this thesis advocates that these stakeholders should utilise existing data observations to project regional climate change but also focus on individual island, climate change projections.

Localised Pacific island, climate change disruption risks are anticipated by the scientific projections and historic field data observations of Dronkers et al. (1990), Gero et al. (2013), Australia Bureau of Meteorology (2014) and IPCC (2015) to affect thermal gradients and atmospheric pressures, influencing local projected winds, currents, subsequent sedimentation rates, coastal erosion rates and vulnerability to shoreline exposure, tidal and wave energy for stakeholders. Localised temperature changes for individual Pacific islands influences coastal upwelling, local PH salinity and rates of ocean acidification. Changes in local sea level may affect changes in scour, sediment and degree of dissipated wave energy/ocean swell. Changes in sea surface temperature are affected by localised vertical stratification of the water column/interconnecting links of an ocean–land–atmosphere model system from Hay (2011), Correro et al. (2011) and Kunreuther et al. (2014).

Downscaling climate change projections as a method, to individual Pacific islands, enables maritime supply chain stakeholders to consider how divergences in the above climate change disruption risk projections between islands, might require adapting business operations, commodity production, risk management, training and other adaptation strategies, to differ for each stakeholder, maritime supply chain stage, island, commodity and climate change risk. Other projected climate change disruption risks including changes in the duration, intensity and frequency of precipitation, storms, tsunamis, cyclones and landslides affecting physical risk exposure are influenced by local changes in precipitation, sea level rise, temperature and other variables for countries such as the Cook Islands, Fiji and Vanuatu (SPC 2015; SPREP 2014; WMO 2015). This paper thus improves upon sources e.g. Walsh et al. (2013) which only project the economic impact of climate change at a regional level, ignoring that local and region specific impact climate change parameters are influenced by local climate, geography, coastal ecosystem protection, economic activity, emissions produced and legal coastal protection.

Proposing a Tool and Screening Criteria to Resolve Stakeholder Uncertainty

Pacific supply chain stakeholders therefore require a tool such as this paper's use of Pacific Climate Change Futures, along with climate change projections that consider a range of scenarios and time horizons aiding effective decision making, when planning to adapt businesses as a reliable information gathering method. Additionally, this tool approach is flexible enough to aid stakeholder adaptation solutions e.g. revising technical design standards, climateproofing existing infrastructure/equipment, transport and processes to determine the degree of resilience and stress/asset lifespan to determine adaptation and post event recovery/replacement cost, disaster reduction/risk management responses (Alesch et al. 2001; Fletcher et al. 2013; Babister and Ball 2014). This is necessary as risk may be significantly underestimated by stakeholders relying on guidelines e.g. Beca International Consultants (2010) for Kiribati and Ports Australia (2014), whose National Ports strategy considers standards of 50–100 years in design but significantly underestimates risk using a probability of 1:100 years of significant storms. Other port developments generally prepare 20–30 years in advance (Ports Australia and Freight Logistics Council of Western Australia 2014). Therefore, this paper considers accurate projections contain advantages for stakeholders in ascertaining how climate change disruption threats originate and subsequently develop; how impacts can differ across various economic sectors, supply chain stages, stakeholders, countries and even between short, medium and long term time horizons.

To reduce constraints to climate change adaptation, this paper also aims to minimise supply chain concerns in providing projections and screening criteria for those stakeholders citing constraints of resources, time, lack of research expertise, staff, resources, information access and technical barriers. These are identified as key challenges for small island developing states, especially in the Pacific by Forbes and Solomon (1997), Magnan (2014) and Paeniu et al. (2015). Another significant constraint is the limited availability of shared information and stakeholder cooperation across different supply chain stakeholders even when mutually advantageous in lowering costs. Accurate information also assists in identifying the an event's timing, the threats and opportunities presented by climate change, to further indicate the need for an integrated stakeholder, joint risk, cooperation approach, in information, communication and adaptation across entire Pacific maritime supply chains to minimise disruption costs to international trade and economic activity, before, during and after a disruption event.

The purpose of these scenarios is to be able to consider how climate change may specifically affect each selected supply chain stakeholder and determine the particular efficaciousness of certain adaptation solutions to address the potential impact costs/opportunities associated with climate change disruption risks. Improved climate change projections and forecasting methods also can more accurately determine the physical vulnerability and risk exposure of maritime supply chain assets/commodities, from direct and indirect climate change related impacts. For example Savonis and Potter (2012) consider the purpose of climate change projections to identify relevant information/articulate data for stakeholder requirements that incorporate the uncertainty, risk and vulnerability of a potential outcome that focuses on combining several tropical climate risks for intermodal transport. These may assist in enhancing the probability of business viability and the physical survival of entire islands and commodity supply chains, to minimise externality, opportunity, congestion, disruption and delay costs. The above assumptions and scenarios possess the following research advantages identified by this paper:

- *Accuracy*: Climate change projection scenarios, assumptions and underlying baseline data, selection criteria have been updated from 2007 IPCC scenario assumption estimates of previous studies to the most recent (IPCC AR5 2015) estimates, while improvements in technology, climate change observation and projection forecasting capacity improve the validity of these projections.
- *Reputable/Credible*: The scenarios utilise scientific sources internationally recognised, to affirm scenario assumptions and predictions for greater reliability. These provide greater certainty/empirical scientific evidence than dependent stakeholders who underestimate climate change's potential disruption risk for a supply chain.
- *Consistent*: These scenarios retain consistency across a significant number of research sources: World Bank (2012), Victor et al. (2014), SPC (2015), Australia Bureau of Meteorology and CSIRO (2014), Netherlands Environmental Agency (2014), IPCC (2015), and used by Pacific island government stakeholders in adaptation. Relying on the IPCC (2015) report ensures a standardised

methodology avoiding data fragmentation/differences in variables across a range of projected causes, impact costs and potential disruption risks.

- *Comprehensive*: These scenarios and assumptions consider both climate and non climatic interdependent causes or 'drivers of climate change, inter-decadal and inter-annual variability along with multiple climate change related risk variables over 100 years to reduce the level and nature of uncertainty of reliable data quality.
- *Autonomously Verifiable/Reduce Complacency*: Certain studies are based on climate change scenario assumptions but do not independently verify them for consistency/accuracy, further increasing the uncertainty for supply chain stakeholders wishing to swiftly adapt but avoid wasting scarce fiscal, time and other resources. Sources including World Bank (2013) and Wong (2015) further multiply systematic error and uncertainty, maladaptation costs and increase the significant opportunity costs associated with risk underestimation through failing to justify evaluation/scenario selection criteria and underlying theoretical framework.
- *Accessible*: The increased institutional research, information gathering/analysing capacity, technological innovation and skilled professional capacity of developed countries in climate change projections can aid less developed countries with similar constraints to small island Pacific nations through accessible data. Pacific nations can reciprocate through providing field research of sudden, disruption risks to maritime supply chains. This allows countries to benefit without wasting scarce resources in isolated efforts and implement adaptation strategies more swiftly to minimise climate change, impact costs.
- *Relevant*: to the study significance or stated objectives of understanding the potential economic impact of climate change upon Pacific maritime supply chains.
- *Simple/Transparent*: minimising litigation, miscommunication, translation and adaptation costs.
- *Effective*: These data sources provide the basis of significant existing efforts in climate change adaptation for supply chain stakeholders.
- *Equity*: Data/scenarios are openly accessible to all and simple to verify.
- *Robust/Costs*: Providing autonomously verified, international government accepted (IPCC 2015), consistent climate change projections, downscaled to Pacific regional and individual island examples, minimises individual stakeholder research, fiscal, training, business forecasting, administration and adaptation costs. Data needs to be succinct, accessible, and affordable.
- *Flexibility*: The three internationally recognised emission scenario types forecast over short (2030), medium (2055) and long term (2090–2100), global, regional and varying across individual Pacific island nations.
- *Data Availability*: Newly present high spatial-temporal resolution models combined with satellite imagery for individual Pacific nations to improve downscaling from general circulation models to regional scale models improves data quality.
- *Satisfying Stakeholder Requirements.*

- *Practical*: in terms of computational, institutional, informational capacity given Pacific supply chain, organisations and governments' resource constraints.
- *Comparable*: Utilised by myriad stakeholders, these assumptions, scenarios and methodology techniques can be applied to different case studies with a common standard of evaluation.

Sources i.e. Australian Government Department of Energy Efficiency and Climate Change (2013) and SPC (2015) actively propose affected maritime supply chain stakeholder consultation but the validity of findings can be compromised without these stakeholders being able to identify specific climate change data scenario assumptions/impact costs, risks and consequences that directly affect them. These stakeholders wary of maladaptation costs from asymmetrical information over issues of timing, intensity and actual consequences and perceptions of scientific uncertainty, are often risk averse in pursuing a response to the ultimate survival threat of climate change. Through providing specific scenarios and assumptions with outlined research advantages, this paper aims to reduce uncertainty factors that prompt moral hazard and inertia by potentially affected stakeholders. For consistency and the above research advantages, this paper utilises the online tool Pacific Climate Futures allowing a range of scenarios and assumptions, Pacific nations/organisations utilised in preparing their national climate change adaptation policies, given existing familiarity and data availability by government stakeholders.

The tool allows users to project specific climate change, variable risks for various IPCC standard scenarios utilising global, regional and nation specific circulation models and assumptions ascertained by IPCC (2015), the Australia Bureau of Meteorology and CSIRO (2014) and the local meteorological services of each Pacific island nations. The user can specify time horizons and climate assumptions, in a simplified graphical representation, even downscaling to localised impacts if necessary. The tool's quality assurance is based on an established scientific consensus that the majority of global nations, their populations and international organisations have recognised. This paper proposes flexibility advantages of users selecting variables to identify the most relevant, efficient, plausible scenarios and assumptions. Each scenario and assumption has been independently verified based on estimates consistent with these nations, IAPH (2013), the World Meteorological Organisation (WMO) (2014), the IPCC (2015) report and historic data observations from each Pacific island's meteorological service. These aim to provide reliable, relevant, consistent, simple, comparatively accurate projection scenarios, to aid climate change anticipation and adaptation as consistently recommended by established literature including Veitayaki et al. (2007), Marra (2014) and Field et al. (2014).

These projections further indicate the urgency of stakeholders to act to climate change, to minimise these threats as the ultimate risk that threatens the future economic, environmental and physical survival of Pacific maritime supply chains. Kinrade and Justus (2006) state that research needs new tools for diagnosing the probability of climate change to aid the determination of effective risk management. High resolution impact data has already aided the similarly climate risk exposed

Caribbean coastal supply chain (Lorde et al. 2013). This paper possesses further supply chain stakeholder advantages in its data gathering method with specific projections and a theoretical screening framework for these stakeholders to independently obtain access to climate change data and scenario simulations, assessment and adaptation without the need to just rely on research of external consultants/conflicting research studies. Accurate, localised, updated projections enable stakeholders to evaluate the costs/benefits of each adaptation and individual stakeholder climate change possible consequences.

Conclusions

In conclusion, the implications of this paper's research consists in addressing the gaps in established literature in providing a source and set of consistent Pacific climate change scenario projections that maritime supply chain stakeholders can utilise to reduce the potential uncertainty of adapting to climate change. It also establishes a set of screening criteria with which these stakeholders can evaluate the use of diverse climate change methods, climate change data projections, information sources and adaptation solutions, to avoid the expense of maladaptation. In addition the findings of this paper which aims to improve the accuracy of existing climate change impact studies and the reliability of climate change projections/information for existing and future vulnerability-risk assessments and climate change impact studies. The advantages of this research paper, includes its replicability across different climate change projected scenarios and can distinguish across countries, individual supply chain stakeholders, commodities and sectors. Further implications of this paper include accurate climate change projections and information can particularly aid maritime supply chain, stakeholders to aid capacity building, risk and adaptation response prioritisation, (CSR Asia 2011) to direct funding, allocate scarce resources towards enhancing commercial/supply chain climate change adaptation and reduce maladaptation. The main findings of this paper include improving climate change projections can reduce uncertainty over when climate change will occur, where supply chains are physically vulnerable, what types of disruption risks exist, which processes are vulnerable and stakeholder adaptation can be more efficiently incorporated into risk management, cost benefit analysis and other established climate change impact study methods in existing literature.

This paper's findings agree with Codiga and Wager (2011), Hay (2011) and Woodhouse and Lumbroso (2015), who consider the value of providing local and regional scale sea level rise information to assist local governments, aid agencies and all supply chain stakeholders to direct resources, effort and attention in ascertaining risks and prioritising pre and post event adaptation. It further enables supply chain stakeholders to consider potential solution effectiveness (Does it enhance resilience/reduce impact costs?) and efficiency (Does it minimise resources used?) for each potential adaptation solution from physical engineering to ecological rehabilitation to improving information gathering, communication and early warning systems traditionally presented as literature responses to projected climate change. The implications of accurate information can further resolve the frequently cited issue of asymmetrical information to stakeholders such as Simpson et al. (2007) who queries: How do projections translate into risks, opportunities and associated costs relevant to individual stakeholders' specific research locations and businesses?

Although IPCC estimates may still retain uncertainty over future specific emissions and climate change, estimates range from 90–99% probability of occurrence. Woodhouse and Lumbroso (2015), note how the bulk of climate change models in existing research predict similar trends and agree on the problem, despite the projected uncertainty of specific impact costs. This paper proposes the above screening criteria for climate change projections are combined as assessment criteria for any maritime supply chain stakeholder seeking to independently determine the reliability of projected climate change assessment impacts, considering the best performing models with the most likely projections. However, these paper's findings are generalizable to other global stakeholders experiencing similar uncertainty over the reliability projected climate change information, to reduce issues of asymmetrical information, time, financial and other concerns that inhibit prioritising climate change adaptation, minimising potential disruption risks and impact costs, to understand personally how climate change will affect each stakeholder. This paper's findings represent an improvement over previous sources e.g. Hay (2011) and Savonis and Potter (2012) that provide adaptation solutions and methodologies but the assumptions and scenarios are implicit rather than critically examined. The prime research limitations primarily extend around a lack of previous research sources for adapting climate change projections to Pacific maritime supply chains. Another study constraint includes the application of this model to a climate change impact study or a risk-vulnerability assessment for a supply chain case study as a future instrument of research. Global climate change data primarily concentrates on sea level rise, CO_2 emissions and surface temperature as gradual climate change related risks, ignoring precipitation, wind velocity, changes in sedimentation and other gradual risks along with sudden, climate change related disaster risks e.g. storms, tsunamis, droughts, floods and landslides.

This paper's contribution to existing literature is motivated by an absence of previous research on the effects of Pacific climate change on maritime supply chains as another significant constraint towards climate change adaptation. Global stakeholder previously lacked scientifically consistent, revised, updated, relevant and focused scenario projections, to prepare for adaptation. This paper provides global, Pacific regional and local climate change projections, data information sources and the Pacific Climate Change Futures electronic tool for stakeholder adaptation plus screening criteria to assess other sources and improve climate change scenarios/models and assumptions. These aim to reduce the uncertainty associated with the intensity, duration and frequency of climate change related disruption risk events for maritime supply chains and dependent stakeholders that limits the pragmatic use of existing climate change impact literature for policymakers that need to prioritise climate change risks and adapt. This paper therefore aims to improve climate change awareness for potentially affected stakeholders, to endorse prioritising climate change adaptation at minimal disruption cost. The risks and impact costs presented by climate change projections presented by this paper, further indicates the need to adapt via increased research, collaboration, communication, awareness training and physical measures including early warning systems to increase Pacific maritime supply chain stakeholder capacities to exist and enhance resilience.

References

Agrawala, S., Carraro, M., Kingsmill, N., Lanzi, E., Mullan., M., & Prudent-Richard, G. (2011). *Private sector engagement in adaptation to climate change: approaches to managing climate risks*. Organisation For Economic Cooperation and Development, Environment Working Paper 39. Paris, France: Organisation For Economic Cooperation and Development Publishing.

Alesch, D., Holly, J., Mittler, E., & Nagy, R. (2001). *Organizations at risk: What happens when small businesses and not-for-profits encounter natural disasters*. Small Organizations Natural Hazards Project First Year Technical Report. Green Bay, USA: University of Wisconsin.

Asian Development Bank. (2013). *The economics of climate change in the Pacific*. Manila, Philippines: Asian Development Bank Report.

Australian Government Bureau of Meteorology. (2015). *State of the climate report 2014*. Australian Government Bureau of Meteorology Report, Sydney, Australia.

Australian Government Bureau of Meteorology and Commonwealth Scientific and Industrial Research Organisation. (2014). *Climate variability, extremes and change in the western tropical Pacific: New science and updated country reports*. Pacific-Australia Climate Change Science and Adaptation Planning Program Technical Report, Melbourne, Australia.

Australian Government Department of Climate Change and Energy Efficiency. (2013). *Economic framework for analysis of climate change adaptation options*. Australia Government Department of Climate Change Report, Canberra, Australia.

Babister, M. & Ball, J. (2014). *Discussion paper on an interim guideline for considering climate change in rainfall and runoff*. Engineers Australia Discussion Paper. Retrieved April 22, 2015, from http://www.arr.org.au.

Beca International Consultants. (2010). *Kiribati shoreline protection guidelines*. Kiribati Government KAP Report, Tarawa, Kiribati.

Becker, A., Satoshi, I., Fischer, M., & Schwegler, B. (2011). Climate change impacts on international seaports: Knowledge, perceptions, and planning efforts among port administrators. *Springer Journal of Science and Business Media*. Retrieved April 22, 2015, from http://unctad.org/sections/wcmu.

Beerman, M. (2010). Linking corporate climate adaptation strategies with resilience thinking. *Journal of Cleaner Production, 19*, 836–842.

Bojinski, S., Verstraete, M., Peterson, T., Richter, C., Simmons, A., & Zemp, M. (2014). The concept of essential climate variables in support of climate research applications and policy. *Journal of American Meteorology*, 1431–1443.

Business for Social Responsibility (BSR). (2014). *Climate—everyone's business. Climate change implications for transport. Key findings from the intergovernmental panel on climate change fifth assessment report*. Cambridge, UK: Business for Social Responsibility/University of Cambridge Technical Report.

Carandang, J., Banaguas, G., & Flores, M. (2015). Modelling climate change risks for food security in the Philippines. *International Journal of Climate Change Strategies and Management, 7*(4), 499–515.

Codiga, D., & Wager, K. (2011). *Sea-level rise and coastal land use in Hawai'i: A policy tool kit for state and local governments*. University of Hawaii, Centre for Island Climate Adaptation and Policy Report. Retrieved August 12, 2015, from http://icap.seagrant.soest.hawaii.edu/icap-publications.

Collins, M., An, S., Cai, W., Ganachaud, A., Guilyard, E., Jin, F., et al. (2010). The impact of global warming on the tropical Pacific Ocean and El nino. *Journal of Nature Geoscience, 3,* 391–397.

Corporate Social Responsibility (CSR) (2011). 'Climate Change Adaptation: Engaging Business in Asia,' *Corporate Social Responsibility Asia Report,* viewed April 6, 2015, from http://www.csr-asia.com/report

Correro, J., Schwartz, H., & Wenger, J. (2011). *Adapting transportation to the impact of climate change: State of the practise 2011*. Washington, USA: United States Transportation Research Board Report.

Dronkers, J., Gilbert, J., Butler, J., Carey, J., Campbell, J., James, E., et al. (1990). *Strategies for adaptation to sea level rise*. Geneva, Switzerland: IPCC Coastal Zone Management Subgroup Report.

Encyclopaedia Britannica (2008). 'Encyclopaedia Britannica,' New York: Charles Van Doren.

Field, C., Barros, V., Mach, K., Mastrandrea, M., van Aalst, M., & Adger, W., et al. (2014). 'IPCC technical summary,' contribution of working group III to the fifth assessment report of the intergovernmental panel on climate change. In O. Edenhofer, R. Pichs-Madruga, Y. Sokona, E. Farahani, S. Kadner, & K. Seyboth, et al. (Eds.), Cambridge, UK: Cambridge University Press.

Fletcher, C., & Richmond, B. (2010). *Climate change in the federated states of Micronesia: Food and water security, climate risk management and adaptive strategies*. Retrieved September 29, 2015, from http://seagrant.soest.hawaii.edu/sites/default/files/publications.

Fletcher, C., Taylor, B., Rambaldi, A., Harman, B., Heyenga, S., Ganegodage, K., et al. (2013). *Costs and coasts: An empirical assessment of physical and institutional climate adaptation pathways*. Southport, UK: National Climate Change Adaptation Research Facility Report.

Forbes, D., & Solomon, S. (1997). *Approaches to vulnerability assessment on Pacific Island coasts: Examples from Southeast Viti Levu (Fiji) and South Tarawa (Kiribati)*. Paper presented at 3rd SPREP Meeting on Climate Change & Sea Level Rise in the Pacific, ORSTOM, Nouméa, 18–22 August 1997.

Garnaut (2008). *Garnaut climate change review: Impact of climate change in Australia and CGE model inputs*. Maunsell Australia and CSIRO Sustainable Ecosystems Review, Sydney, Australia.

Gero, A., Fletcher, S., Thiessen, J., Willett, J., Rumsey, M., Daly, J., et al. (2013). *Disaster response and climate change in the Pacific*. Sydney, Australia: University of Technology Sydney Report.

Goodrich, D., Church, J., Henderson-Sellers, A., Giffard, R., Noone, K., & Holper, P., et al. (2015). *Future climate change research and observations: GCOS, WCRP and IGBP learning from the IPCC fourth assessment report*. Workshop and Survey Report. Retrieved June 19, 2015, from http://www.researchgate.net/publication.

Hansen, H. (2013). Meeting the climate change challenges in river basin planning: A scenario and model-based approach. *International Journal of Climate Change Strategies and Management, 5*(1), 21–37.

Hay, J. (2011). *Pacific adaptation to climate change: Past approaches and considerations for the future*. Cardno Acil Emerging Markets Report, Australian Department of Climate Change and Energy Efficiency, Canberra, Australia

Hay, J., Mimura, N., Campbell, J., Fifita, S., Koshy, K., McLean, R., et al. (2003). *Climate variability and change and sea-level rise in the Pacific islands region: A resource book for policy and decision makers, educators and other stakeholders*. Tokyo, Japan: Japan Ministry of The Environment.

Hemer, M., Katzfey, J., & Hotan, C. (2011). *The wind-wave climate of the Pacific ocean.* Australian Government Bureau of Meteorology Centre For Australian Weather and Climate Research Report, Canberra, Australia

Inoue, S. (2012). Climate initiatives of the world's ports. In R. Asariotis & H. Benamara (Eds.), *Maritime transport and the climate change challenge.* London, UK: Routledge Taylor and Francis Group Press.

Intergovernmental Panel on Climate Change (IPCC). (2015). Fifth assessment report of the intergovernmental panel on climate change. In O. Edenhofer et al. (Eds.), Cambridge, UK: Cambridge University Press.

International Association of Ports and Harbours. (2013). *WG150: Sustainable ports: A guidance for port authorities.* International Association of Ports and Harbours Report, Tokyo, Japan

International Climate Change Adaptation Initiative. (2013). *Climate change in the Pacific: Scientific assessment and new research volume 1: Regional overview.* Australian Department of Climate Change Report. Retrieved April 14, 2015, http://www.environment.gov.au/climate-change.

Jia, M., Dawson, J., Twilley, B., & Hu, G. (2015). *Pacific sea level monitoring project, CGPS coordinate time series analysis report.* Canberra, Australia: Geoscience Australia.

Kinrade, P., & Justus, M. (2006). *Impacts of climate change on settlements in the western port region climate change risks and adaptation.* Australian Government Department of Climate Change and Energy Efficiency Report. Melbourne, Australia: Broadleaf Press.

Knapp, S., Bordeau, E., & Falco, E. (2011). *Climate change adaptation and sustainable design at the port of New York and New Jersey.* New York, USA: Port Authority of New York and New Jersey Report.

Kong, D., Setunge, S., Molyneux, T., Zhang, G., & Law, D. (2013). Structural resilience of core port infrastructure in a changing climate. National Climate Change Adaptation Research Facility Report. Southport, UK: Griffith University.

Koshy, K. (2008). *Modelling climate change impacts on Viti Levu (Fiji) and Aitutaki (Cook Islands). A final report submitted to assessments of impacts and adaptations to climate change (AIACC).* Washington, USA: University of South Pacific Report, International Start Secretariat.

Krämer, I., Borenäs, K., Daschkeit, A., Filies, C., Haller, I., Holger, J.,H., et al. (2013). *'Climate Change Impacts on Infrastructure in the Baltic Sea Region,' Sectoral Impact Assessments for the Baltic Sea Region,'* Climate Change Impacts on Biodiversity, Fisheries, Coastal Infrastructure and Tourism Coastline Reports, viewed June 24, 2015.

Krey, V., Masera, O., Blanford, G., Bruckner, T., Cooke, R., & Vanden, K., et al. (2014). Annex II: Metrics & Methodology. In O. Edenhofer, R. Pichs-Madruga, Y. Sokona, E. Farahani, S. Kadner, & K. Seyboth, et al. (Eds.), *Climate change 2014: Mitigation of climate change. Contribution of working group III to the fifth assessment report of the intergovernmental panel on climate change.* Cambridge, UK: Cambridge University Press.

Kunreuther, H., Gupta, S., Bosetti, V., Cooke, V., Dutt, M., Ha-Duong, H., et al. (2014). Integrated risk and uncertainty assessment of climate change response policies in climate change 2014: Mitigation of climate change. In O. Edenhofer, R. Pichs-Madruga, Y. Sokona, E. Farahani, S. Kadner, K. Seyboth, A. Adler, I. Baum, S. Brunner, P. Eickemeier, B. Kriemann, J. Savolainen, S. Schlömer, C. von Stechow, T. Zwickel, & J. Minx (Eds.), *Contribution of working group III to the fifth assessment report of the intergovernmental panel on climate change.* Cambridge, UK: Cambridge University Press.

Lutz, J., Volkholz, J., & Gerstengarbe, F. (2013). Climate projections for Southern Africa using complementary methods. *International Journal of Climate Change Strategies and Management, 5*(2), 130–151.

Lorde, T., Gomes, C., Alleyne, D., & Phillips, W. (2013). 'An Assessment Of The Economic And Social Impact Of Climate Change Upon The Maritime And Coastal Sector in the Caribbean,' Australia Aid and Economic Commission for Latin America and the Caribbean Report. viewed May 7, 2015. from http://www.cepal.org

Magnan, A. (2014). Avoiding maladaptation to climate change: Towards guiding principles. *Sapiens Journal, 7*(1). Retrieved July 30, 2015, from http://sapiens.revues.org/1680.

Marra, J. (2014). Enhancing capacity for adaptation to climate change and variability in the Pacific small Island developing states. USAID and National Oceanographic-Atmosphere Administration, Pacific Small Island Developing States Final Report, Washington.

Matear, R., Chamberlain, M., Sun, C., & Feng, M. (2014). Climate change projection for the western tropical Pacific ocean using a high resolution ocean model: Implications for tuna fisheries. *Journal of Deep-Sea Research II.* Retrieved September 11, 2015, from http://www.per.marine.csiro.au.

Metternicht, G., Sabelli, A., & Spensley, J. (2014). Climate change vulnerability, impact and adaptation assessment. *International Journal of Climate Change Strategies and Management, 6*(4), 442–476.

Mori, N., Shimura, T., Yasuda, T., & Mase, H. (2013). Multi-model climate projections of ocean surface variables under different climate scenarios. Future change of waves, sea level and wind. *Journal of Ocean Engineering, 71,* 122–129.

Moser, S., Williams, S., & Boesch, D. (2012). Wicked challenges at land's end: Managing coastal vulnerability under climate change. *Annual Review of Environmental Resources, 37,* 57–78.

Nagy, G., Seijo, L., Verocai, J., & Bidegain, M. (2014). Stakeholders' climate perception and adaptation in coastal Uruguay. *International Journal of Climate Change Strategies and Management, 6*(1), 63–68.

Netherlands Environmental Assessment Agency, (2014). *'Climate Change 2014 Synthesis Report,'* Amsterdam: Netherlands Environmental Assessment Agency.

Pacific Catastrophe Risk Assessment and Financing Initiative. (2013). *Better risk information for smarter developments catastrophe risk assessment methodology.* New York, USA: World Bank Press.

Pacific Climate Change Science Program.PACCAP (2014). *Climate change in the Pacific: Vanuatu country report.* Pacific Climate Change Science Program Review. Retrieved May 7, 2015, from http://www.pacificclimatechangescience.org.

Pacific Islands Climate Change Cooperative. (2015). *Projections of climate change for Hawai'i and other Pacific Islands: Both wetter and drier?'* Pacific Islands Climate Change Cooperative. Retrieved September 26, 2015, from http://piccc.net.

Pacific Islands Forum Secretariat. (2012). *Pacific climate change and disaster risk management education framework.* Paper presented at Forum Education Minister's Meeting, Port Vila, Vanuatu.

Paeniu, L., Viliamu, L., des-Combes, J., N'Yeurt, D., Korovulavula, A., & Isoa, T., et al. (2015). *Coastal protection: Best practices from the Pacific.* University of South Pacific, Pacific Centre for Environment and Sustainable Development. Suva, Fiji: (PACE-SD) Press.

Ports Australia. (2014). *Leading practises in port master planning. Approaches and future opportunities.* Ports Australia Report. Retrieved July 9, 2015, from http://www.portsaustralia.com.au.

Ports Australia and Freight Logistics Council of Western Australia. (2014). *Leading practises in port and supply chain protection. Current practises and future opportunities.* Ports Australia Report. Retrieved July 9, 2015, from http://www.portsaustralia.com.au/assets/Publications.

Regmi, M., & Hanaoka, S. (2009). Impact of Climate Change on Transport and Adaptation in Asia, Proceedings of Eastern Asian Society for Transportation Studies. Vol 7, viewed May 29, 2005, from http://www.researchgate.net

Savonis, M., & Potter, M. (Eds.). (2012). *Impacts of climate change and variability on transportation systems and infrastructure: Gulf coast study, phase I.* US Climate Change Science Programme and Sub Committee on Global Change Synthesis and Assessment Report, Washington, USA.

Secretariat of the Pacific Community (SPC). (2015). *Pacific resilience programme regional environmental and social management framework.* Secretariat of the Pacific Community and Pacific Islands Forum Secretariat Review Joint Report, Suva, Fiji.

Secretariat of the Pacific Regional Environmental Programme (SPREP). (2014). *Mainstreaming climate change into development in the Pacific: A regional guide*. Suva, Fiji: SPREP Report.

Secretariat of the Pacific Regional Environmental Programme and CSIRO. (2011). *Food security in east timor and the Pacific and its vulnerability to climate change*. Australian Government Department of Climate Change and Energy Efficiency Report, Canberra, Australia.

Secretariat of the Pacific Regional Environmental Programme and United Nations Development Programme. (2015). *Decision support tools for climate-resilient coastal development: A case study from the cook islands Pacific adaptation to climate change (PACC) project*. Suva, Fiji: Secretariat of the Pacific Regional Environmental Programme Report.

Sexton, J., Everingham, Y., & Timbal, T. (2015). Harvest disruption projections for the Australian sugar industry. *International Journal of Climate Change Strategies and Management, 7*(1), 41–57.

Simpson, M., Scott, D., Harrison, M., Silver, N., O'Keeffe, E., Harrison, S., et al. (2007). *Quantification and magnitude of losses and damages resulting from the impacts of climate change: Modelling the transformational impacts and costs of sea level rise in the Caribbean*. Georgetown, Barbados: United Nations Development Programme Report.

Thomas, V., Albert, J., & Perez, R. (2013). *Climate-related disasters in Asia and the Pacific*. Manila, Philippines: Asian Development Bank Report.

Thorpe, J., & Fennel, S. (2012). *Climate change risk and supply chain responsibility. How should companies respond when extreme weather affects small-scale producers in their supply chains?* Oxfam Discussion Paper. Retrieved May 9, 2015, from https://www.oxfam.org.

Veitayaki, J. Manoa, P., & Resture, A. (2007). *Addressing climate change and sea level rise in the Pacific Islands*. Kagoshima University Research Centre for the Pacific Islands Occasional Papers. Retrieved October 29, 2015, from http://www.spc.int.

Victor, D., Zhou, D., Ahmed, M., Dadhich, P., Olivier, J., & Rogner, H., et al. (2014). 'Introductory chapter in climate change 2014: Mitigation of climate change.' Contribution of working group III to the fifth assessment report of the intergovernmental panel on climate change. In O. Edenhofer, R. Pichs-Madruga, Y. Sokona, E. Farahani, S. Kadner, K. Seyboth, A. Adler, I. Baum, S. Brunner, P. Eickemeier, B. Kriemann, J. Savolainen, S. Schlömer, C. von Stechow, T. Zwickel & J. Minx (Eds.), Cambridge, UK: Cambridge University Press.

Wallingford, H. (2014). *Future climate for africa scoping report, the use of climate services for decision making in the ports sector*. Wallingford, UK: HR Wallingford Report.

Walsh, C., Roberts, D., Dawson, R., Hall, J., Nickson, A., & Hounsome, R. (2013). Experiences of integrated assessment of climate impacts, adaptation and mitigation modelling in London and Durban. *Journal of Environment and Urbanisation, 25*(2), 361–380.

Wong, K. (2015). Mitigation and adaptation responses to sea level rises. *The Open Hydrology Journal, 9*, 24–27.

Woolhouse, G., & Lumbroso, D. (2015). Using climate information to achieve long term development objectives for African ports. *Journal of Climate and Development Knowledge*. Retrieved May 29, 2015, from http://cdkn.org.

World Bank. (2013). *Acting on climate change and disaster risk for the Pacific*. New York, USA: World Bank Report.

World Bank. (2012). *Turn down the heat: Why a* 4 °C *world must be avoided*. Potsdam Institute For Climate Impact Research and Climate Analytics Report. Washington, USA: World Bank Press.

World Meteorological Organisation. (2015). *World meteorological organisation statement on the status of the global climate in 2014*. Retrieved May 27, 2015, from https://www.wmo.int.

Author Biography

Jack Dyer is a maritime economist who graduated with a BSc (Honours) degree in Economics with Econometrics from the University of Kent, Canterbury UK in 2012 and a Master of Commerce in Maritime Studies from the University of KwaZulu-Natal, Durban South Africa in 2015. Jack Dyer has served as a Lecturer/Tutor in International Trade at the University of KwaZulu-Natal, a Lecturer in Shipping and Maritime Affairs at the South African Maritime School and Transport College and a consultant for various stakeholders including South Africa's port pricing, customs modernisation, Durban, Mombasa and Rio de Janeiro's proposed port expansions, African maritime education, aquaculture cruise industry, cabotage, and space economy. In 2015, Jack Dyer was awarded a University of Tasmania PHD scholarship to study the potential economic impact of climate change on Pacific ports, shipping and maritime supply chains, currently undertaking in addition to serving as a part time Lecturer there in Data Analytical Methods and involved in a book based on his Master's degree: 'Are Durban's and Other Proposed Port Expansions Really necessary?".

Chapter 13
Climate Change Adaptation in Pacific Countries: Fostering Resilience Through Gender Equality

Cecilia Aipira, Allanah Kidd and Kate Morioka

Introduction

This paper presents findings from research undertaken by United Nations (UN) Women in 2015 to explore the interlinkages between gender and climate change across the Pacific region. The purpose of the study was to provide evidence-based information on the gender impacts of climate change and disasters, to identify how gender inequalities can be addressed through climate change adaptation (CCA), and how women's empowerment can be strengthened to improve the effectiveness of CCA activities in Pacific Island countries (PICs).

The study assessed five spheres of influence: policy; institutional frameworks; implementation and practice; knowledge and data; and participation and leadership (Fig. 13.1). In each area, the extent to which gender equality aspects have been integrated in climate change activities in the Pacific was analysed, and best practice examples were gathered from the field to inform more effective CCA in the future.

Objectives

The objectives of this research paper were as follows:

1. To illustrate the importance of gender responsive CCA

C. Aipira (✉) · A. Kidd · K. Morioka
UN Women Pacific, Level 3 Kadavu House, Victoria Parade, Suva, Fiji Islands
e-mail: cecilia.aipira@unwomen.org

A. Kidd
e-mail: allanah.m.kidd@gmail.com

W. Leal Filho (ed.), *Climate Change Adaptation in Pacific Countries*,
Climate Change Management, DOI 10.1007/978-3-319-50094-2_13

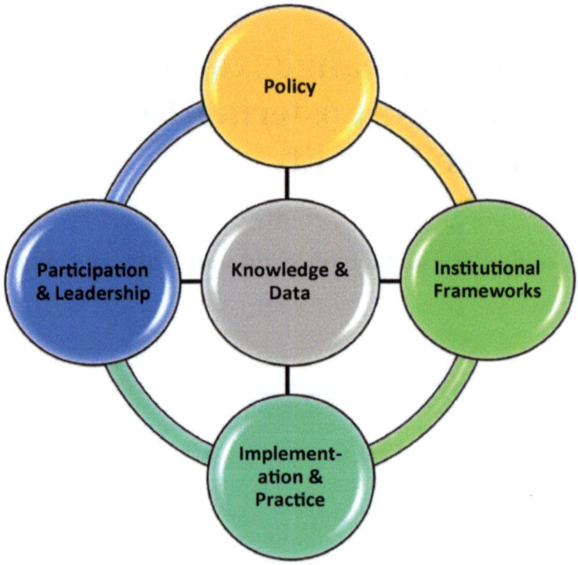

Fig. 13.1 Research framework

2. To evaluate regional capacity for gender responsive climate change action through current status of policy, institutional arrangements, implementation and practice, knowledge and women's participation
3. To recommend measures for meaningful incorporation of gender equality in future CCA policy, planning, implementation and monitoring.

Rationale

The Pacific region is highly vulnerable to climate change. Flooding, drought, tropical cyclones, rising sea level, salinisation and extreme heat are common. Over the past three years, the Pacific region has experienced seven major climate-related disasters.[1] The region combines high exposure to frequent and damaging natural hazards with low capacity to manage the resulting risks. As a result, the 2014 World Risk Report identified Vanuatu, Tonga, Solomon Islands and Papua New Guinea in the top 15 most at-risk countries in the world based on their exposure to natural hazards and societal vulnerability (UNU and Alliance Development Works 2014).

[1]Cyclone Evan: Samoa and Fiji 2013, Cyclone Lusi: Vanuatu 2014, Cyclone Ian: Tonga 2014; flash flooding: Solomon Islands 2014; Cyclone Pam: Vanuatu and Tuvalu 2015, Cyclone Maysak: Federated States of Micronesia 2015; Cyclone Winston: Tonga and Fiji 2016.

Numerous studies indicate that the effects of climate change are not gender-neutral (Enarson and Fordham 2001; Enarson 2000; Peterson 1997). Women are disproportionately affected due to pre-existing inequalities.

Firstly, women and girls are at a higher risk of physical impacts of disasters and extreme climatic events. Such risks include higher fatalities (Table 13.1) and injuries among women and girls. Women also suffer secondary impacts due to pre-existing and incredibly high rates of gender-based violence (GBV) in the Pacific, which increase post-disaster and in times of stress. After two cyclones hit Vanuatu in 2011, the Vanuatu Women's Centre in Tanna recorded a 300% increase in reported violence cases (Kilsby et al. 2012; Asian Development Bank 2014).

Secondly, women, unlike men, have a higher dependence on natural resources. The majority of women in the Pacific work in agriculture, often at subsistence level which increases their exposure to climate change. In Tuvalu women undertake 78% of subsistence agriculture (ILO 2009). In Papua New Guinea, it's 80% (Morioka 2012). Because this is also the sector hardest hit by climate change, women's resilience is limited. As agriculture becomes less productive and water and fuel harder to find, women must spend more time on subsistence production and household activities, rather than activities that are leisure- or income-related. This 'time poverty', together with unequal access to resources, means women are less able to adapt to and cope with climate change, and are unable to develop a reliable and independent source of income. For example, following the devastating tropical cyclone Pam in Vanuatu in 2015, women spent long hours foraging for food in order to provide for their families and voluntarily reduced their own food intake to stretch resources as far as possible. Cyclone Ian in Tonga in 2014 destroyed the pandanus tree crop. The weaving of mats and baskets from pandanus leaves is an integral part of women's culture in the Pacific and a source of income. Because it takes up to two years for leaves to grow long enough for harvesting, women's ability to generate income was affected long after the cyclone (Kingdom of Tonga 2014).

Despite these inequalities, women also play an important role in climate change adaptation. For example, Pacific women can use traditional knowledge to preserve

Table 13.1 Female mortality in Asia-Pacific disasters

Year	Disaster/country	Female mortality (%)
1991	Cyclone 0B2/Bangladesh	90
2004	Tsunami/Aceh Indonesia	77
2004	Tsunami/Tamil Nadu Indonesia	73
2008	Cyclone Nargis/Myanmar	61
2009	Tsunami/Samoa, Tonga	70
2014	Flash floods/Solomon Islands	96*
2016	Cyclone Winston	50

*Women/children

and store food and seeds ahead of approaching storms, floods or drought, which can carry their families through the recovery months. They hold critical knowledge on where or how to find clean water, which crops to grow in a flood or a drought season and how to survive through climate extremes. They also play a pivotal role in managing household finances and investing their savings in education, health, livelihood and other activities that assist their families to adapt and respond to climate effects (Morioka 2012).

As the above examples show, the gender implications of climate change are significant. Such considerations must be integrated in the design of disaster risk reduction (DRR) and CCA measures, with particular attention to the linkages between climate-related disasters, sustainable development and women's inequality (Nellemann et al. 2011). Recent global agreements are finally recognising these linkages. The Paris Agreement on Climate Change, set out at the UN Framework Convention on Climate Change (UNFCCC) Conference of the Parties (COP21) calls for a "*country-driven, gender-responsive, participatory and fully transparent approach to foster climate resilience and reduce vulnerability*". Similarly, the Sendai Global Framework for DRR calls for *"a gender, age, disability and cultural perspective in all policies and practices;* and emphasises that *"women and their participation are critical to effectively managing disaster risk and ... implementing gender-sensitive disaster risk reduction policies, plans and programmes"*. The priorities of the Sendai Framework and the outcomes of COP 21 offer new impetus to substantively start tackling the gender dimensions of climate change.

Research Methodology

Based on the outcomes of a desk review, and for reasons of sub-regional representation, three countries were selected as case studies: the Republic of Marshall Islands (RMI) in Micronesia, Vanuatu in Melanesia and Samoa in Polynesia. Research comprised high level analysis of national policy framework, semi-structured interviews and focus group discussions with relevant stakeholders from government ministries, development agencies, civil society organisations, nongovernmental organisations (NGOs) and staff from UN Women sub-regional offices. The questions and discussion focused on the five predetermined areas of influence (Fig. 13.1) to assess the extent to which CCA in the Pacific is gender inclusive.

Further desk analysis was undertaken to assess project level implementation across the Pacific. Publically available documents for all climate change projects in the region funded by the Global Environment Facility (GEF) in the period 2010–2015 were reviewed and assessed for gender analysis, technical expertise and gender responsive budgeting and monitoring.

Limitations

It is notable that the project review outlined above largely assessed projects at the design stage, rather than during implementation. Therefore it is possible that in some cases, projects designed in a gender-sensitive way were not implemented as such, that a gender-responsive approach was taken but not articulated in project documents, or that projects that were designed gender blind had gender considerations retro-fitted at a later stage.

It is also important to note that, while the findings of this study are intended to inform future climate change work across the Pacific region, not all PICs are equally represented or assessed. Fieldwork was undertaken in a representative country from Micronesia, Melanesia and Polynesia in an effort to obtain sufficient representation of the Pacific as a whole, and examples from other countries uncovered during the research have also been cited.

In addition, time constraints in the field limited the number and breadth of stakeholders consulted in each study country. Time constraints also resulted in only publically available documents being consulted during analysis of projects. Further analysis would be possible if project managers were consulted.

Findings

Country Case Studies

Republic of the Marshall Islands

RMI's National Disaster Management Plan (1997) mentions women as being vulnerable to disaster. The National Climate Change Policy (2011) includes an objective on promoting gender-specific adaptation responses. In both cases, no specific actions are proposed. However, gender-specific actions do transfer to the National Action Plan (JNAP) on Disaster Risk Management (DRM 2007). Likewise, the proposed Gender Equality Policy (2015-draft) includes more substantial outcomes on CCA and DRM.

When it comes to Marshallese institutional arrangements, the Secretary of the Ministry of Internal Affairs (MIA), the national women's machinery, sits on both the National Disaster Committee and the National Climate Change Committee. NGOs have a key role in service delivery and implementation of government policies. For example, the Women United Together for Marshall Islands (WUTMI) organisation works closely with MIA and is also a member of the National Disaster Committee. During drought in RMI in 2013, WUTMI played a key role in the delivery, monitoring and evaluation of food package distribution to the outer islands. Focus group discussions found that other NGOs are yet to fully grasp the link between gender and vulnerability and resilience to climate change.

While gender-responsive climate change action is encouraged in RMI through the JNAP, it lacks dedicated funding or consistent interventions. Neither MIA nor

WUTMI have the necessary financial, human or technical resources to drive gender mainstreaming in the climate change and DRR sectors. For this reason, no sex and age disaggregated data (SADD) was available in relation to climate change. The research also found failings such as atoll development plans—that had been prepared with extensive consultation and planning exercises with men, women and youth at community level, assisted by WUTMI—that had not been endorsed by the government. This not only delays and weakens the influence of the action plans, it erodes the community's belief that consultation exercises will result in meaningful action.

Samoa

Samoa's national Climate Change Policy (2008) was found to be completely gender blind, and it's National Policy for Women merely with an indicator on '*measurable understanding by women and girls on issues of climate change adaptation…*' However, both documents are under review and the second generation of policies promises to be more gender responsive. For example, the new draft DRM strategy has substantive analysis of gender issues in relation to disasters, steps for capacity development of its stakeholders and a proposed stand-alone 'Gender and DRM Policy'.

Despite the uncertain policy environment, real opportunities exist for Samoa to translate gender into implementation and practice. The Disaster Management Office works closely with the Ministry of Women, Community and Social Development (MWCSD), and both are taking proactive steps to empower women by making sure they are participating and contributing to community discussions on DRR. MWCSD will be well-equipped to mobilise women's village committees to engage them in other climate change and DRR work following these recent community development projects.

Gender mainstreaming can be seen in Samoa in the formulation and implementation of some projects such as the economy-wide integration of CCA and DRM to reduce climate vulnerability of communities, implemented by the Government of Samoa with technical support from UNDP. This project makes gender equality an integral part of its project design, aims to identify and strengthen the livelihoods of women through livelihood diversification and directly allocates $500,000 to the implementation of these activities. Further investigation showed these positive steps to be largely attributed to the presence of a gender and climate change specialist in the project formulation team. A similar scenario was behind gender and protection issues being addressed in Samoa's Post Disaster Needs Analysis (PDNA) following 2012 Cyclone Evan, a departure from other, gender blind PDNAs from this period. In that case, a GenCap Gender Specialist was responsible for conducting gender analysis (Government of Samoa 2013). Despite obvious benefits in terms of knowledge and data gained from the approaches above, there remains no systematic collection and analysis of SADD in relation to climate change in Samoa.

Vanuatu

At the time of consultation, the Vanuatu Ministry of Climate Change was in the process of formulating Vanuatu's first joint policy on climate change and DRR (2015a). This document recognises the vulnerability of women and people with disabilities to climate change and disasters, following stakeholder consultation on the draft policy which highlighted the need for gender and social inclusion to be a standalone objective. Vanuatu's national policy on gender equality (2015a) highlights mainstreaming of gender issues into other sectors e.g. agriculture and health, as a priority area, with specific strategies for strengthening gender responsiveness in the climate change and DRR sectors. The needs of women are also recognised in sectoral climate change policies for agriculture, livestock, forestry and energy.

Regarding institutional arrangements, Vanuatu's National Advisory Board on Climate Change (NAB) functions as a coordination body for all climate change and DRR programmes. It includes directors from all government ministries and the Climate Action Network (VCAN) which represents the voices of marginalised groups. VCAN, along with other CSOs, was directly involved in drafting of the Climate Change and DRR Policy. The Department of Women's Affairs (DWA) advocates for greater focus on gender and climate change through its representation on the NAB, and liaises regularly with the Ministry of Climate Change. Vanuatu's national delegation to COP19 was led by the female director of DWA and the selection process for the delegation ensured equal gender representation. At the COP, the delegation presented to the UNFCCC on *Gender Balance of UNFCCC Bodies*, stressing the integral role of women in CCA and mitigation, and the need for their engagement in decision and policy making on climate change.

The level of gender responsiveness in climate change action in Vanuatu is mixed. Gender assessments were found to feature more strongly in NGO-led CCA programmes, for example Live and Learn's food security project (Live and Learn 2010) and CARE's baseline assessments on Tanna Island, collected for the purpose of informing future gender analyses and to monitor the effectiveness of CARE's activities, but also used to inform its post-Cyclone Pam relief efforts in 2015. In contrast, the Climate Change Ministry, NDMO and DWA do not have systematic procedures for gathering SADD. As a requirement of some donor-funded projects (e.g. UN, World Bank, GIZ) these institutions record the number of women and men participating in project activities but no other details are systematically collated. Furthermore, it seems there is no platform for sharing information such as that collected by NGOs in order to inform the decisions of others. Vanuatu has an active and effective humanitarian cluster group dedicated to gender and protection chaired by DWA and NGOs CARE and Save the Children (Government of Vanuatu 2015b). This represents one possible pathway for such sharing of information, but would only apply in relation to disaster response, rather than wider climate change adaptation.

Project Review

Analysis of climate change mitigation and adaptation projects funded by the Global Environment Facility in the Pacific in the period 2010–2015 showed that overall, the consideration of gender in project design and implementation to date has been

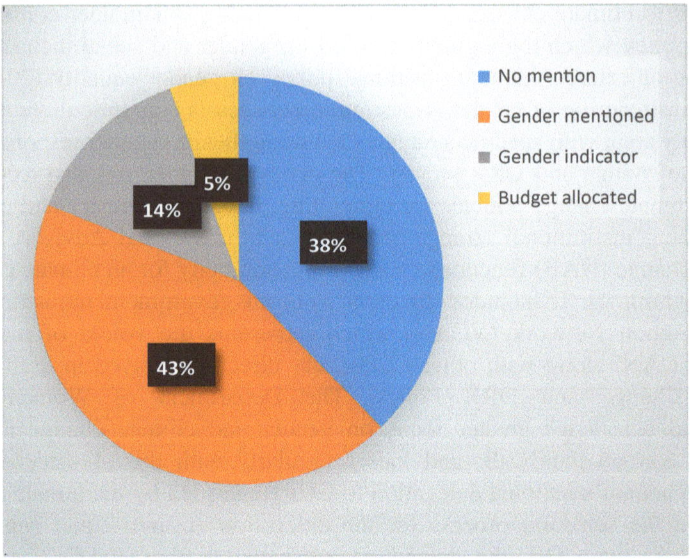

Fig. 13.2 Pacific GEF projects 2010–2015 gender analysis

limited (Fig. 13.2). Of 42 project documents, only two made obvious mention of dedicated staff resource. 38% made no mention of gender considerations and 43% made at least passing mention of gender but often in the context of even gender spread during consultation, rather than meaningful involvement of women in project design and implementation.

In contrast, at the local level, better progress has been made on smaller scale projects. Of seven community projects funded by the GEF Small Grants Programme in RMI, three are entirely led and managed by women's groups. This includes the introduction of smokeless stoves in nine outer islands to reduce women's exposure to respiratory diseases, an organic farming project in Wotje atoll and the installation of solar panels in Ebon. In Samoa, 25% of grants awarded through the GEF Small Grants Programme office are projects initiated and managed by women, while others have been awarded to village councils. While many of these projects described above are mitigation, rather than adaptation projects, it is evidence that that women are not only participating but are taking a proactive role in local communities. Interestingly, consultations with members of the Vanuatu NAB found that GEF small grants projects implemented at the community level are in fact more effectively managed by women than men, especially in budgeting and execution of project activities.

Discussion

The findings above are analysed in more depth below, based on the five spheres of the framework. Further examples from other Pacific Islands are provided, as uncovered during the literature review.

Policy

Gender-responsiveness of national climate change policies differed widely between PICs. While there is already some integration of climate change in national gender policies, and of gender equality in climate change policies in the country case studies, Vanuatu being a good example of this, other PICs have failed to connect these sectors completely. Therefore from the highest level, there is a failure to recognise women's different needs and priorities, relative vulnerability, and the potential leadership role of women in developing and implementing climate change strategies.

With international frameworks now recognising the need for gender responsive CCA, and with this observed inconsistent reference and integration of gender issues in climate change policy, the first step for all PICs must be to ensure establishment of enabling climate change policies with gender specific provisions at the national level to ensure the needs and participation of women are taken into account.

In regards to national gender policies, it was observed that in some instances, climate change is prioritised as a standalone policy outcome e.g. Cook Islands (2012), while other cases take the approach of mainstreaming gender across all government sectoral policies including the environmental and climate change sectors (Government of Tuvalu 2014). Both of these approaches to incorporating climate change considerations into gender policy can be effective. Rather, effectiveness is dependent on the context and political dynamics in each country and the level of implementation of policy statements in each case. Recognising the link between gender and climate change adaptation in national policy, while essential, has little meaning if not translated into practice.

Institutional Framework

In order to achieve the goals set by climate change and gender policies, establishment of conducive institutional frameworks for implementation is necessary. Gender mainstreaming in climate change adaptation is only possible with strategic collaboration between ministries, NGOs, CSOs, and communities.

In most PICs, the mandate for gender equality falls to the ministry responsible for women's affairs. These ministries usually have very limited political advocacy, technical capacity and financial resources for policy and programme delivery (SPC 2012). But with effective cross-sectoral coordination, in particular between the gender department and climate change departments of government, policy commitments can be turned into on-the-ground action. As an example, the Solomon Islands Climate Change Policy is supported by specific provisions that articulate the involvement of women in operations and decisions: the MWYC is included in the National Climate Change Council and its thematic working groups are articulated in the Climate Change Policy (Government of Solomon Islands 2012).

Other PICs (e.g. Vanuatu) achieve collaboration through disaster response 'cluster' systems. This offers potential for mainstreaming gender policy into climate change more generally by starting with disaster response.

The study also highlighted the critical role of NGOs in putting policy into action. Some NGOs have institutionalised minimum criteria on gender equality as part of their programming practice: e.g. the Red Cross ensures gender parity in all community disaster committees and SADD in risk assessments, and CARE's programmes are all required to report SADD, attain gender balance for consultations and promote women's leadership in decision making. Without WUTMI during the 2013 drought in RMI, it would have been impossible to provide disaster relief to remote and widespread communities in a fair and equitable manner.

Implementation and Practice

Often in the Pacific, despite policy commitment, institutional responsibilities and gender responsive project design documents, the goal of gender equality and empowerment of women is not achieved on the ground due to lack of technical capacity, financial resources and evaluation criteria (SPC 2012). Policy must therefore be supported by tangible actions for implementation, including allocated resources for addressing women's needs and building their capacity.

The review of GEF projects showed that awareness of the issue is growing. Generally, more recent projects were more likely to have considered gender equality in project design and in some cases, gender considerations have been retrofitted to existing project design. However, overall, gender issues were poorly considered in design, implementation and monitoring phases. Although gender sensitivity is one criterion used to assess applications for GEF grants, grantees are not required to evaluate and report on the outcome as part of the grant acquittal process. This means there are no gender-specific indicators or outcomes, making it difficult to tell whether the project has met women's needs or contributed to their overall social and economic empowerment. It provides no mechanism for projects to sustain their focus on gender issues. The GEF Small Grants Programme Team has recognised this gap and is working with gender experts on its Technical Advisory Group to develop specific indicators for improving monitoring and reporting on gender outcomes across the region.

Similar findings were evident in UN Women research in Tonga where national climate change and DRR projects were assessed on their gender-sensitivity using the Inter-Agency Standing Committee (IASC) Gender Marker, a coding system developed for tracking gender allocations of humanitarian interventions. It revealed nearly half of all projects were completely gender blind and another quarter had limited or purely cosmetic reference to gender or social vulnerability (Kingdom of Tonga 2015).

Like climate change policies, climate change project design documents may feature gender rhetoric that is not followed by gender-responsive implementation and practice. For example, a gender assessment of the Tuvalu 'NAPA1/+' Project showed that while the National Development Plan made explicit commitment to "*promote gender equity and expand the role of women in development*", in reality only one of 15 Members of Parliament was female, and members of the Project Board and Technical Working Committees were predominantly men (Bernard 2013). While project design may fully integrate gender considerations, in reality, constraints such as the absence of a clear organisational intent or a dedicated budget can wane conscious efforts to make projects gender-responsive. Adequate resourcing and gender indicators for monitoring throughout project implementation should be conditional for project approval.

Knowledge Management

The advocacy and knowledge sphere is at the heart of the mainstreaming framework as it is integral to all spheres. It must inform climate change and DRR policy reform, organisational change, CCA project implementation, and community-driven risk assessments. Increased awareness of the gender impacts of climate change among all members of society as well as increased substantiated knowledge of how to address gender equality in these thematic areas is fundamental in promoting and sustaining equitable resilience.

A review of the Sendai Framework on DRR's predecessor, the Hyogo Framework, showed gender disaggregated data to be available in only 14% of Asia-Pacific countries, with a complete absence of gender disaggregated data in most Pacific countries (UNISDR 2013). Due to extremely limited collection and use of SADD in the Pacific, current data are insufficient for informing decision making. The lack of data means that proper analysis on the immediate and long-term social impacts of climate change is not done well, and similarly the impact of climate change adaptation initiatives cannot be quantified without baseline data.

The presence of a gender specialist can make a substantial difference to data availability and quality. For example, gender and protection issues were extensively addressed in both the Cyclone Evan PDNA (Government of Samoa 2013) and the Cyclone Pam PDNA (Government of Vanuatu 2015c). Both assessments made gender-specific recommendations such as support to help men and women find alternative livelihood options, training and resources for women farmers, addressing land ownership issues to assist female headed households who lost their homes but have no rights to land, increasing access to reproductive and maternal health services, and GBV counselling and outreach. In both cases, there were gender specialists with responsibility for conducting gender analysis. Similarly, in cases where climate change projects have a gender advisor on the project team, or a nominated gender focal point, clear benefits to women can be seen in the form of gender indicators that are tracked throughout project implementation to ensure that clear benefits to women are achieved.

Another observation was the severe lack of inter-agency, cross-organisational or cross-sectoral sharing of data, limiting opportunities for this information to be fully utilised and analysed for improving policy and programming work. For example, the government stakeholders consulted in Vanuatu for this study noted that NGOs were collecting data on the gender impacts of climate change and disasters but there was no process for sharing this with relevant government agencies. The stakeholders suggested that if such data was shared, it may be useful in informing policy and programming work, although this is also reliant on adequate technical capacity in conducting gender analysis being available within the agency. Similar observations were made during the Tonga Climate Finance and Risk Governance Assessment (CFRGA) (Kingdom of Tonga 2015).

Women's Participation and Leadership

Participation and leadership of women is important in climate change adaptation for two reasons: women's higher vulnerability due to gender inequality and their traditional roles in families and communities must inform planning and decision making, but also, women demonstrate tremendous capacity for promoting and leading climate change adaptation. This study uncovered that in some PICs, women are leading, initiating and implementing climate change and DRR projects in communities more so than men. This is certainly the case for the UN GEF Small Grants Programme in RMI and Samoa. However, with no gender-specific indicators or outcomes, it is difficult to track to what level these projects are contributing to gender equality and the empowerment of women.

CCA programmes must ensure their interventions advance, rather than reverse, equality between men and women. Where they do, the challenge now is to make women's participation visible and to promote their leadership at all levels. Moving beyond seeing women as victims of climate change and rather as agents of change will require a transformation of ingrained gender norms and cultural perceptions in the Pacific.

Next Steps

While this study has illustrated the situation in the Pacific at the macro level, assumptions and generalisations have been made based on the three case studies and policies and projects that were available at the time of writing. It must be recognised that while there are very clear conclusions that relate to the region as a whole, there are significant variations between the fourteen Pacific countries. In order for the recommendations below to be most useful, the research would benefit from similar analysis being undertaken at the national, or even subnational level in order to identify priority actions that are more specific to the country or community context.

Conclusion

The findings of this paper emphasise that the concept of gender equality is still embryonic in climate change adaptation in the Pacific Islands. While recognised in theory (i.e. policy and project design), it is not often well supported by action and practice, nor well measured and monitored.

The study observed a notable lack of mainstreaming of gender in national policies, and therefore most likely a lacking general awareness of the linkage between gender equality, climate change and DRR in the Pacific. Despite overwhelming evidence that climate change and disasters disproportionately affect women, gender equality is not often well integrated into climate change policy, nor climate change into gender policy. Where climate change and gender equality are well mainstreamed into national policy, corresponding institutional frameworks that are well-resourced, coordinated and informed are required, along with adequate technical capacity within government agencies, NGOs and project teams, in order to achieve this in practice. At all levels and in all countries, SADD related to climate change, natural disasters and sustainable development were found to be non-existent or limited, which constrains proper assessment to inform and improve policies, project design, risk assessments and programming, and ongoing measurement of progress.

Through case studies and success stories, this study identified clear recommendations for improving gender equality in climate change adaptation, as outlined below:

(1) National climate change and disaster management **policy** should clearly recognise gender principles and the importance of an inclusive and consultative approach. Likewise, Pacific gender equality policies should recognise climate change as not only a significant risk to women and their livelihoods, but also as an opportunity to empower women and achieve gender equality.
(2) Strengthened **collaboration** and positive relations between government and NGOs are important to effectively embed gender analysis into climate change

policies, laws and projects. This may be achieved by leveraging active networks on outer islands (e.g. women's councils in villages) to ensure remote communities are represented and engaged in national dialogue on climate change and DRR and to enhance the gender responsiveness of national climate change projects.

(3) Networks or communities of practice will be effective for **sharing knowledge**, advising and raising awareness of the need for gender responsive climate change action. Such networks should engage a wide range of stakeholders to inform planning and develop consistent messaging about gender equality and climate change.

(4) Gender equality outcomes must be included in CCA assessment criteria to ensure gender analysis is integrated into the design of all climate change policies, plans and projects. Adequate **resourcing** and specific **gender indicators** for monitoring throughout project implementation should be conditional for project approval, to ensure grant recipients monitor, evaluate and report how their projects contribute towards gender equality outcomes. This in turn may be shared with stakeholders through a community of practice (refer (3)).

(5) Women's involvement is not only essential during consultation. Participation of women in climate change processes and dialogue needs to be further strengthened towards active **engagement and leadership**.

It is clear that participation and skills of women and girls will lead to more effective, and much needed, climate change adaptation in the Pacific. The recommendations above aim to improve gender equality through adaptation to climate change, and therefore, sustainable development overall. Underlying social norms and institutional constraints will continue to hamper the Pacific Islands' progress towards fully achieving gender equality and women's empowerment in CCA, but the best practice examples identified through this study are evidence that these challenges can be overcome.

References

Asian Development Bank. (2014). *Gender-based violence in Asia and the Pacific*. Retrieved May 29, 2016 from http://www.adb.org/news/infographics/gender-based-violence-asia-and-pacific.

Bernard, K. (2013). *Tuvalu gender assessment of NAPA1/+Project*. Suva: UNDP.

Enarson, E. (2000). Gender and natural disasters: Working paper no. 1. In Focus Programme on Crisis Response and Reconstruction. Geneva: ILO.

Enarson, E., & Fordham, M. (2001). From women's needs to women's rights in disasters. *Environmental Hazards* 3(3), 16–139. Cambridge: Taylor and Francis.

Government of Samoa. (2008). *National policy on combating climate change*. Apia, Samoa: Government of Samoa.

Government of Samoa. (2013). *Post disaster needs assessment cyclone Evan 2012*. Apia, Samoa: Government of Samoa.

Government of Solomon Islands. (2012). *National Climate Change Policy 2012–2017*. Honiara: Government of the Solomon Islands.

Government of the Cook Islands. (2012). *National policy on gender equality and women's empowerment*. Rarotonga, Cook Islands: Office of the Prime Minister.

Government of the Republic of the Marshall Islands. (2007). *National action plan on disaster risk management 2008–2018*. Majuro: Government of RMI.

Government of the Republic of the Marshall Islands. (2011). *National climate change policy framework*. Majuro: Government of RMI.

Government of the Republic of the Marshall Islands National Disaster Management Committee. (1997). *National disaster management plan*. Majuro: Government of RMI.

Government of Tuvalu. (2014). *National gender policy 2014–2016*. Funafuti: Government of Tuvalu.

Government of Vanuatu. (2015a). *National gender equality policy 2015–2019*. Port Vila: Government of Vanuatu.

Government of Vanuatu. (2015b). *Climate change and disaster risk reduction policy 2016–2030*. Port Vila: Government of Vanuatu.

Government of Vanuatu. (2015c). *Post disaster needs assessment: Tropical cyclone Pam, March 2015*. Port Vila: Government of Vanuatu.

International Labour Organisation (ILO). (2009). *Decent work country programme: Tuvalu*. Funafuti: Government of Tuvalu.

Kilsby, D., & Rosenbaum, H. (2012). *Scoping of key issues in gender, climate change and disaster risk management to inform proposed UN women five-year, five-country Pacific programme*. Internal briefing document, Suva: UN Women.

Kingdom of Tonga. (2014). *Safety and protection assessment of tropical cyclone Ian—Ha'apai*. Government of Tonga: Tonga. Nuku'alofa.

Kingdom of Tonga. (2015). *Climate financing and risk governance assessment*. Unpublished.

Live and Learn. (2010). *Navigating false seasons: A research of aspirations and perceptions toward climate change adaptation for food security in Western Pacific Island communities*. Honiara: Live and Learn Environmental Education.

Morioka, K. (2012). *A climate for change: Understanding women's vulnerability and adaptive capacity to climate change from ActionAid's rights-based approach—case studies from Papua New Guinea and Solomon Islands*. Sydney: Action Aid.

Nellemann, C., Verma, R., & Hislop, L. (2011). *Women at the frontline of climate change: Gender risks and hopes, a rapid response assessment*. Norway: UNEP, GRID-Arendal.

Peterson, K. (1997). From the field: Gender issues in disaster response and recovery. *Natural Hazards Observer* 21(5), 3–4. Boulder: University of Colorado.

Secretariat for the Pacific Community (SPC). (2012). *Regional stocktake of the gender mainstreaming capacity of PICs*. Retrieved May 22, 2016 from http://sids-l.iisd.org/news/spc-publishes-gender-mainstreaming-assessments/.

UNISDR. (2013). *Hyogo framework for action on DRR 2011–2013*. Asia Pacific Synthesis Report. Suva: United Nations Office for Disaster Risk Reduction.

United Nations University and Alliance Development Works. (2014). *World risk report 2014*. Bonn, Germany: Alliance Development Works and UNU Institute for Environment and Human Security. Retrieved May 29, 2016 from http://www.worldriskreport.com/uploads/media/WorldRiskReport_2014_online-II_01.pdf.

Chapter 14
Urban Resilience to Climate-Related Disasters: Emerging Lessons from Resilience Policy and Practice in Coastal Tourism Cities

Elnaz Torabi, Aysin Dedekorkut-Howes and Michael Howes

Introduction

Climate change entails rising sea levels and temperatures, as well as increasing the duration, intensity and/or frequency of extreme weather events such as floods (Intergovernmental Panel on Climate Change [IPCC] 2007, 2012; Climate Commission 2013). Nearly a quarter of the world's population lives within 100 km of the coast and this is likely to rise to half by 2030 (Adger et al. 2005). Tourism in many coastal cities is an important part of the economy due to the proximity to water and its amenities. Yet this dynamic global industry is vulnerable to the impacts of climate change on coastal and marine ecosystems, as well as communities, utilities and infrastructure (Becken and Hay 2007; United Nations Environment Programme [UNEP] 2008). The Pacific Region is an important case in point with tourism being the major sector and source of foreign exchange earnings (Baker 2013). Samoa, Vanuatu, the Solomon Islands and Tuvalu, for example, are heavily dependent on tourism (UNEP 2006), but these islands are also highly vulnerable to climate change (IPCC 2007; Keener et al. 2013). Local governments therefore need to better understand vulnerability and resilience in order to integrate climate change adaptation and disaster risk reduction with spatial planning (Bajracharya et al. 2011). This paper focuses on resilience planning in coastal tourism cities, using Australia's Gold Coast and Sunshine Coast as examples. It aims to explore the root causes of urban vulnerability and their implication for enhancing resilience to climate-related disasters in coastal tourism cities.

E. Torabi (✉) · A. Dedekorkut-Howes · M. Howes
Cities Research Institute, Griffith University,
Gold Coast, QLD 4222, Australia
e-mail: elnaz.torabi@griffithuni.edu.au

W. Leal Filho (ed.), *Climate Change Adaptation in Pacific Countries*,
Climate Change Management, DOI 10.1007/978-3-319-50094-2_14

The Need for this Research

Australia is highly vulnerable to climate-related disasters with 85% of its population living on the coast (Department of Climate Change [DCC] 2009; Australian Academy of Science 2015). Within Australia the state of Queensland is particularly vulnerable (IPCC 2007, 2012). The 2011 Queensland floods, for example, led to 38 deaths, the inundation of tens of thousands of buildings, and insurance pay-outs of A$2.4 billion (Productivity Commission 2014). The Gold Coast and the Sunshine Coast are both located in the South East Queensland region of this State, which has been identified as a climate change 'vulnerability hotspot' due to its coastal location, geography, local climate and the distribution of population (IPCC 2007; Burton 2014; IPCC 2014a). These cities are built on low-lying flood-prone land. They are similar in their biophysical and socioeconomic characteristics but have different settlement patterns as can be seen in Fig. 14.1 (DCC 2009; Infrastructure Australia Major Cities Unit 2010; IPCC 2014a).

At the local level, responses to climate-related disasters are heavily influenced by land-use planning and disaster risk reduction policies (Shaw and Theobald 2011; Howes et al. 2013; Howes 2015). While the international focus has been on mitigation, there is an increasing need for adaptation to the unavoidable impacts of climate change, but local governments who are the main stakeholders face some major challenges (United Nations Framework Convention on Climate Change 2007; IPCC 2014a; 2014b). Unfortunately politics has driven a decline in climate change policy in Australia and Queensland over the last few years that has significantly affected the case studies (Howes and Dedekorkut-Howes 2013).

Developing a more appropriate response requires an understanding of the underlying drivers of urban vulnerability and resilience as a first step (Berkes 2007; Romero-Lankao and Dodman 2011). Factors such as exposure, sensitivity, lack of coping, and adaptive capacity appear regularly in the relevant research literature (Cutter et al. 2008; United Nations University 2014). It is clear that urban

Fig. 14.1 Low-lying settlements on the Gold Coast and the Sunshine Coast; **a** Surfers Paradise, Gold Coast (Courtesy of Skyepics.com.au, May 2013) **b** Twin Waters, Sunshine Coast (Courtesy of Skyepics.com.au, January 2010)

vulnerability has both biophysical and socioeconomic dimensions. The biophysical dimension includes factors such as: location, geography, sea-level rise, the built environment, infrastructure, and natural resources. Socioeconomic vulnerability is an indication of the inability to cope with stresses (Romero-Lankao and Qin 2011), and this in turn is influenced by factors such as: population growth, demography, governance, the economy, wealth, education, health, and social capital.

Methods

This research undertakes a qualitative approach to understanding urban vulnerability in coastal tourism cities by using empirical evidence. The overall research strategy is a comparative analysis of two coastal tourism cities in Australia that sit on the edge of the Pacific Ocean: the Gold Coast and the Sunshine Coast. Comparative case study helps with an in-depth investigation of the research question within a real life context to provide more detail and expand on qualitative findings (Yin 2003). The research proceeded in a series of steps. First, a comprehensive literature review was conducted on the concept of urban resilience that identified the core themes and defined key terms. Second, the case studies were selected based on their similar vulnerability to climate change but their significantly different local government responses. Third, thirty seven semi-structured interviews were conducted in 2015. Respondents included local and state government officials, researchers, representatives from non-government organisations, the insurance industry, and consulting firms (see Appendix). The interviews were analysed thematically with NVivo and the findings were compared to what emerged from the initial literature review (Kvale and Brinkmann 2009). The main impediments to this study included the limitation of available data due to the political sensitivity of the topic of climate change and the unwillingness of some respondents to disclose more detail. Moreover, due to time constraints we did not include resident and tourist communities' perception of their vulnerability or resilience and this study focused mainly on higher level stakeholders influencing decision-making.

Results

The key themes that emerged from the research show that the risks associated with climate-related disasters include elements of: the nature of extreme weather events; the exposure of communities and settlements; sensitivity; and, capacity building and resilience. We discuss the biophysical and socioeconomic vulnerabilities under each of these themes below.

The Nature of Extreme Weather Events

Extreme weather events are having an increasing impact on coastal populations (Centre for Research on the Epidemiology of Disasters 2012). Most of our interviewees, while acknowledging the vulnerability of inland areas, thought that coastal settlements were more vulnerable due to the nature of additional extreme weather events such as coastal erosion, storm surges, and cyclones. Respondent 17 also discussed the difference in the nature of flooding within the cities: "the exposure of a flood height can be greater in hinterland catchments which means that the roads can be flooded more regularly whereas in coastal catchments the extent of flooding can be more significant". Flash flooding is a major risk factor for the Gold and the Sunshine Coasts, however, as Respondent 33 stated: "we don't, here in Queensland, understand flash flood risk very well. I don't think people consider that with where they settle, it is an affordability issue and a lifestyle issue". In short, the nature of extreme weather events means that coastal communities are exposed to additional kinds of risk that are less well understood.

Exposure

The research literature indicated that exposure to hazards is an important contributor to vulnerability (Adger 2006). Urban systems with people, infrastructure, services and assets are located in areas that are exposed to the risk of climate-related disasters (IPCC 2012). The biophysical vulnerability of coastal areas is a function of geography, population, and development patterns. This can exacerbate the pressure to develop low-lying waterfront land, as noted by Respondent 10:

> Things get more intense more quickly, everyone wants that view, everyone wants that breeze, there is a lot of competition [...] the land is very highly valued and it is quite dramatic how that clusters around the coast and often the coast also has very high recreational values and very high habitat values and all that is occurring on top of each other, so it is a great place to learn about general urban planning.

This desire for proximity to water has resulted in development patterns that increase exposure. Respondent 7 gave the example of Brisbane floods in 2011 and noted that:

> What you should do is build the houses totally differently, we should retreat from the river bank, we should allow the river bank to be empty, which is probably fine in theory but [...] settlement patterns today don't allow for that, what people want to do is [...] to be right by the river because the river is an open space.

In coastal cities many of the tourism-related activities take place on the water's edge, resulting in the development of assets on exposed land. Respondent 30 discussed the implications of this for spatial planning:

> Things like caravan parks or holiday parks are often proposed to be located in areas that wouldn't normally be considered for normal residential development but because it is a special type of development without permanent residents then there is often a push to have that sort of development in these low-lying and where you call vulnerable areas.

The risks are exacerbated by actions that reduce natural elements of resilience such as filling in wetlands, the removal of mangroves, etc. This has been done on a significant scale on the Gold Coast and Respondent 27 described its exposure: "in particular its location adjacent to floodplains and the amount of waterways which allow storm tide and flood to sort of reach further inland than it might otherwise". So it is clear that exposure is due not only to the geography of these regions, but also to their population growth and development decisions.

Sensitivity

Susceptibility (in disaster risk management) and sensitivity (in climate change adaptation) are terms used in the literature to indicate the intrinsic and context-related conditions that can make a system collapse during a hazardous event (IPCC 2012). Sensitivity is understood as the characteristics that describe the weakness of an exposed system (Birkmann 2007). The natural sensitivity of coastal areas, the old age of infrastructure, and an economy that has an overreliance on tourism, are the main drivers of susceptibility in coastal tourism cities as identified by the interviewees. Respondent 9 considered the coastal zone itself to be exacerbating the vulnerability of settlements as it is: "a very dynamic zone naturally with coastal processes […] due to geomorphology".

Another important issue is the old age of infrastructure such as sewage, storm-water drainage, and roads in highly exposed areas. Old infrastructure networks are flooded more frequently as they are usually constructed to serve smaller populations or they are built to outdated design standards (Queensland Floods Commission of Inquiry 2012). Coastal tourism cities usually start as small resort towns or retirement destinations for nearby communities in low-lying areas without any consideration of future climate, so their supporting infrastructure can be problematic.

An economy that has an overreliance on tourism is another aspect of sensitivity in coastal tourism cities. In 2011 the annual earnings of the Gold Coast and the Sunshine Coast from tourism were A$7,914 and A$8,330 per capita respectively (Tourism Tropical North Queensland 2011) with tourism-related industries such as retail, health care and social assistance, construction, accommodation and food services accounting for the highest percentages of total employment among other industries in both areas (Australian Bureau of Statistics [ABS] 2011). The desire to be close to water coupled with the impacts of climate change on the beach as the main tourism attraction can have unavoidable consequences for the city's economy. Respondent 12 underscored this issue especially on the Gold Coast:

> [In] 2012 when cyclone Oswald came in and there was that massive amount of erosion that took place there was a lot of concern from tourist operators that […] unless council pumps or replaces sand […] they would be facing a downturn, less business. So I guess that there is that vulnerability where climate has the capacity to reduce the qualities that make a place a tourist place.

Climate-related disasters can also lead to reputational economic risk for tourism cities. Overall then, the sensitivity of coastal cities is a factor of geography, ageing infrastructure, and the dependence of the economy on tourism.

Capacity Building and Resilience

The vulnerability of a system is inversely proportional to its resilience and capacity building (Vogel et al. 2007). These capacities include: *the capacity to anticipate risk*; *the capacity to respond*; and *the capacity to recover and change* as important contributors to resilience (Brooks et al. 2005; Gallopín 2006; IPCC 2012). The ***capacity to anticipate risks*** in coastal tourism cities is undermined by the transient tourist population because of their lack of familiarity with local hazards or emergency management arrangements. Respondent 13 stated that: "They [visitors] don't have support mechanisms, they don't have people who give them local knowledge so that leaves them quite exposed". An example to this was the aftermath of Boxing Day tsunami of 2004 where tourists, oblivious to dangers, went back to the beaches to watch the retreating waters. Respondent 3 mentioned that: "You are trying to save human life in this rather than just infrastructure, […] we have so many international visitors that I don't think we do nearly enough for people who don't speak English". Respondent 14 suggested:

> If you have permanent residents who are used to natural disasters and know what to do when they happen then you will expect those communities to be able to better cope. If there is a lot of people who don't know what to expect and don't know what to do when they do happen, for sure there is a social vulnerability there, certainly, but I think that would be specific to climate-related disasters that don't have long warning times.

Some respondents discussed the nature of resident population themselves being very transitory, less socially connected and therefore lacking coping capacities to deal with disasters. In fact, 34.8% of Gold Coast's population and 24.8% of Sunshine Coast's population were born overseas while internal migration rates in each local government area were 38.3 and 39.7% respectively (ABS 2011). Respondent 22 stated that: "I think just the fact that you have not got that broader population engaged in the social networks where the rest of the community is, must make those cities more vulnerable". Respondents also discussed the ability of the permanent population of hinterland towns in predicting events compared to coastal communities as they have lived in the area longer and thus are more aware of the risks including historical flood levels. Moreover, the Sunshine Coast has an older population with 31.5% of population aged over 55 compared to the Gold Coast's

25.8% (ABS 2011). Respondent 18 suggested this to be important in local residents' awareness of the risks: "they know what to do, they can predict the amount of rainfall we are getting, how high the river is".

The ***capacity to respond*** and accommodate the impacts of climate-related disasters lies in part in different physical measures from applying building codes and construction of seawalls, offshore reefs, dams, and levees to nourishment of the beach. As noted by Respondent 26: "they [the local government] are spending a lot of money every year on rejuvenating, nourishing the sand […] but that has an impact, nobody wants to go there and just see the black coffee rug, you know, you want to go and see a nice beach". The Gold Coast's approach to accommodating risks has been highly reliant on engineering measures. The Sunshine Coast community has imposed strict development regulations relevant to a development-free buffer, building height, and density. The nature of the most resilient settlement patterns, however, has not been agreed. Respondent 10 criticised development of nearby locations such as the Byron Bay and the Sunshine Coast for expanding further along the coast as opposed to Gold Coast's compact form: "we create coastal tourism amenity here, perhaps that is protecting other parts of the coast […] we have got 600,000 people within our coast, 36 km long. That is a huge tax base to do what is necessary to respond to climate change".

Interviewees discussed the vulnerability of three major settlement types on the Gold Coast and the Sunshine Coast: the beachfront settlements, canal estates, and suburban/hinterland settlements. The major difference between the two case studies is in their beachfront development with the Gold Coast being mainly high-rise as opposed to Sunshine Coast's low to medium-density beachfront development. A major mitigation infrastructure for Gold Coast's exposed shoreline is the city's seawall (A-line) that runs between the beach and dunes. Despite its importance in accommodating climate risks, the interviewees raised concerns about the integrity of the A-line due to shifting responsibilities between the local council and homeowners and lack of a full assessment by the local council. Respondent 4 also pointed to the asset life of the A-line saying that "it has about 50 years of life in terms of the value of A-line […] but beyond that we don't really have a strategy. I don't think we are prepared for anything beyond 50 years in an infrastructure sense".

Canal estates are vulnerable to the impacts of climate-related disasters, especially interactions of regional flooding with coastal events such as storm surge. On the Gold Coast some interviewees considered canals to have higher capacity to cope with disasters due to both the engineering involved in their construction and the impact of city's other extensive mitigation structures. The interviewees identified transport systems as the most important factor undermining the coping capacity of all settlements on the Gold Coast and the Sunshine Coast. While the lack of redundancy in the transport network (one way in and one way out) in canal estates can result in isolation of communities during emergencies, Respondent 4 discussed the scarcity of connecting bridges on the Gold Coast: "we have far more water's edge than Venice and Amsterdam put together and we have canals along the river but […] we have 1% of the bridges that they have". This also has implications on the limited east-west connection across the city as noted by the respondents.

Another critical issue is the isolation of communities during disasters due to complete or partial inundation of critical transport routes that are not designed to accommodate future impacts of climate change.

The development-free buffer is important in accommodating risks and reducing exposure to climate hazards, especially on the Sunshine Coast. Respondents also discussed the quality of the buildings and the city infrastructure. Upgrading the existing built structure of the city is an important step in building the capacity to respond to climate change. This includes interventions to the buildings such as raising structures on stilts or raising floor levels using hydraulic lifts, etc. However, these types of adaptive measures can only have impact if they are considered in the context of the lifecycle of buildings and the horizon of climate impacts. Respondent 7 discussed this in the context of 2011 Brisbane floods: "The difficulty with that is the frequency of those events is so long like over 30 years so if you did, if the council decided to put the houses on hydraulic lifts, in 2013 you put it in but then 2020 you sell the house and the next person wouldn't maintain it so when the flood came…". This highlights the importance of the concept of asset life in highly exposed coastal tourism cities and the need to consider the viability of alternatives such as short-term or ephemeral buildings.

Risk transfer and insurance are also important in shaping the city's capacity to respond. The insurance industry plays an important part in the formation of resilient settlement patterns as noted by Respondent 4:

> The insurance people are going to determine where development occurs or doesn't occur because it will become unviable for some property owners to maintain properties to hang on to them, whereas local councils don't have the courage to say that 'sorry you are just in an area which is just not going to be feasible for redevelopment, we are not going to protect it'.

Respondent 5 discussed this in the context of cross-subsidisation and increases in insurance premiums due to high risks associated with some parts of the city that affect everyone else:

> There is a wish not to disrupt the property market. So you can't have the [exposed] property with an insurance policy forty times bigger than somebody at the hilltop perfectly safe from storm surge so they even it up. So we are all paying for these increased risks […] when the insurance premiums go up they don't assign the blame to climate change.

The ***capacity to recover and change*** in coastal tourism cities is highly influenced by the political climate and governance mechanisms that promote adaptation planning as well as encouraging information sharing and community engagement. An important issue for long-term adaptation of these cities is accommodating climate change impacts while still continuing to attract tourists. Respondent 15 underscored this issue:

> How would the tourism activities on the beach change if the beach was gone and you are left with a sea wall? […] what would that do to tourism? Now if you look at the European context, they have evolved to accommodate that. And there is probably just as many people to go to see the water as if there was a beach there. So you would have to have a shift from a focus away from the beach to that land-water interface. And that is a real potential outcome of climate change.

The biophysical and socioeconomic context of coastal tourism cities highlight the need for governance mechanisms that recognise and address all elements of vulnerability. However, as noted by Respondent 8: "governments don't take the time or the effort to inform resident population of those risks anyway and that is because of the politics of it […] that has implications in terms of property values and the opportunity for politicians to get re-elected and those types of issues which come into play a great deal in local governance". Acknowledging the local council's cautious approach to information sharing, Respondent 21 highlighted the big challenges as: "the need to go through engagement but also to not instil fear about the potential risks". These concerns then limit the ways that local governments approach community engagement and what information they release. The local political climate in this sense is a fundamental underpinning of exploiting the benefits of climate change as suggested by Respondent 4: "yet it could be one of the ways to actually create new industries. It is such a brilliant thing for our city". The interviewees underscored the role of academia and collaborations with industries and local governments as an important contributor to adaptive capacity. An example to this is the collaboration between the University of the Sunshine Coast and the local council in research on community resilience.

Interviewees also discussed the amount of money spent in the Planning and Environment Court to ensure good planning outcomes. Respondent 8 considered the settlements above the 1-in-100 year flood event to be the most resilient and discussed the court rulings for development on floodplain as: "it is entirely appropriate for a unit development or for a high-rise development to be constructed in the Gold Coast even though in a 1-in-100 year flood event there will be no flood free access […] as long as they have got on-site refuges". Respondent 25 criticised the approval of a low socioeconomic caravan park development on the Sunshine Coast on the condition that they have a community centre with a second floor above the major flood levels with a helicopter pad and enough food at that floor. He notes: "that just seems to be that […] OK it is going to flood, here is your boat, good luck! […] it is not a solution. It is very short-term". Similarly there are development approvals on the Gold Coast where the disaster management response is to use rowing boats. Respondent 3 called this "a complete nonsense" and further discussed the shelter-in-place strategy as a more appropriate approach especially considering the nature of the flooding: "the water stream, if it is not running fast it is not dangerous. So I can get anything I need to that place […] in a different scenario, if in that case the water were going through at 8 metres per second or something, different issue. Then you can't satisfactorily get there". Respondent 26 also highlighted the need to consider asset life and the population turnover in these types of developments: "most of the people who buy those units will be young couples without children […] eventually as that building gets older and older the tenants will be aged so it will be cheaper so older people would move in eventually you will end up with an older building with a lot of retired people in there who are going to have a lot of trouble getting down the stairway let alone getting into a boat to escape". So these cases confirm that capacity building and resilience involves anticipation, response, recovery and adaptation.

Discussion

With climate change there is a need for a comprehensive understanding of the nature of extreme weather events to be factored into urban and regional planning. This is particularly important for coastal settlements that face additional risks and where historical data are no longer a good predictor for the future. The development of the Gold Coast on the edge of water, its rapid growth, and population density increase its exposure to climate-related disasters relative to the Sunshine Coast. This, coupled with sensitivities such as overreliance on the tourism economy, contributes to vulnerability. Resilience, on the other hand, lies in the capacity to anticipate, respond, recover and adapt. The itinerant nature of the community in coastal tourism cities call for a redesign of disaster planning and management. A major difference between the Gold Coast and the Sunshine Coast is in the nature of their approach to accommodating the risks of climate-related disasters on their built environment. While planning on the Sunshine Coast has focused on preventive measures and allowed for appropriate development-free buffers and building setbacks from the water, the Gold Coast's approach is in defensive engineering solutions. Although the compact form of the Gold Coast provides a strong tax-base to cope with the climate change impacts, as opposed to lower-density and more dispersed nature of the settlements on the Sunshine Coast, with future climate change the coping capacity of both approaches will be challenged. There is a need to focus on the viability of such decisions under future climate change scenarios. Moreover, the engineering measures are expensive and can jeopardise the coping capacity of the city in the long run, an issue especially important in the context of less developed tourism-dependent nations in the world such as small island nations. This underscores the need for enhancing more transformative and adaptive approaches to capacity building and planning for climate change in coastal tourism cities. The key to this lies in taking more proactive approaches to design and planning of the built environment as well as tailored governance responses. Taking longer planning horizons by implementing concepts such as the asset life of the buildings or acceptance of a range of possible design benchmarks instead of one single threshold (such as the 1-in-100 year flood) are examples of this. The transport network of the city is an important piece of infrastructure in this sense because of its potential to shape future settlement patterns while ensuring effective disaster management. The local government, universities, and the insurance industry play a very important role in this context in sharing information and risks of climate-related disasters. There is also a need for a better communication of climate risks with the public without instilling fear.

Our research also showed that while vulnerabilities of coastal tourism cities are heavily influenced by their biophysical characteristics such as their exposed locations, it is equally important to consider socioeconomic factors. There is a need to shift the focus from reactive disaster management solutions to more proactive risk reduction and adaptation approaches in addressing climate change impacts on both the natural and built environment. This can be done in part by integrating short-term disaster management and land-use planning approaches with long-term climate change adaptation responses. The comparison between these two case studies with similar socioeconomic and biophysical context underscored the role of local government planning in capacity building and climate adaptation. This has important implications for other local governments throughout the Pacific Region as it shows the power of local governments in making a difference and building resilience.

Conclusion

With climate change there will be an increase in the duration, intensity and/or frequency of some types of climate-related disasters. The majority of the population in the Pacific Region lives in coastal areas that are often dependent on tourism. However, the vulnerability of this sector to the impacts of climate-related disasters coupled with the biophysical and socioeconomic vulnerabilities of coastal cities exacerbates the risks. The examples of Australia's Gold Coast and Sunshine Coast can be useful for other local governments in the region. Our findings suggest that vulnerability of coastal tourism cities includes elements of: the nature of extreme weather events; the exposure of communities and settlements; sensitivity; and capacity building, and that resilience is underpinned mainly by the capacity to recover and change. While the Gold Coast's approach to managing risk was heavily based on engineering solutions, the Sunshine Coast took a more adaptive approach by allowing wider setbacks and greater community engagement. These findings highlight the need to a more comprehensive understanding of vulnerability in planning for urban resilience in coastal tourism cities. Future research on urban resilience should incorporate a more in-depth consideration and analysis of the underlying socioeconomic and biophysical contributors to vulnerability that in turn drives urban resilience to climate-related disasters. Inclusion of a more diverse range of stakeholders including the community and triangulating qualitative findings with quantitative methods could benefit future research in this area. There is also a need for more international comparisons between coastal tourism cities across the world, including the Pacific Region, to better understand urban vulnerability and resilience.

Appendix: Interviewee List

Respondent	Gold Coast	Respondent	Sunshine Coast	Respondent	Other
1	Academic researcher	16	Academic researcher	26	Environmental law consultant
2	Property consultant	17	Local government official	27	Planning consultant
3	Former local government official	18	Local government official	28	Environment consultant
4	Architect	19	Local government official	29	Intergovernmental forum
5	Non-government organisation	20	Non-government organisation	30	Engineering consultant
6	Non-government organisation	21	Academic researcher	31	State government official
7	Local government official	22	Local government official	32	State government official
8	Local government official	23	Academic researcher	33	State authority
9	Academic researcher	24	Non-government organisation	34	State authority
10	Local government official	25	Planning consultant	35	Non-government organisation
11	Local government official			36	State government official
12	Local government official			37	Insurance industry
13	Academic researcher				
14	Planning consultant				
15	Academic researcher				

References

Adger, W. N. (2006). Vulnerability. *Global Environmental Change, 16*(3), 268–281.

Adger, W. N., Hughes, T. P., Folke, C., Carpenter, S. R., & Rockström, J. (2005). Social-ecological resilience to coastal disasters. *Science, 309*(5737), 1036–1039.

Australian Academy of Science. (2015). *The science of climate change: Questions and answers*. Canberra, Australia: Australian Academy of Science.

Australian Bureau of Statistics (ABS). (2011). *Data by region—census 2011*. Retrieved February 7, 2014 from http://stat.abs.gov.au/itt/r.jsp?databyregion#/.

Bajracharya, B., Childs, I., & Hastings, P. (2011). Climate change adaptation through land use planning and disaster management: Local government perspectives from Queensland. In *17th Pacific Rim Real Estate Society Conference*, Gold Coast, Australia.

Baker, D. M. (2013). Understanding the economic impact of tourism in the Asian Pacific region using the tourism satellite account (TSA). *International Journal of Business and Social Science, 4*, 3.

Becken, S., & Hay, J. E. (2007). *Tourism and climate change: Risks and opportunities clevedon*. Buffalo, Toronto: Channel View Publications.
Berkes, F. (2007). Understanding uncertainty and reducing vulnerability: Lessons from resilience thinking. *Natural Hazards, 41*(2), 283–295.
Birkmann, J. (2007). Risk and vulnerability indicators at different scales: Applicability, usefulness and policy implications. *Environmental Hazards, 7*(1), 20–31.
Brooks, N., Adger, W. N., & Kelly, P. M. (2005). The determinants of vulnerability and adaptive capacity at the national level and the implications for adaptation. *Global Environmental Change, 15*(2), 151–163.
Burton, P. (Ed.). (2014). *Responding to climate change: Lessons from an Australian hotspot*. Melbourne, Australia: CSIRO Publishing.
Centre for Research on the Epidemiology of Disasters. (2012). *Number of Natural Disasters Reported 1900–2012*. Retrieved May 12, 2013, from http://www.emdat.be/natural-disasters-trends/.
Climate Commission. (2013). *The critical decade: Extreme weather events*. Canberra, Australia: The Climate Commission Secretariat.
Cutter, S. L., Barnes, L., Berry, M., Burton, C., Evans, E., Tate, E., et al. (2008). A place-based model for understanding community resilience to natural disasters. *Global Environmental Change, 18*(4), 598–606.
Department of Climate Change (DCC). (2009). *Climate change risks to Australia's coast, a first pass national assessment*. Canberra, Australia: Commonwealth of Australia.
Gallopín, G. C. (2006). Linkages between vulnerability, resilience, and adaptive capacity. *Global Environmental Change, 16*(3), 293–303.
Howes, M. (2015). Disaster risk management and climate change adaptation: A new approach. In J. P. Palutikof, S. L. Boulter, J. Bernett, & D. Rissik (Eds.), *Applied studies in climate adaptation*. Oxford: Wiley.
Howes, M., & Dedekorkut-Howes, A. (2013). The paradox of paradise: Declining government responses to the increasing risks of climate change on the Gold Coast. In *State of Australian Cities Conference*. Sydney, Australia.
Howes, M., Grant-Smith, D., Reis, K., Tangney, P., Bosomworth, K., Heazle, M., Mcevoy, D., & Burton, P. (2013). *Rethinking disaster risk management and climate change adaptation*. Gold Coast, Australia: The National Climate Change Adaptation Research Facility.
Infrastructure Australia Major Cities Unit. (2010). *State of Australian cities*. Canberra, Australia: Commonwealth of Australia.
Intergovernmental Panel on Climate Change (IPCC). (2007). Climate change 2007: Impacts, adaptation and vulnerability: Working group II contribution to the fourth assessment report of the IPCC intergovernmental panel on climate change. In M. L. Parry, O. F. Canziani, J. P. Palutikof, P. J. van der Linden & C. E. Hanson (Eds.). Cambridge, United Kingdom and New York, USA: Cambridge University Press.
IPCC. (2012). Managing the risks of extreme events and disasters to advance climate change adaptation. A special report of working groups I and II of the intergovernmental panel on climate change. In C. B. Field, V. Barros, T. F. Stocker, D. Qin, D. J. Dokken, K. L. Ebi, et al. (Eds.). Cambridge, United Kingdom and New York, USA: Cambridge University Press.
IPCC. (2014a). Climate change 2014: Impacts, adaptation, and vulnerability. Part A: Global and sectoral aspects. Contribution of working group II to the fifth assessment report of the intergovernmental panel on climate change. In C. B. Field, V. R. Barros, D. J. Dokken, K. J. Mach, M. D. Mastrandrea, T. E. Bilir, et al. (Eds.). Cambridge, United Kingdom and New York, USA: Cambridge University Press.
IPCC. (2014b). Climate change 2014: Mitigation of climate change. Contribution of working group III to the fifth assessment report of the intergovernmental panel on climate change. In O. Edenhofer, R. Pichs-Madruga, Y. Sokona, E. Farahani, S. Kadner, K. Seyboth, (Eds.). Cambridge, Unied Kingdom and New York, USA: Cambridge University Press.

Keener, V., Marra, J. J., Finucane, M. L., Spooner, D., & Smith, M. H. (Eds.). (2013). *Climate change and Pacific Islands: Indicators and impacts: Report for the 2012 Pacific Islands regional climate assessment*. Washington, DC: Island press.
Kvale, S., & Brinkmann, S. (2009). *Interviews: Learning the craft of qualitative research interviewing*. Los Angeles, USA: Sage Publications.
Productivity Commission. (2014). *Natural disaster funding*. Inquiry Report no. 74. Canberra, Australia: Commonwealth of Australia.
Queensland Floods Commission of Inquiry (QFCI). (2012). *Final Report*. Brisbane, Australia: QFCI.
Romero-Lankao, P., & Dodman, D. (2011). Cities in transition: Transforming urban centers from hotbeds of GHG emissions and vulnerability to seedbeds of sustainability and resilience: Introduction and editorial overview. *Current Opinion in Environmental Sustainability, 3*(3), 113–120.
Romero-Lankao, P., & Qin, H. (2011). Conceptualizing urban vulnerability to global climate and environmental change. *Current opinion in environmental sustainability, 3*(3), 142–149.
Shaw, K., & Theobald, K. (2011). Resilient local government and climate change interventions in the UK. *Local Environment, 16*(1), 1–15.
Tourism Tropical North Queensland. (2011). *Tourism: Driving the TNQ economy*. Cairns, Australia: Tourism Tropical North Queensland.
United Nations Environment Programme (UNEP). (2006). *Trade on human terms: Transforming trade for human development in Asia and the Pacific, Asia-Pacific human development report*. Colombo, Sri Lanka: UNDP.
UNEP. (2008). *Disaster risk management for coastal tourism destinations responding to climate change*. Paris, France: UNEP.
United Nations Framework Convention on Climate Change (UNFCC). (2007). *Climate change: Impacts, vulnerabilities and adaptation in developing countries*. Bonn, Germany: UNFCC.
United Nations University. (2014). *World risk report*. Germany: Alliance Development Works and UNU-EHS.
Vogel, C., Moser, S. C., Kasperson, R. E., & Dabelko, G. D. (2007). Linking vulnerability, adaptation, and resilience science to practice: Pathways, players, and partnerships. *Global Environmental Change, 17*(3–4), 349–364.
Yin, R. K. (2003). *Case study research design and methods* (3rd ed.). Thousand Oakes, CA: Sage Publications.

Chapter 15
Coastal Erosion Monitoring on Ouvea Island (New Caledonia): Involving the Local Community in Climate Change Adaptation

M. Le Duff, P. Dumas, O. Cohen and M. Allenbach

Introduction

Adapting to the current and future effects of global warming on coastal areas, particularly in the Pacific, is the challenge all countries will have to respond to in the 21st century. Some population are already having to ask tough questions and make urgent and sometimes painful decisions requiring them to give up their land under growing pressure from marine and radiative forcing and coastal erosion processes. While this dramatic situation requires immediate responses, the approach needs to be careful and measured, because although current developments in these territories can be linked to the climate change observed throughout the world (IPCC 2013, 2014), neither local practices that could exacerbate coastal erosion, i.e. the particular dynamics of certain hazards, nor the vulnerability factors specific to each society can be ignored. Also Pacific island territories' geomorphological characteristics involve many different situations in which the timing and urgency of responses vary widely.

There are many different types of local pressure on sandy shorelines caused by factors as varied as a community's social and economic circumstances, a country's

M. Le Duff (✉) · P. Dumas
CNEP: Centre for New Studies on the Pacific,
University of New Caledonia (UNC), Noumea, New Caledonia
e-mail: matthieu.leduff@yahoo.fr

M. Le Duff · M. Allenbach
LIVE: Island Life and Environment Laboratory, University of New Caledonia (UNC),
Noumea, New Caledonia

M. Allenbach
Labex CORAIL, University of New Caledonia (UNC), Noumea, New Caledonia

O. Cohen
Oceanography and Geoscience Laboratory (UMR 8187),
Université du Littoral de la Côte d'Opale (ULCO), Calais, France

W. Leal Filho (ed.), *Climate Change Adaptation in Pacific Countries*,
Climate Change Management, DOI 10.1007/978-3-319-50094-2_15

political stability, economic development strategies or even the choices made by individual communities or government bodies when developing land and building infrastructure (Dumas 2004). Such local pressure emphasises how ambiguous people's relationship is with the shoreline, whether as a leisure area, home or resource, but the interest shown in it goes no further than the uses it is put to and ignores the fact that the mechanisms that govern its dynamics and the impact of human practices on those processes make it a perpetually readjusting area. For decades, sandy shores have been used for quarrying sediment in sometimes large quantities for private and public building works. Ever more building projects that disrupt coastal sediment transport are carried out on the shoreline and economic development requires increasingly heavy use of the seafront. As the shoreline recedes with disastrous consequences for coastal communities, the responses all too often ignore the basic principles of coastal dynamics, exacerbating an already difficult situation (Allenbach 2013; Allenbach and Hoibian 2006; Allenbach et al. 2014). The seawalls and other "non-compliant" dikes that are constantly being erected as a response have become a serious issue and projects built directly on the backshore only serve to destabilise the whole system.

Defining an adaptation strategy is hindered by the many issues related to the paucity of available data on these areas. The missing data include historical, social, political and cultural aspects as well as information on local coastal dynamics, which are the *sine qua non* of any initiative for these perpetually moving grounds (Jeanson and Dolique 2011). Both the general public and government bodies are also unaware of the present and past consequences of their practices and methods when using the shoreline. They always, and almost dogmatically, cast the blame on external causes related to climate change. There is no doubt that global warming has had and will have effects on these areas, but laying the whole blame on it boils down to ignoring part of the problem and preventing effective strategies from being implemented. Before defining any adaptation policies, a much-needed education, information and awareness campaign on the broad principles of coastal dynamics and practices that can artificially accelerate the processes in some areas also must urgently be carried out for both the communities and governing authorities.

The main purpose of this paper is to describe the methods used to raise people's awareness and involve them in coastal management.

In New Caledonia, the island of Ouvea in the Loyalty Islands (cf Fig. 15.1a) has been affected by erosion. Local communities are concerned about this and it has been confirmed by a few technical reports (Garcin and Vendée-Leclerc 2014, 2015; Le Duff et al. 2015). Since 2009, the local authorities, namely the Loyalty Islands Province, the municipal council and traditional leaders, in liaison with environmental groups and representatives of the island's other traditional leaders, have formed a task force and have been attempting to understand why the shoreline is receding as well as discussing the most appropriate response strategy (Goarant et al. 2015). Existing hazard-management or other planning frameworks for reducing public exposure to coastal hazards are not operational in New Caledonia, owing to the territory's special status. As 26% of land is customary land and, as such, is managed by traditional authorities and deemed inalienable, non-transferrable, held in absolute ownership and immune

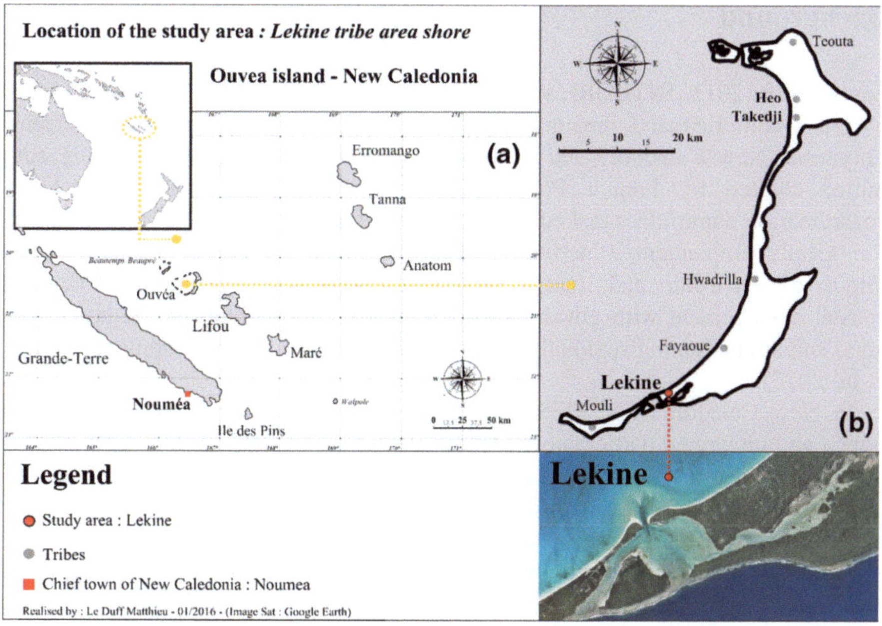

Fig. 15.1 Location of the study area: the shore of the Lekine tribal area on Ouvea, New Caledonia

from attachment, land-use planning regulations in force in mainland France do not apply to it (Merle 1999).

The question is, therefore, how to define an effective climate-change adaptation and vulnerability-reduction strategy for local communities.

This paper presents the methodological approach developed on Ouvea Island (Fig. 15.1a) as part of a regional development programme entitled INTEGRE (Pacific Territories Initiative for Regional Management of the Environment)[1] funded by the European Union (EU) and managed by the Secretariat of the Pacific Community (SPC). The first part of the paper provides more detailed background information on the study and the INTEGRE programme. It then focuses on the methodological approach developed to not only impart "expert" knowledge to the "communities", but also to jointly construct a coastal monitoring process by overhauling the governance method and placing it under the Integrated Coastal Management (ICM) framework.

[1]Together with funding from the French Ministry of Overseas Territories (MOM), further activities are being developed on another site on Northern Ouvea.

Background

Launched in 2013, INTEGRE is a 4-year programme being carried out in various overseas countries and territories (OCTs), namely Wallis and Futuna, French Polynesia, New Caledonia and Pitcairn. It is driven by a regional steering committee chaired by French Polynesia and a network of territorial technical co-ordination committees tasked with providing guidance and technical monitoring for locally implemented activities. The programme's priority objectives are improving environmental management and development, promoting integrated coastal management with government authorities, communities and regional partners, and strengthening regional co-operation in sustainable development (Goarant et al. 2015).

In New Caledonia, INTEGRE provides support for provincial environmental policy and integrated participatory management of World-Heritage properties.[2] The programme also aims at facilitating discussions on the territory's climate-change adaptation strategies. It is against this background that this study for developing a participatory coastal-erosion observatory is being carried out.

The study area is on Ouvea (Fig. 15.1a), located north of the Tropic of Capricorn between latitudes 20° 20′ and 20° 80′ south and longitudes 166° 10′ and 166° 80′ east. Ouvea is a fairly low island rising to 43 m above sea level with a surface area of 132 km^2, 60% of which lies below 10 m (Le Duff et al. 2016). The site studied under INTEGRE is located at Lekine tribal area (Fig. 15.1b) on the island's southern end.

Ouvea has a small population of 3374 according to the latest census (ISEE 2014), i.e. 26 inhabitants/sq. km, making up 18.44% of the Loyalty Island Province (PIL) population. So human pressure on the natural environment is fairly low, although there are significant issues for coastal areas. Firstly, most of Ouvea's population[3] lives on the coast and, secondly, one of the main industries is tourism[4] operating essentially in coastal areas.

With Ouvea islanders being so exposed to marine and radiative forcing, the anthropological dimension of the approach also needs to be considered. Kanaks, New Caledonia's indigenous population, are the majority ethnic group on Ouvea[5] and land is the basic cultural cornerstone for them (Herrenschmidt 2004) with groups building their identities around it. Their relationship with the land must be seen in the light of three main factors, namely cultural, political and historical considerations (Fig. 15.2).

[2]The Ouvea Lagoon was UNESCO World-Heritage listed in 2008.

[3]44% of the population, 68% of environmentally-classified facilities and 67% of public access buildings (PABs) on Ouvea are located less than 500 m from the shoreline and lower than 10 m above sea-level.

[4]90% of the island's tourist facilities are located less than 500 m from the shoreline and lower than 10 m above sea level (Le Duff 2012, 2013).

[5]97.5% of the Ouvea population is Kanak (ISEE 1996).

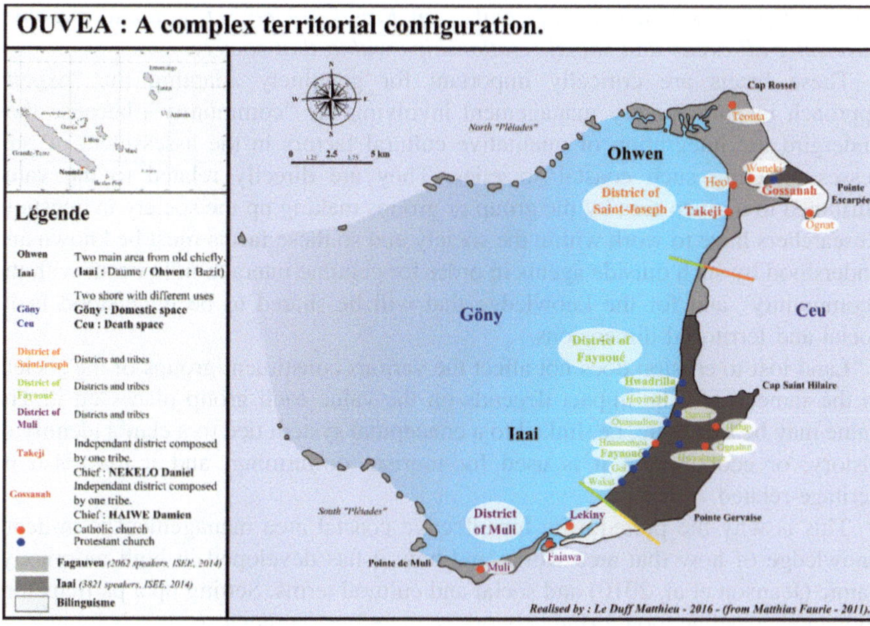

Fig. 15.2 A territorial configuration marked by history and complex balances of power

Cultural because *the bond with the land is a clan's bodily and spiritual relationship with the area where its home mound is located … and the different stages in its history. In a broader sense, the link with the land is about the emotional bonds that hold family/clan and land together* (Customary Senate, 2014). To take a Kanak's land away from him is to strip him of his name, social ties, position and role in custom and everything that defines him as a man in his society.

Political because the independence cause is rooted in land claims and the restitution of land grabbed during colonisation (Merle 1999; Leblic 1991). On a smaller scale, it is also about traditional intra- or inter-clan politics that lead to land claims and clashes within the community itself. Groups build their society and territory, in other words their identity, based on a number of locations within a network or what Herrenschmidt (2004) called a "*topotype*". "*Topotype*" management is complex and based on strategic policy that changes according to needs, opportunities and relationships at any given time with other groups.

Historical because the currently-known social and spatial organisation must be examined in the light of changes all brought about at the same time when religion and colonisation arrived (Fig. 15.2). "Traditional" Kanak culture only takes on meaning when thought of as a dynamic process interacting with all the endogenous and exogenous factors that the communities have dealt with over time. As emphasised by Leblic (1991) there is no such thing as a static, unchanging society: *What Kanaks today claim as … their "custom" is not identical to what existed prior to colonisation… some aspects endured while others changed. It is the result of this historical process that currently makes up Kanak society's traditional structure governing all areas and social relationships within that society.*

These facets are critically important for genuinely adapting the "expert" approach in participatory management involving the "community", because they undergird the integration of qualitative cultural factors in the assessment of processes affecting such coastal societies. They are directly related to the value attributed to a given area by the group or groups making up the society in question. Researchers have to work within the society and so these facets must be known and understood by such outside agents in order for genuine interaction to occur with the "community" and for the knowledge that will be shared to be understood in its social and territorial dimensions.

Land lost to erosion does not affect the various constituent groups of the society in the same way. The impact depends on the value each group places on it. The value may be symbolic, i.e. linked to a conceptual system tied to a clan's identity or history; or economic if it is used for tourism or farming; and it may also be heritage-related.

This is why the prerequisite for effective coastal area management is in-depth knowledge of how that area works and how it has developed in both morphodynamic (Jeanson et al. 2010) and social and cultural terms. Setting up a participatory shore and coastline monitoring network is, therefore, a crucial stage in helping local communities, particularly traditional authorities who are at the forefront of coastal management in the areas studied, acquire certain skills and improve their knowledge of this environment.

Methodology

A network of geodetic marks was set up in order to construct topographic profiles (Fig. 15.3). The marks are used as "profile heads" and georeferenced with the Global Navigation Satellite System (GNSS) based on a known levelling point and linked to New Caledonia's co-ordinate system (RGNC 91-93 Lambert NC). The mean vertical error margin when fine-tuning the marks was estimated at ± 1.5 cm and horizontally at ± 0.8 cm. The marks were set up beyond the backshore to prevent the reference points from being destroyed in extreme episodes, the idea also being to have a permanent network. In the Lekine area, the network is made up of 20 evenly distributed marks, i.e. one every 200 m (Fig. 15.3).

Using the geodetic mark network, cross-sections can be made in any given direction. Measurements are carried out regularly every quarter and spot measurements can also be made, e.g. after marine and radiative forcing, as was the case after cyclones Pam (March 2015) and Solo (April 2015). The areas studied can then be described by comparing the profiles.

In addition to acquiring data for improving our understanding of the physical processes at work on Ouvea Lagoon's sandy shores, the monitoring also serve to involve the task force members mentioned above as well as any community

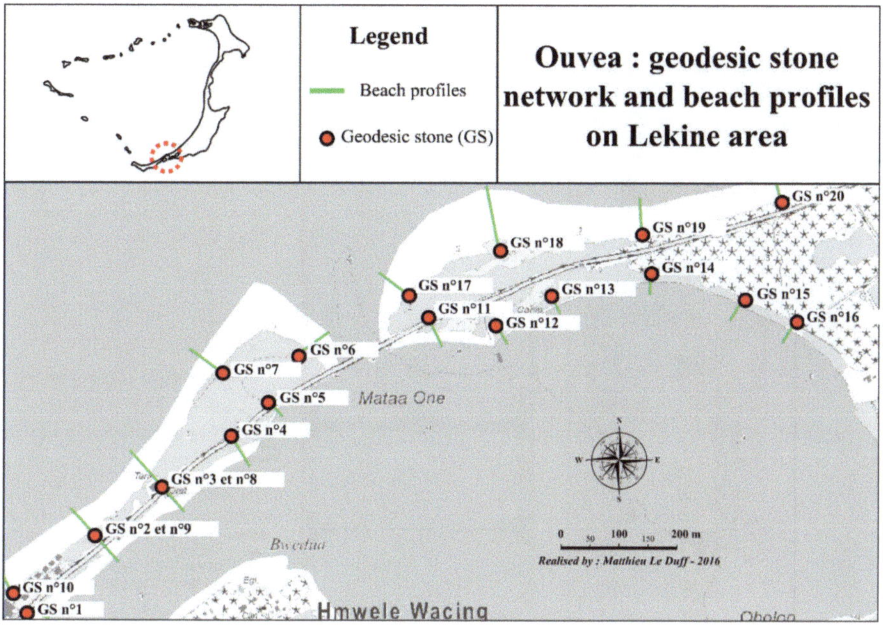

Fig. 15.3 A marker network forming the basis of participatory implementation

members who wish to contribute to the project. The aim of the participatory observatory is to disseminate general scientific knowledge in the community regarding hazards that could affect their coasts, such as erosion, subsidence and sea-level rise, and a number of awareness activities have been organised throughout the project. Transferring technical and analytical skills to traditional stakeholders is also an integral part of the project, as it is a mandatory requirement for making this kind of initiative sustainable.

In order to achieve this aim, we opted for a special measuring instrument (Fig. 15.4) inspired by the Emery (1961) method as adopted and improved by Troadec (1991, 2002) and used for research into coral-beach water and sediment dynamics on Reunion Island (Troadec 2002; Cordier 2007) as well as for a thesis on integrated coastal management in the Comoros by Sinane (2013). This "topometer" (Sinane 2010, 2013) is a simple instrument (Fig. 15.4) that provides highly satisfactory results based on the general principles of field surveying for measuring the difference in elevation between two points. It is an appropriate solution for this type of participatory experiment, as it is very cheap and can be used to involve local communities in data acquisition and coastal dynamics observation and, in the process, impart vital knowledge for gradually, but sustainably, building a genuine hazard-mitigation culture while at the same time giving local communities the opportunity to make their own building choices in a more relevant manner. The aim is to develop an integrated coastal management method that enables local communities to strengthen their climate-change adaptation capacities.

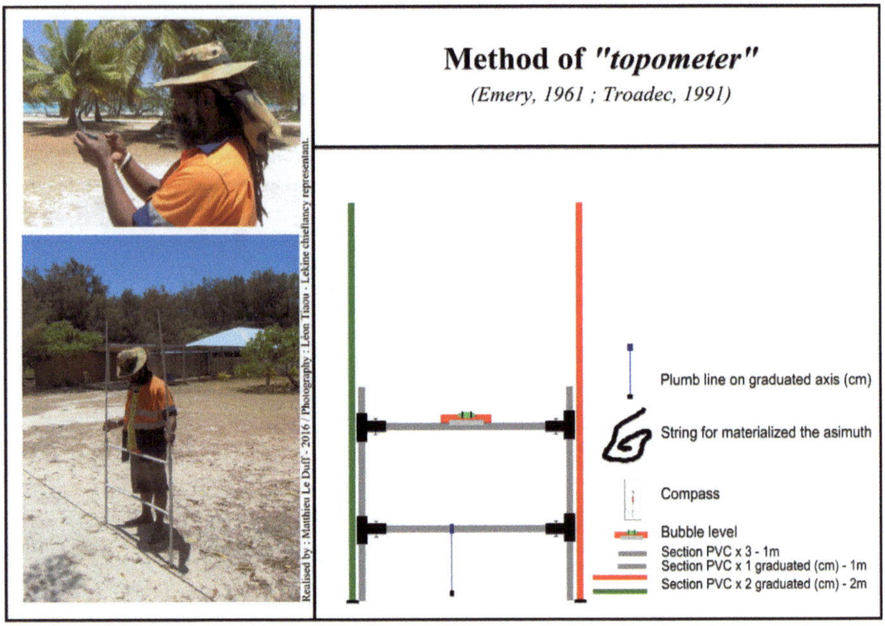

Fig. 15.4 The data acquisition instrument, i.e. a version of the "topometer" (Emery 1961; Troadec 1991)

The version of the "topometer" we proposed is made up of a 1 m/1 m PVC frame (Fig. 15.4) with hollow vertical bars into which fairly wide centimetre-graduated 2 m PVC tubes are inserted to serve as target rods. A horizontal level is fitted to the upper crossbar and the opposite crossbar is also graduated and fitted with a plumb-line for determining the exact location of grade changes in a section. By using a compass and standing directly above a marker, the initially-defined bearing can be found, as it is perpendicular to the shoreline. The bearing is then indicated with a cord placed on the ground. The measuring pace is defined by the "topometer" size, in this case every meter. Each measurement is recorded on a field log-sheet and then entered into the database and processed using Profiler 3.1 software (Cohen 2014).

Limitations and Constraints

The topometer-acquired data were then compared with GNSS surveys based on a similar measuring protocol (Fig. 15.5a). For the same bearings and measuring pace, the topometer data correlated closely with the GNSS survey data. The topometer's vertical error margin can be estimated at ±0.5 cm (Sinane 2013) for each measurement. The method's main drawback is that occasional reading errors are aggregated. The longer the profile, the higher the risk of compounding errors.

Such discrepancies in measurements, particularly in the lower parts of profiles (Fig. 15.5b), have been noted by other studies using this type of instrument (Emery 1961; Delgado and Lloyd 2004; Sinane 2010, 2013). To reduce errors related to the rod's subsiding into the sand, a larger leg was added to the ends of each target rod to keep them from sinking in. Errors arising from operator readings or rounding-off can be reduced by using a finer graduation, such as half-centimetres or millimetres. Once again, however, this deviation trend in profiles as compared with reference profiles (GNSS) is more pronounced in some, though not all, of the longest profiles (Fig. 15.5a) and, when it does occur, only very slightly affects the relevant profiles' morphology (Fig. 15.5b), so qualitative comparisons between the morphotypes can be maintained (Sinane 2010, 2013).

Obviously, this type of instrument is not as accurate as GNSS nor can data obtained with it be as thoroughly processed, but it is nevertheless highly appropriate, because it is easy to use and can be readily reproduced by the communities themselves in participatory observatories. The lack of any data at all on coastal areas is particularly detrimental to communities and research organisations do not have sufficient funds to conduct major field surveys. Involving communities in producing data not only helps them break away from the fence-sitting, ossifying and disempowering mind-set they are sometimes caught up in but also allows them to become active players in reducing their climate-change vulnerability and knowledge and skills that are useful and relevant for day-to-day use in coastal areas can be sustainably passed on to them.

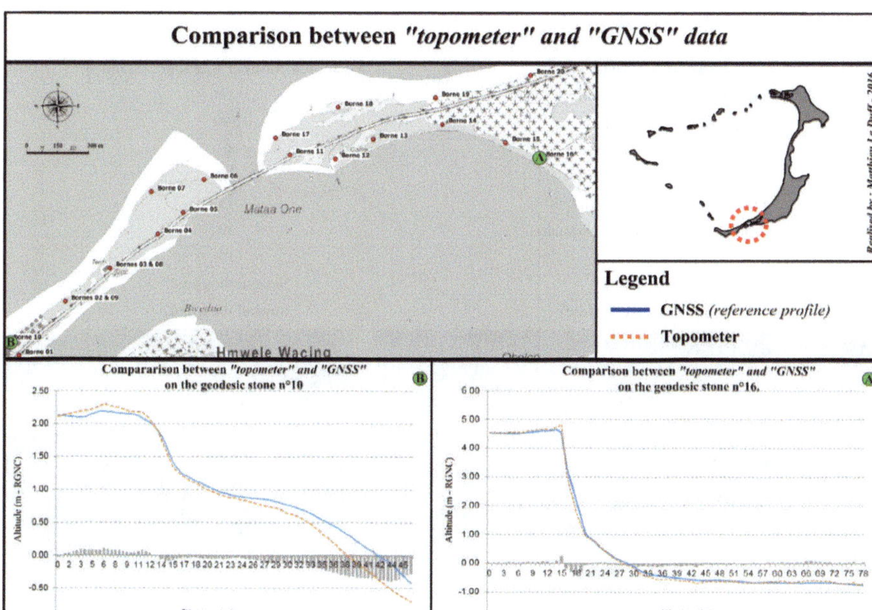

Fig. 15.5 Comparisons between GNSS and "topometer" data in different profiles

Developing and Disseminating Acquired Knowledge

Against a backdrop of economic development in which the coast's attractiveness to tourists is a genuine asset for communities (Dumas 2004, 2009), knowing about sediment transport is a prerequisite for effectively managing such development, not only for sustainable environmentally-sound development, but also for the long-term prospects of financial investment in tourist infrastructure building. As such, studying coastal dynamics is part and parcel of integrated coastal management.

The monitoring network or future observatory's role will also be to define an interface for liaising between the local, government, traditional and scientific stakeholders that is conducive to discussion and understanding areas where not just physical but also social and cultural dynamics are intertwined. The observatory, as a platform for disseminating "expert" knowledge in the community and schools, will be a means of passing on the acquired knowledge through awareness activities such as conferences held in tribal areas or at tourist, church or community events, local press articles and radio shows and by developing teaching aids and holding introductory workshops on the issues. The workshops would be both for coastal-area professionals and users in the broadest sense and involve local associations, such as environmental groups, and church networks (Sunday schools and conventions, etc.)

During our research, we organised several activities (Fig. 15.6), such as public meetings and workshops that were opportunities to reach out to several hundred people who are directly affected by the receding shoreline. They are gradually becoming more aware of how the current dynamics work, particularly in terms of

Awareness Activity	Estimated Number of People Reached
Guillaume Douarre Junior Secondary School	25
Protestant Convention	500
Public Meeting (Heo)	80
Public Meeting (Hwadrilla)	80
Public Meeting at the town hall	15
World Heritage Forum (Gossanah)	50
Study group	10
Radio show (Djido)	Several thousand?
TV show (news on NC1ere)	Several thousand?

Fig. 15.6 Table of activities held and a photograph taken during the workshop held in the Hwadrilla tribal area (April 2015 as part of the R2C3 research programme)

timing, and the need to lay groundwork immediately for discussions that will guide development in tribal areas in the years to come. The situation on Ouvea is very different from that in the Micronesian atolls in many respects, as there is as yet no urgent need to displace communities but rather a need to examine past and present physical, social and cultural dynamics. Also, the mistakes committed in previous building projects need to be corrected as a matter of priority and questions asked about individual as well collective practices that affect the coast.

Conclusion

In terms of climate-change adaptation, there are many advantages to a research-action approach to coastal erosion monitoring that leads to developing an observer network.

The main results are

At the governance level

- The transfer of knowledge and sharing information between "experts", "communities" and managers drive change in the governance approach.
- The community and traditional stakeholders are identified and acknowledged as key players in the management process without casting doubt on government responsibilities.
- The observer network also links up all the local stakeholders, paving the way for planning awareness, information and education activities on the erosion processes currently affecting the island's shores. Disseminating the knowledge to the widest possible audience is a key lever for ensuring the management options adapted by the community and managers remain sustainable in the long term.
- The empowerment of local people take to strengthen their ability to adapt.

At the scientific level

- The approach is a way of offsetting the lack of data that so cruelly hinders our understanding of processes brought about by global warming.
- It is also vital to set up a database in the medium to long-term on selected sites to ensure the sustainability of the measures taken.
- While it is true that the suggested data acquisition protocol is simple and the information it produces must be treated with caution because of its limitations.
- It is, nevertheless, cheap and easy to use, opening up opportunities for developing whole series of measures in many geographical areas where the lack of data currently stands in the way of any substantial and truly relevant discussion.

This approach to integrated coastal management (ICM) is a little unusual, as it no longer aims to just form partnerships with government authorities or sometimes self-styled "official" representatives of civil society such as certain NGOs (Lavaud 2011). Unfortunately, local citizen groups and association networks are very seldom involved in that type of approach (Meur-Ferec 2007). Genuinely involving communities and traditional authorities in coastal management decision-making, analysis and management processes, especially for long periods extending beyond one-off research programmes, is something of a novelty.

Perspectives
The central issue of keeping such a system sustainable remains unaddressed, however, as an observatory has no real meaning unless it has a long lifespan

- Setting up an association on Ouvea, as planned, could be a first step in the right direction.
- The second step would be to integrate this type of participatory approach into a territorial observatory. His main objective could be making it possible to centralise the data and then process, analyse and provide feedback on such data to local communities and managers.
- A longer-term vision sustainability could involve the membership in a coastal surveillance network of international scope.

The method's simplicity and adaptability have already caught the attention of World Heritage management committees on Ouen Island and the Isle of Pines, New Caledonia, where erosion affects populated coastal areas.

In conclusion, against a backdrop of climate change, ensuring the sustainability of cross-sectoral structures that combine scientific methods and local stakeholder involvement through jointly developed strategies could be an area that needs to be given priority in a territory such as New Caledonia or further afield in Small Island Developing States (SIDS) in the Pacific and beyond in order to understand the local implications of this global disruption and provide better solutions.

Acknowledgements We would like to acknowledge the traditional authorities of Saint Joseph (Heo), Takedji, Mouli and Gossanah districts as well as the people and government authorities of Ouvea for their contribution to this project. We would particularly like to thank Mr. Fayssen Wea, Mr. Auguste Omniwack, Mr. Léon Tiaou, Mr. Zéphyrin Touhoulou, Mr. Alibi Ouaignepe, Mr. Jacques Adjougniope, Mr. Wasasea Adjougniope, Mr. André Wea, Mr. Jean-Baptiste Dao, Mr. Benoit Tangopi, Mr. Jean-Luck Majele, Mr. Raymond Ouckewen, Mr. André Houquet, Mrs. Céline Touet, Mrs. Yolaine Bouteiller, Mr. Luen Iopue and Mr. George Kakue. We are also grateful to the Ouvea municipal authorities and their departments, the Loyalty Islands Province, ASBO association, Bomene Tapu GDPL, Iaai Traditional Area, SPC, French Ministery of Overseas Territories (MOM) and the European Union (EU).

References

Allenbach, M. (2013). Changement climatique et migration dans le pacifique insulaire, Ed. Outre-terre, n° 35–36, pp. 505–526. ISSN: 1636-3671. doi:10.3917/oute.035.0505.

Allenbach, M. & Hoibian, T. (2006). Enjeux, spécificités et perspectives de la gestion intégrée du littoral dans l'espace insulaire français du sud-ouest Pacifique, Ed. Paralia, *IXèmes Journées Nationales Génie Civil – Génie Côtier, 12–14 septembre 2006, Brest,* pp. 265–289.

Allenbach, M., Hoibian, T., Touraivane, T., Bantos, S., Worliczek, E., Malau, A. et al. (2014). Les actions de l'IFRECOR relatives au changement climatique en milieu récifal. Application à la gestion intégrée du littoral des îles Wallis et Futuna (Pacifique Sud-Ouest). Ed. Paralia, *XIIIeme journées Nationales Génie Côtier – Génie Civil, Dunkerque, 2–4 juillet 2014*, pp. 791–798. doi:10.5150/jngcgc.2014.087.

Cordier, E. (2007). Dynamique hydrosédimentaire du récif frangeant de l'Hermitage/La Saline (La Réunion): Processus physiques et flux sédimentaires. Ph.D. thesis in physical oceanography, sedimentology, under the direction of Coudray J and Millet B, U.F.R. Science and Technology, University of La Réunion, 208 pp.

Cohen, O. (2014). Profiler 3.1 XL, un logiciel gratuit pour la construction et l'analyse de profils topographiques dans Microsoft Excel. 061 Ed. Paralia, *XIIIeme Journées Nationales Génie Civil- Génie Côtier, Dunkerque, 02–04 juillet 2014*, pp. 557–564. doi:10.5150/jngcgc.2014.

Delgado, I., & Lloyd, G. (2004). A simple low cost method for one person beach profiling. *Journal of Coastal Research, 20*(4), 1246–1252. West Palm Beach (Florida). ISSN: 0749-0208.

Dumas, P. (2009). La gestion du littoral en Nouvelle-Calédonie. In *De la Nouvelle-Calédonie au Pacifique*. Eléments de recherches en Lettres, Langues et Sciences Humaines. Coll. Portes Océanes. Ed. L'Harmattan, pp. 53–78.

Dumas P. (2004). Caractérisation des littoraux insulaires: approche géographique par télédétection et SIG pour une gestion intégrée. Application en Nouvelle-Calédonie. Ph.D. thesis in Geography, University of Orléans, 404 pp.

Emery, K.-O. (1961). A simple method of measuring beach profiles. *Limnology and oceanography, 6*(1), 90–93. doi:10.4319/lo.1961.6.1.0090.

Garcin, M., & Vendée-Leclerc, M. (2014). Observatoire du littoral de Nouvelle-Calédonie – Rapport préliminaire: observations, état des lieux et constats. Ed. BRGM, Rapport BRGM/RP-63235-FR, 125 pp.

Garcin, M. & Vendée-Leclerc, M. (2015). Observatoire du littoral de Nouvelle-Calédonie – Bilan des activités 2014. Ed. BRGM, Rapport BRGM/RP-64824-FR, 145 pp.

Goarant, A.-C., Bouteiller, Y. & Le Duff, M. (2015). La gestion intégrée des zones côtières: pour une meilleure gestion de l'environnement au bénéfice des populations. Présentation dans le cadre du séminaire de l'UICN «Solutions fondées sur la nature et changements climatiques», Paris le 14 avril 2015.

Herrenschmidt, J.-B. (2004) *Territoires coutumiers et projet de développement en Mélanésie du sud (îles Loyauté, Vanuatu, Fidji)*. Ph.D. thesis in Geography, University of Paris IV, 656 pp.

IPCC. (2013). Summary for policymakers. In T. F. Stocker, D. Qin, G.-K. Plattner, M. Tignor, S. K. Allen, J. Boschung, A. Nauels, Y. Xia, V. Bex & P. M. Midgley (Eds.), *Climate Change 2013: The Physical Science Basis.* Contribution of Working Group I to the Fifth Assessment Report of the Intergovernmental Panel on Climate Change. Cambridge, New York: Cambridge University Press.

IPCC. (2014). Summary for policymakers. In O. Edenhofer, R. Pichs-Madruga, Y. Sokona, E. Farahani, S. Kadner, K. Seyboth, A. Adler, I. Baum, S. Brunner, P. Eickemeier, B. Kriemann, J. Savolainen, S. Schlömer, C. von Stechow, T. Zwickel & J. C. Minx (Eds.), *Climate Change 2014: Mitigation of Climate Change.* Contribution of Working Group III to the Fifth Assessment Report of the Intergovernmental Panel on Climate Change (32 pp.). Cambridge, New York: Cambridge University Press.

ISEE. (1996). Structure de la population et évolutions. L'évolution de la population par communauté d'appartenance. http://www.isee.nc/population/recensement/communautes.

ISEE. (2014). Structure de la population et évolutions. La population légale en 2014. http://www.isee.nc/population/recensement/structure-de-la-population-et-evolutions.

Jeanson, M. & Dolique, F. (2011). An observatory for monitoring the coastal dynamics under extreme meteo-marine conditions. In *Symposium Vulnerability of coastal ecosystem to global change and extreme events*, Biarritz, France, 18–21 October 2011, Abstracts and Proceedings, 38 pp.

Jeanson, M., Dolique, F., & Anthony, E. (2010). Un réseau de surveillance des littoraux face au changement climatique en milieu insulaire tropical: l'exemple de Mayotte. *VertigO, 10*(3), 22. doi:10.4000/vertigo.10512.

Merle I. (1999) La construction d'un droit foncier colonial: De la propriété collective à la constitution des réserves en Nouvelle-Calédonie, *Enquête* (pp. 97–126). doi:10.4000/enquete.1571.

Meur-Ferec, C. (2007). La GIZC à l'épreuve du terrain: premier enseignements d'une expérience française. *Développement durable et territoires* [En ligne], Varia, mis en ligne le 07 janvier 2007. doi:10.4000/developpementdurable.4471.

Lavaud, J. (2011). La taxe de bienfaisance. *Multitudes,2011/4*(47), 92–96. doi:10.3917/mult.047.0092.

Leblic, I. (1991). Représentations du foncier en Nouvelle-Calédonie et identité culturelle kanak. *Evolution de ces représentations liée à la revendication identitaire et aux processus de développement économique. 3e journées scientifiques de la Société d'écologie humaine, Perceptions et représentations de l'environnement, Novembre 1991*, Aix-en-Provence, France, 25 pp.

Le Duff, M. (2012). *Rapport sur le risque Tsunami à Ouvéa: état des lieux et préconisations.* Rapport administratif interne SAIL (Subdivision Administrative des Iles Loyauté), juillet 2012, Lifou, 29 pp.

Le Duff, M. (2013). *Rapport de synthèse: prévention du risque tsunami aux îles Loyauté.* Rapport administratif interne SAIL (Subdivision Administrative des Iles Loyauté), octobre 2013, Lifou, 26 pp.

Le Duff, M., Allenbach, M., Dumas, P. & Cohen O. (2015) Shoreline evolution in Ouvea island (New Caledonia). In *A diachronic approach using (past) long term data. A method for a middle term in field measurements for the future.* Bridging Information Gaps By Creating Smarter Maps, 16th–19th November 2015, USP Japan ICT Auditorium, Suva, Fiji Islands.

Le Duff, M., Dumas, P., Cohen, O. & Allenbach, M. (2016). Observation et suivi de l'érosion côtière à Ouvéa (Nouvelle-Calédonie): exemple d'une démarche participative. In *Colloque LITEAU: Observation et recherche en appui aux politiques du littoral et de la mer/Observation and research in support of ocean and coastal policies.* Affiliation: Centre de l'IFREMER, Brest, France, janvier 2016. doi:10.13140/RG.2.1.3701.3526.

Sinane, K. M. (2013). Les littoraux des Comores, dynamique d'un système anthropisé: le cas de l'île d'Anjouan. Ph.D. thesis in Geography, University of La Réunion, 296 pp.

Sinane, K. M., David, G., Pennober, G. & Troadec, R. (2010). Fragilisation et modification des formations littorales meubles sur l'île d'Anjouan (Comores): quand l'érosion d'origine anthropique se conjugue au changement climatique. *VertigO - la revue électronique en sciences de l'environnement, 10*(3). doi:10.4000/vertigo.10528.

Troadec, R. (1991). Courantologie et sédimentologie des baies de Saint-Paul et de la Possession à l'île de La Réunion. Ph.D. thesis in Geography, University of Aix-Marseille II, 217 pp.

Troadec, R., et al. (2002). Suivi de l'évolution des profils des plages coralliennes de la Réunion 1998–2001. Internal report - LSTUR, University of La Réunion, 20 pp.

Part III
Information, Communication, Education and Training on Climate Change

Chapter 16
Towards a Mega-Pacific Islands Education Curriculum for Climate Adaptation Blending Traditional Knowledge in Modern Curriculum

Mohamed Walid

The Devil at Lonburas
There was once a devil (named Ias) who lived in a cave outside Lonburas Village. Every day he came to the village when food (lap-lap) was being prepared and consumed huge amounts, including the tables on which it was served. The villagers
decided to rid themselves of the devil, so questioned him to try to discover what scared him. Ias revealed his fear to be
saltwater, so the villagers went down to the bay and collected saltwater in a special kastom leaf, brought it back to the Kastom Man in Lonburas, who poured it into a hole in a black palm tree. The Kastom Man sang a song that made the saltwater
rise up inside the tree and urge the ocean to come and join it. When the song ended the villagers heard the sound of a wave rushing towards them and felt a strong wind. Alarmed, the
devil awoke and asked the villagers what the noise was. The villagers joined their Kastom Man singing the song again and again until the saltwater reached Lonburas. The devil climbed a banyan tree in fear, but the Kastom Man sang the song one last time, and then the ocean swallowed the devil. Then the villagers retrieved the devil from the ocean, cut him, and retrieved all the
tables he had swallowed.
(A traditional Pacific Island story narrated by a friend in Palau; found also in text in Walshe and Nunn 2012).

M. Walid (✉)
London School of Economics, Houghton Street, London WC2A 2AE, UK
e-mail: M.Walid-Lotfy@lse.ac.uk

W. Leal Filho (ed.), *Climate Change Adaptation in Pacific Countries*,
Climate Change Management, DOI 10.1007/978-3-319-50094-2_16

Introduction

Naturally occurring climate change has affected the Pacific Islands for centuries. More recently, human activity also has contributed to climate change (Cambers and Diamond 2010). Consequently, these islands face potential destruction (UNOHRLLS 2009), as described in the Fourth Assessment Report of the Intergovernmental Panel on Climate Change (2007). Tuvalu, the fourth smallest nation in the world, is expected to be the first nation to be submerged in the ocean as a result of global climate change (Levine 2002).

Pacific Islanders have found sustainable ways to adapt to environmental changes and innovative approaches to climate change adaptation (UNESCO 2013). They have longstanding indigenous knowledge in environmental resource management systems and traditions of story-telling, oral history and art symbolic representations that can help researchers develop culturally appropriate strategies for climate adaptation. For example, Campbell (2006) documented positive traditional responses to natural disasters. Möhlendick (2009) and Percival (2008) dissected aspects of traditional knowledge in relation to climate change. In Māori communities, preliterate knowledge has been used to address contemporary challenges when modern ways fail (New Zealand Library 2014).

The United Nations Educational, Scientific, and Cultural Organization (UNESCO) recognizes the link between traditional knowledge and effective learning for sustainable living (2007). Compiling and sharing this knowledge is beneficial to students and the wider community. Principle 4 of the Pacific Islands Framework Environment Programme calls to educate, train, and raise awareness in the respective Pacific Islands (Secretariat of the Pacific Community 2010). Equally, article 6 of the United Nations Framework Convention on Climate Change states that climate change education should instil "awareness and understanding at a young age," as it is "ultimately the best way to change behaviors and attitudes" and that it should be "oriented to the local context and prioritize passing traditional knowledge and practices to learners" (UNESCO 2011, p. 5).

To that end, this paper looks into previous and current education curricula related to climate change. In doing so, it evaluates how the traditional learning methods on these islands are used and can be integrated in a Pacific Islands-wide climate adaptation curriculum. Ideally, principles underlining this Pacific-wide curriculum are to be inclusive to the views of the Pacific population. It should give autonomy to teachers and local communities so that they can adapt it to local contexts. It should promote creative problem-solving in students in order to be empowered by climate change and not against it (Lawler 2011; UNICEF 2013). The intended curriculum should raise environmental awareness among Pacific youth in a manner that appeals to them (Minowa 2015). Finally, the curriculum should use local communication styles and local languages in disseminating the education information.

Climate Change: Traditional Knowledge, Modern Science, and Perceptions

Discussions about how climate change affects Pacific Island nations usually focuses only on its imminence and intensity. However, Pacific Islanders have been adapting to natural changes for centuries (Vize 2012). They perceive climate change as a natural part of life. For example, the indigenous Ni-Vanuatu cultures consider natural disasters to be rather normal or at least occasionally expected (Galipaud 2002). The 1999 tsunami that struck Pentecost Island in Vanuatu reached heights of more than six meters yet caused only five fatalities in a population of 300, because most residents of Baie Martelli knew to leave, according to local custom (*kastom*). Interviews with 55 persons from the devastated area indicated that community leaders and others could sense the coming event (Walshe and Nunn 2012). International organizations admit to this fact as well. Director of PAHO/World Health Organization's (WHO) Emergency Preparedness and Disaster Relief programme poetically says, "Who knows more about their lands and their environment than the people who live closest to it?" (ReliefWeb 2015, p. 1).

Barnett and Campbell (2010) identify a gap between "inside and outside knowledge." The inside knowledge of preliterate communities is transmitted using traditional learning methods, such as story-telling (Walshe and Nunn 2012). Outside knowledge, such as that from international research projects, usually lacks information about the local environment and ways of life (Baba 2008; UNESCO 2015). Yet, these outside organizations often influence curriculum design. Additionally, despite the demise of political control on the Pacific Islands, international aid organizations and locals often disagree about what is relevant and necessary information to be taughted among the Pacific population. This gap of knowledge and tensions created therefrom help explain why climate adaptation discussions in the region focus on generic vulnerabilities rather than specific regional issues.

To bridge this gap, the focus should be on assembling knowledge that is unique to each island, so that local communities can better understand pending disasters and prepare themselves for climate change in a sustainable manner that respects local traditions and customs. This traditional knowledge can make an even more substantial contribution to environmental management than non-indigenous, science-based externally imposed systems which may not address the unique needs of these islands (Davies 2015; McAdoo et al. 2009).

Global climate change poses a serious threat to this region (Dakaica 2005). Thus, it is imperative that future generations understand climate change in a way that is meaningful for this region. In the 2014 UNESCO project, *Sharing Perceptions of Adaptation, Resilience, and Climate Knowledge* (SPARCK), researchers worked with grassroots communities to gather information about climate change perceptions. In Samoa, 65% of teachers who were surveyed thought capacity-building workshops would be an effective method for teaching about global climate change. In Vanuatu, teachers preferred combining traditional knowledge with climate change science.

Whereas 98% of Samoan teachers thought that climate change education was of critical importance, only 48% of Vanuatu teachers thought so (UNESCO 2014a). In Fiji, 80% of communities were willing to take immediate action to counter climate change; in Vanuatu, 73% were willing and in Samoa, 64% (ibid). Respondents to the survey expressed concern that their ways of life would be negatively impacted by climate change. Their responses emphasized a point to consider; that is, climate education should improve adaptation skills rather than abstract knowledge (Lawler 2011).

The 2014 UNESCO project showcases the importance of evidence-based science for measuring Pacific Islanders' perceptions of climate change. However, it does not investigate reasons for the varied responses and disparate education levels across the islands. The lack of high-quality education in general and climate change education in particular might explain the relatively poor coping strategies in these areas (GIZ 2016). Nevertheless, the variations in perception regarding climate education seem to be grounded in the communities' understanding of climate change.

Challenges and Opportunities for a Pacific Islands Climate Adaptation Curriculum

Having explored the relationship between traditional knowledge, modern science, and perceptions of Pacific Islanders towards climate adaptation education, additional obstacles hinder the implementation of climate education in this region. First, climate change topics have not been effectively integrated into their school curricula (Hartmann et al. 2010). In Vanuatu, teachers state that climate change education is largely absent from the school curriculum. In Tuvalu, climate education is taught only in science and geography (UNESCO 2015). Even topics that are pertinent to climate change do not draw connections to local climate change (UNESCO 2014a). That is, greenhouse effects are discussed in the abstract, rather than drawing on the regional history of climate change that would allow students to visualize the accumulated, local effects of climate change. Additionally, teachers note that climate change must be integrated at the junior level in schools to increase learning opportunities.

Building on the first point, children's perceptions and voices should be included. As one study indicates, children are able to perceive climate change better than their parents (Notaras 2011). This study might require further inquiry as to the extent of how children comprehend the significance of climate change. Nevertheless, what stands out more confidentially is that children are aware of their environment given its direct influence on their daily activities as in playing or going to school.

Third, this region lacks adequate teacher training on climate change. UNESCO identified teacher training as the "priority of priorities" in its Education for Sustainable Development agenda (UNESCO 2011). Educational resources are scare, but teachers can provide alternative learning activities, such as escorting students on field trips to study local climate change impacts. Such activities may increase awareness on the impact of climate change in the wider community of children (UNESCO 2014b). These trips would also expand the realm of climate education beyond the classroom. The fact that until recently around 40% of school children in the Pacific Islands did not finish basic education showcases the opportunity as well as the necessity to target the learning of climate change outside of school (Young 2011).

Fourth, climate change education competes with educational agendas from organizations that prefer to focus on other educational topics (e.g., health education). These agendas draw attention away from climate change education (Vize 2012). Coordination between the various educational organizations must allow climate change education to be incorporated into the curriculum. A cooperative, multidisciplinary approach can combine climate change course work with other topics. For example, the University of the South Pacific offered a summer course in 2011 that combined human rights with climate change (Vize 2012).

Fifth, language differences can present a communication barrier regarding perceptions on climate change. For example, in Fiji, highly technical information on climate change is usually delivered in English, which is not most Fijian's first language. Delivering such information in the Fijian language, iTaukei, and using traditional story-telling techniques in each local language would help to disseminate climate change science more effectively (UNESCO 2014a). In Tonga, which has a strong oral history, traditional media can be integrated with scientific content to inform and educate (Taylor 1995). For example, a story that teaches about volcanic eruption could be presented as a story in which a sorcerer ignites a volcano magically. In Palau, the legend of the giant Uab, who ate so much that the island on which he lived started to sink, could be used (PBS Learning Media 2016). Story-telling can be used to teach scientific and health principles, including issues of obesity, for example, which is prevalent among Pacific Islanders and which has a direct and indirect relation to climate change (WHO 2010).

Finally, uniting the community and the classroom is important in the discussion of climate change. The Sparck project, as shown in Table 16.1, demonstrates how dialogue between communities and teachers leads to innovative activities that could be implemented in climate change curriculum across the Pacific (UNESCO 2014a). The Pacific Institute also uses community-based participatory approaches to bridge various segments of local society for climate education purposes (Pacific Institute 2016).

Table 16.1 Climate education innovative activities for students resulting from community-teacher discussions

Community	Teachers	Resulting student activities
Deforestation	Reforestation	Tree planting
Water tanks	Water purification	Water conservation
Solar system	Distillation	Energy conservation

Note Data compiled from the Sparck project (UNESCO 2014a)

Traditional Knowledge: Pedagogical Medium and Thematic Content

Most communities across the region agree that integrating climate change into the educational curriculum is important (UNESCO 2014a). This paper advocates for an integrated curriculum, designed specifically for the Pacific Island nations, that compiles and disseminates traditional and scientific knowledge using a variety of media formats (e.g., electronic, books, etc.). Cost-benefit analysis to this culturally aware and resourceful cross-Pacific curriculum begets lower transactional cost. For instance, cost of printing would be shared among countries/communities/schools etc. across the Pacific or in between some of them sharing the very same climate issues—as in the case of low atoll islands, facing similar threats that could be taught to children across this cluster of islands. By virtue of this, students, parents, teachers and the wider communities would learn about other communities facing similar as well as island-specific threats. For example, three weeks before the November 1999 tsunami in Baie Martelli, Vanuatu, the National Disaster Management Office in Vanuatu showed villagers a video about the 1998 Aitape tsunami in Papua New Guinea. This video, along with local knowledge, contributed to the low fatality rate in Vanuatu after the 1999 tsunami (Walshe and Nunn 2012).

On that note, worth-reiterating, although the Pacific Islands share many common features, they often have unique ecosystems (FAO 2008). The localization of climate change curriculum must account for these differences. For instance, Vanuatu is located in the southwestern Pacific Ocean is vulnerable to most tropical cyclones. It cannot be treated like Kiribati, which suffers some other different ecological challenges (UNICEF 2011a). Traditional, local knowledge is thus a key component in the design and implementation of the curriculum. *Traditional knowledge as a pedagogical medium* of education communication can be helpful for tailoring a Pacific Islands curriculum that celebrates indigenous wisdom (Smith and Jones 2008). The curriculum would incorporate each island's culture, geography, and ecosystems using *traditional knowledge as thematic content.* Thematic content includes stories, traditions, and other local knowledge. For example, participants from Vanuatu suggested blending traditional knowledge about climate change and scientific data into the curriculum as part of a series of workshops organized by UNESCO under the Education for Sustainable Development initiative (UNESCO 2014c). In Kiribati, participants wanted to focus on developing capacities of teachers to explain coastal erosion and loss of biodiversity to kids (ibid). These local issues would be incorporated in sections of the curriculum that are specific to the respective island.

Pacific Islands Climate Adaptation Education Attempts

A few fragmented attempts have been made to merge climate change adaptation education into the Pacific Islands curricula. Three significant ones are discussed in this section. The first was the 2001 Scientific Educational Resources and

Experience Associated with the Deployment of Argo Drifting Floats in the South Pacific Ocean (SEREAD) project. It generated awareness among Pacific communities, teachers, and students on topics such as sea-level rise and its local impacts (Partnership for Observation of Global Oceans 2001). However, a major criticism of this project is that it was limited to secondary school students. Since SEREAD, few programmes have offered regional inclusiveness.

The second attempt was the climate change educational resource, *Learning about Climate Change the Pacific Way*, launched in 2014 by the Secretariat of the Pacific Community and the German Deutsche Gesellschaft für Internationale Zusammenarbeit. The Pacific Islands Climate Education Partnership (2016) initiated the *Learning about Climate Change the Pacific Way* project in the Pacific islands, using the following guidelines:

- Local knowledge offers time-honored wisdom about climate change.
- Scientific knowledge should inform adaptation and mitigation actions in concert with Pacific Island cultural values.
- Traditional Pacific Island cultural values that identify with and respect the natural environment should be central to climate change adaptation.
- Formal and community-based educational systems should be included to enhance understanding of climate change and decisions about climate adaptation strategies.

Nevertheless, the German-sponsored project has had trouble balancing local and regional context. The curriculum was produced for Fiji, Kiribati, Samoa, Vanuatu, Tuvalu, and Tonga, using 16 pictures to illustrate the causes of climate change, along with ways to adapt and mitigate its impacts on these six islands (Secretariat of the Pacific Community 2011). The eighth picture is specific to the island in-focus; all of the other pictures are applicable to all of the six islands, thus representing all Pacific Islands as one imaginary island called "Pasifika" that lacks any specific details about individual island ecosystems. Although the curriculum accounts for the power of tradition and uses pictures to communicate ideas directly, it omits traditional pedagogical methods of communication, such as story-telling. The project also does not cover all grades. Additionally, as this project relies heavily on top-down communication through government documents, it does not consider equally voices from the local communities. Last, it uses a web-based platform, which is not accessible to all islanders. Nevertheless, the Deputy Director General of the Secretariat of the Pacific Regional Environment Programme (SPREP) commended the programme and the fact that workshop participants were from different islands (SPREP 2014). He urged participants to share their newly acquired knowledge with their home islands. Unfortunately, without a deeper appreciation of the local contexts of each island and without the inclusion of all students, this project may not obtain its intended outcome.

Third, SPREP developed the 2013 initiative, *Children Take Action—A Climate Change Story*. Unlike the first two projects, *Children Take Action—A Climate Change Story* is not regional. It uses stories instead of pictures to teach children the basics of climate science in a simple and appealing manner. The stories are collected in a book that was made specifically for Kiribati and thus written in both the local Te-Kiribati language and in English (SPREP 2013). The dual languages ensure effective communication in a local language and help to teach English, as in the following excerpt:

> Jone didn't know what climate change was and asked his grandfather to explain. Grandpa told Jone that the Earth's temperature is becoming hotter.
>
> "My temperature gets hot when I am unwell," said Jone.
>
> "Yes!" said Grandpa. "The Earth is becoming unwell too. There is less food for the birds and the fish. That is why they are leaving our island."
>
> "What is making the Earth sick?" Jone asked.
>
> "We are," said Grandpa. "Gases from our cars, buses and factories are making the Earth too hot."
>
> "People are driving more cars and building more factories. So the Earth is getting hotter and hotter."
>
> "Just like putting too many blankets on me!" said Jone (SPREP 2012, p. 5).

This paper calls for a more inclusive curriculum for the Pacific Islands Region. At their 2011 meeting about climate change education in formal and non-formal education, the Pacific Islands' heads of education agreed that current endeavours have not yet fulfilled this goal (UNESCO 2012). On the short run, the curriculum should combine these three education efforts, as well as others, in a coordinated Pacific Islands education management network. On the long run, a wholly new Pacific Islands curriculum should be developed.

Methodology

This research paper relies on meta-analysis of previous studies and literature review, as well as email communication with informants from the Pacific Islands Region. The study uses qualitative text-based analysis of primary and secondary materials. In tandem, it provides contextual explanations to quantitatively-conducted studies. Quantitatively, the study uses an ATLAS coding system. It focuses on the conjuncture of two major dimensions in previous and current education programmes: traditional knowledge and climate change and adaptation education.

Limitations

This study has several limitations, which can be classified into two categories: methodological limitations and limitations related to developing and implementing the proposed curriculum. As for the first category, the study was primarily based on a literature review and a meta-analysis of previous studies. However, current documentation of traditional knowledge regarding climate adaptation is limited, and time constraints did not allow for field research. On the second category, attempts to develop Pacific climate adaptation curricula are quite recent. As a result, the chance to evaluate any such climate change curriculum with more accuracy is limited. Developing and implementing the proposed Pacific Islands curriculum is difficult because of the wide dispersal of islands. Additionally, the need for climate adaptation education varies significantly from one locality to another, which increases the cost of developing relevant curricula. Lastly, the study was not able to elaborate fully or equally to the different façades of traditional knowledge given the broadness and depth of traditional knowledge in these nations.

Discussion

Considerable variation exists in climate adaptation education of the Pacific Islands region. Internationally, the focus is on integrating climate adaptation education into aid and development packages that are provided to the region. Regionally, the Pacific Islands Climate Education Partnership promotes climate adaptation education, though political efforts seem slow to include climate change education in major strategies. For example, the *Pacific Culture and Education Strategy* (2010–2015) does not cite climate change education. This less robust commitment in the region is another reason why current Pacific Island education attempts have not yet been implemented (UNESCO 2015). Nationally, governments seem to be less attentive to the urge of explicitly incorporating climate adaptation education into their educational strategies. While for instance in Tuvalu, the 2012–2016 National Strategic Action Plan for Climate Change and Disaster Risk Management was made effective, its implementation has been less clear (UNESCO 2015). Two reasons could explain that. First, Climate-vulnerable islands, such as Vanuatu and Palau, focus in principal on different education goals, such as the Millennium Development Goal of universal primary education (Palau Ministry of Education 2006; Vanuatu Ministry of Education 2006). This is where funds flow more. Secondly, there is less integration of climate education throughout the entire government. Instead, usually Ministries of Education would stand alone responsible for climate education (UNESCO 2015). Locally, communities seem to care about climate education; nevertheless, if their voices, which represent traditional local knowledge, are not directly integrated into curricular projects, international interventions will be far less effective in engaging and appealing to local communities. Hence, an effective Pacific Islands climate change adaptation curriculum must account for the different levels of developmental and sociocultural variances in the Pacific Islands region to promote positive behavioural changes toward climate adaptation (UNESCO 2014c).

The UNESCO-led Sparck project shows that not all community members believe that climate change is imminent (UNESCO 2014a). Local media, such as news agencies can narrow this gap of knowledge between the scientifically literate and climate change skeptics by transmitting climate change-related information in ways that the public can comprehend and relate to. For example, the media might explain how climate change affects tourism in Fiji or how ocean acidification negatively affects fisheries in Palau. Climate education also should be presented as a "smartship" quest, rather than a hardship: Instead of portraying the region as always prone to natural disasters, climate education should be promoted as a means to a smart and better way of living. When the taro leaf blight wiped out the lucrative taro trade in Samoa in 1993, climate adaptation education got recognition as a means to manage the disaster and innovate ways to revitalize the economy. A Samoan climate change toolkit has since been implemented nationally (Peace Climate Change Portal 2012). After the success of a 2012 science fair, themed "Adapting to Climate Change," the science fair with the theme of climate change became compulsory for primary and secondary schools in Samoa (ibid). In this way, threats posed by climate change turn into opportunities for innovation.

With the goal of innovation in mind, this paper advocates incorporating regional goals and local communication methods into the Pacific Islands climate adaptation curriculum. It promotes creative problem-solving and explores entertainment options to engage students from kindergarten through 12th grade, as well as adults. It emphasizes learning through action and interaction, inside and outside of the classroom, and integrating modern scientific methods with traditional knowledge. The UNESCO Sandwatch project is a good example. It presents an opportunity for students to discuss the threats facing beaches and ways to deal with them. Since its launch in 2001 in the Caribbean, about 40 nations have joined, including nations of the Pacific Islands (Sandwatch Foundation 2011).

Despite the range of divergent opinions regarding climate change and adaptation, there is a consensus that preparedness education allows people to cope with global climate change and natural disasters. This education is essential for economic, social, and cultural survival. Implementing a climate adaptation curriculum that is tailored to the Pacific Islands region is a viable prospect. To echo the words of Davidson Hepburn, former president of the General Conference of UNESCO, "Education plays an essential role in increasing the adaptation capacity of communities and nations with regards to climate change" (UNESCO 2012, p. 7). Major initiatives in the Pacific Islands that address sustainable education, such as the Education for Sustainable Development, have flourished as a result of the enormous impact of climate change (Hartmann et al. 2010; Vize 2012). Nonetheless, that climate change adaptation education is not yet recognized explicitly and systematically enough as a vital component of sustainable living, at least on par with other climate adaption mechanisms, indicates that more has be done (GIZ 2016).

Conclusions

This study stresses the necessity of preserving traditional knowledge and heritage in climate adaption education. It is not meant to dismantle the use of modern ways of addressing climate threats, yet it invites its reader to think of the balance in integrating traditional means of education in the Pacific Islands Region with modern science. Thanks to this traditional knowledge, locals in Vanuatu were able to survive the 2015 Tropical Cyclone Pam (Minowa 2015).

Further, the research highlights several lessons and future prospects for academia and policy makers. Although climate change is imminent, international efforts to develop a climate adaption curriculum that is specific to the Pacific Islands Region and its unique local knowledge base are recent developments. The current curricula still do not incorporate enough pedagogical methods (e.g., story-telling, arts, field trips). Thus, there is a risk of losing this valuable knowledge, especially as social, cultural, and economic changes occur throughout the region (Campbell 2006). Future opportunities can be harnessed if the gap between inside and outside knowledge shrinks. Henceforth, this study recommends that the climate adaption curriculum of the Pacific Islands region should be a cooperative project, rather than one that international humanitarian and development organizations impose. The cooperative project should begin with a Pacific-wide ethnographic research about the respective islands (inter- and intra-communities); their perceptions of climate change; detailed transcriptions of adaptation methods and local suggestions for addressing climate change. This includes children and youth in general sharing their views. The product of this Pacific Islands-wide research should be diffused into formal and informal education systems (AusAid 2008).

The design and development of the proposed Pacific Islands curriculum should encourage ways to bring formal education into informal settings in the communities. Of particular importance is that Pacific Islanders perceive climate change differently, compared to other regions around the world. Thus, the paper calls for climate adaptation *with* rather than *against* climate change. This approach seems more valid, given the proximity and frequency of climate threats there to the Islanders, as opposed to those from nations with less climate turbulence. As with previous education attempts, however, implementation costs are high, and is it challenging to find ways to incorporate such information into school curricula, particularly if teachers are not familiar with the material or if communities are skeptical.

This broader approach requires that policymakers revisit and revamp "business-as-usual" model that targets schools in a way that is disconnected from the community, as called for by New Zealand foreign minister Murray McCully (Young 2011). They need to consider the entire community as the school. People's natural habitat should be conceived of as the primary area of effective learning, especially in the less urbanized areas of the Pacific Islands. Otherwise, regardless of the finances invested in projects, kids will either be excluded from climate adaptation education or not effectively approached. Future curricula should include materials that are presented in the native languages of these islands (UNICEF 2013). Moreover, education should focus on humanitarian action in disaster preparedness and risk reduction. Climate adaptation education should permeate all climate-related policies of international agencies and local governments(UNESCO 2007). Indeed, to do so, education should not be treated as a separate component from other policy interventions, especially when it comes to efforts such as child protection (UNICEF 2014). Eventually, by reframing education as a tool and medium that is valid across all policy domains, these education initiatives can leverage the region's rich and diverse traditional knowledge for long-term climate adaptation.

The study finds that few international organizations and academic institutions dedicate resources to climate adaptation education in the Pacific Islands region, leaving a significant gap in the knowledge base between climate change science and traditional means of adaptation (ESCAP 2014). The suggested framework for a regional curriculum in the Pacific Islands represents an opportunity for more tangible academic and policy collaboration. As Dr. Farnafi Larkin of Western Samoa is reported to have remarked, "The only thing Pacific islanders have in common is the water around them" (Baba 2008, p. 125). Water is certainly a shared resource of the islands, but they also have precious traditional knowledge that has tangible and intangible benefits for the well-being of the population (UNESCO 2013).

References

AusAid. (2008). *The gendered dimensions of disaster risk management and adaptation to climate change*. Suva, Fiji: UNDP Pacific Center.

Baba, U. N. (2008). Decolonising framings in Pacific research: Indigenous Fijian Vanua research framework as an organic response. *Alternative an International Journal of Indigenous People, 4*, 2.

Barnett, J., & Campbell, J. (2010). *Climate change and small island states: Power, knowledge, and the South Pacific*. Washington, US: Earthscan.

Cambers, G., & Diamond, P. (2010). *Sandwatch: Adapting to climate change and educating for sustainable development*. Paris, France: UNESCO.

Campbell, J. (2006). *Traditional disaster reduction in Pacific Island communities, GNS Scientific Report, 038*. Institute of Geological and Nuclear Sciences Ltd.: New Zealand.

Dakaica, S. (2005). *The challenges of education for sustainable development and planning in small island developing countries of the South Pacific—a case study of the University of the South Pacific*. Melbourne, Australia: The Australia and New Zealand Association of Planning Schools.

Davies, K. (2015). *Kastom, climate change and intergenerational democracy: Experiences from Vanuatu*. New York, US: Springer.

ESCAP. (2014). *Climate change and migration issues in the Pacific*. Pacific Office, Suva, Fiji: UNESCAP.

FAO. (2008). Climate change and food security in Pacific Island countries, Rome, Italy.

Galipaud, J. C. (2002). Under the volcano: Ni-Vanuatu and their environment. In R. Torrence & J. Grattan (Eds.), *Natural disasters and cultural change*. London, UK: Routledge.

GIZ. (2016). *Coping with climate change in the Pacific Island region*. German Federal Ministry for Economic Cooperation and Development. Retrieved March 26, 2016, from https://www.giz.de/en/worldwide/14200.html.

Hartmann, H., Sabass, H., Tagivakaini, S., & Ries, F. (2010). Climate change education in the Pacific Island states, Report of the SPC-GIZ Preparatory Mission.

Hay, J. (2009). Assessment of implementation of the Pacific Islands framework for action on climate change. SPREP. Apia, Samoa. Retrieved April 3, 2016, from https://sprep.org/climate_change/PYCC/documents/HayReport_to_PCCR_2009.pdf.

Intergovernmental Panel on Climate Change. (2007). *Climate change, impacts, adaptation, and vulnerability*. Contribution of Working Group II to the Fourth Assessment Report of the Intergovernmental Panel on Climate Change. M. L. Parry, O. F. Canziani, J. P. Palutikof, P. J. van der Linden, & C. E. Hanson (Eds.). Cambridge, UK: Cambridge University Press.

Lawler, J. (2011). *Children's vulnerability to climate change and disaster impacts in east Asia and the Pacific*. UNICEF: Bangkok, Thailand.

Levine, M. (2002). Tuvalu Toodle-oo. Outside Online News. Retrieved February 11, 2016, from http://www.outsideonline.com/1821146/tuvalu-toodle-oo.

McAdoo, B. G., Moore, A., & Baumwoll, J. (2009). Indigenous knowledge and the near field population response during the 2007 Solomon Islands Tsunami. *Natural Hazards, 48*(1), 73–82.

Minowa, T. (2015). Local knowledge saves lives in Vanuatu. UNISDR. Retrieved April 8, 2015, from https://www.unisdr.org/archive/45870.

Möhlendick, K. (2009). *Traditional Samoan methods that can help to adapt to climate change with a focus on food security*. Apia: Unpublished report prepared for UNESCO.

New Zealand Library. (2014). Preliterate Māori knowledge frameworks as postliteracy strategies. *Information Management Journal, 54,* 4.

Notaras, M. (2011). Climate change is not a child. Tokyo, Japan: United Nations University. Retrieved April 10, 2016, from http://ourworld.unu.edu/en/climate-change-is-not-childs-play.

Pacific Institute. (2016). Community resilience on climate change. Oakland, USA. Retrieved March 25, 2016, from http://pacinst.org/issues/climate-change-vulnerability-and-resilience/community-resilience-to-climate-change/.

Palau Ministry of Education. (2006). Education master plan 2006–2016. Republic of Palau. Retrieved March 16, 2016, from http://www.unesco.org/education/edurights/media/docs/e51733290f3523016b8384e8a0ec6da32de9fcff.pdf.

Partnership for Observation of the Global Oceans. (2001). Project SEREAR launched in the Pacific. Suva, Fiji. Retrieved March 19, 2016, from http://www.ocean-partners.org/project-seread-launched-in-the-pacific.

PBS Learning Media. (2016). Climate change and the Pacific Islands. Retrieved February 26, 2016, from http://www.pbslearningmedia.org/collection/pacific/.

Peace Climate Change Portal. (2012). Education: Samoa profile. Secretariat of the Pacific Regional Environment Programme. Apia, Samoa. Retrieved March 2, 2016, from http://www.pacificclimatechange.net/index.php/education.

Pacific Islands Climate Education Partnership (PCEP). (2016). PCEP principles. *Tangient LLC*. Retrieved March 3, 2016, from https://pcep.wikispaces.com.

Percival, G. S. (2008). *An assessment of indigenous environmental knowledge (IEK) in the Pacific region to improve resilience to environmental change*. New South Wales, Australia.

Pacific Platform for Disaster Risk Management. (2012). Joint outcome statement from the 4th session of the Pacific platform for disaster risk management and Pacific regional water and sanitation consultations. Retrieved March 10, 2016, from http://www.pacificdisaster.net/pdnadmin/data/original/PPDRM2012_Final_Outcome_Statement_Eng.pdf.

ReliefWeb. (2015). Combining indigenous knowledge with scientific expertise can help mitigate disaster risks. OHCA. Retrieved April 10, 2016, from http://reliefweb.int/report/world/combining-indigenous-knowledge-scientific-expertise-can-help-mitigate-disaster-risks.

Sandwatch Foundation. (2011). The results of the Sandwatch 2014 photo competition "A Snapshot in Time–Beaches Past and Present" are in! Retrieved February 20, 2016, from http://www.sandwatch.ca.

Smith, A., & Jones, K. (2008). Cultural landscape in the Pacific Islands, the international council on monuments and sites (ICOMOS), Paris, France.

Secretariat of the Pacific Community. (2010). SPC/GIZ coping with climate change in the Pacific Island region. Suva, Fiji. Retrieved March 3, 2016, from http://www.spc.int/lrd/spcgiz-coping-with-climate-change-in-the-pacific-island-regionq.

Secretariat of the Pacific Community. (2011). Learning about climate change the Pacific way. Suva, Fiji. Retrieved February 24, 2016, from http://www.spc.int/cc-project/.

SPREP. (2013). The children take action–Kiribati launches children's story book in te-Kiribati. Suva, Fiji. Retrieved April 13, 2016, from https://www.sprep.org/climate-change/the-children-take-action-kiribati-launches-childrens-story-book-in-te-kiribati.

SPREP. (2014). Climate change education resource launched in Samoa. Suva, Fiji. Retrieved March 8, 2016, from https://www.sprep.org/news/climate-change-education-resource-launched-in-samoa.

Taylor, P. W. (1995). Myths, legends and volcanic activity–an example from northern Tonga. *Journal of the Ploynesian Society, 104*(3), 323–346.

United Nations Educational, Scientific, and Cultural Organization (UNESCO). (2006). Pacific education for sustainable development framework. Apia, Samoa.

UNESCO. (2007). UNESCO in the Pacific, a strategic plan for UNESCO in the Pacific states, 2008–2013. Apia, Samoa.

UNESCO. (2011). Teacher training in climate change education is in its infancy. Retrieved March 11, 2016, from http://www.unesco.org/new/en/education/themes/leading-the-international-agenda/education-for-sustainable-development/climate-change-education/dynamic-content-single-view/news/teacher_training_in_climate_change_education_is_in_its_infancy/#.Vve2o8dB3PB.

UNESCO. (2012). Report and recommendations on climate change, education for sustainable development in small island developing states. Apia, Samoa.

UNESCO. (2013). Traditional knowledge for adapting to climate change, safeguarding intanglible cultural heritage in the Pacific. Apia, Samoa.

UNESCO. (2014a). *Understanding community perceptions about climate change in the Pacific*. Apia, Samoa: SPARCK.

UNESCO. (2014b). Report and recommendations on climate change education for sustainable development in Asia and the Pacific. Apia, Samoa. Retrieved February 20, 2016, from http://unesdoc.unesco.org/images/0023/002345/234563e.pdf.

UNESCO. (2014c). Pacific trainers and educators introduced to climate change education inside and outside the classroom. Apia, Samoa. Retrieved March 13, 2016, from http://www.unesco.org/new/en/media-services/single-view/news/pacific_trainers_and_educators_introduced_to_climate_change_education_inside_and_outside_the_classroom/#.VvrhVcdB3PA.

UNESCO. (2015). Putting climate change education into practice. Apia, Samoa. Retrieved April 13, 2016, from http://unesdoc.unesco.org/images/0023/002330/233083e.pdf.

UNICEF. (2013). Climate change and environmental education. Apia, Samoa. Retrieved January 15, 2016, from http://www.unicef.org/education/bege_61668.html.

UNICEF. (2014). Humanitarian action and climate change in the Pacific Island countries. Apia, Samoa. Retrieved April 7, 2016, from http://www.unicef.org/pacificislands/overview_10818.html.

United Nations Office of the High Representative for the Least Developed Countries, Landlocked Developing Countries and Small Island Developing States (UNOHRLLS). (2009). The impact of climate change on the development prospects of the least developed countries and small island developing states. New York, US.

Vanuatu Ministry of Education. (2006). Vanutau education sector strategy 2007–2016. Republic of Vanuatu. Retrieved March 16, 2016, from http://planipolis.iiep.unesco.org/upload/Vanuatu/Vanuatu-Education_sector_strategy.pdf.

Vize, S. (2012). Using education to bring climate change adaptation to Pacific, communities. *Journal of Education for Sustainable Development, 6*(2), 219–315. Washington, DC: Sage publications.

Walshe, R. A., & Nunn, P. (2012). Integration of indigenous knowledge and disaster risk reduction: A case study from Baie Martelli, Pentecost Island, Vanuatu. *International Journal of Disaster Risk Science, 3*(4), 185–194.

WHO. (2010). Pacific Islanders pay heavy price for abandoning traditional diet. *Bulletin of the World Health Organization, 88,* 481–560. Retrieved March 15, 2016, from http://www.who.int/bulletin/volumes/88/7/10-010710/en/.

Young, A. (2011). Horror Pacific Islands education stats. *New Zealand Herald.* Retrieved April 4, 2016, from http://www.nzherald.co.nz/nz/news/article.cfm?c_id=1&objectid=10749535.

Author Biography

Mohamed Walid is a researcher and consultant with five master degrees spanning international development in education, behavioral science, international law, culture and security, and international negotiations and policy-making. He was very gladly and gratefully selected by Dr. Nuria Castells at the United Nations Joint Inspection Unit (JIU) to work on the JIU Report on recommendations to the UN General Assembly on Small Islands Developing States, released in 2015 (JIU/REP/2015/2). Mr. Walid is serving as a Board Member on Education for Global Peace at the University of California, Irvine. He is currently working on his PhD.

Chapter 17
Dynamic Adaptive Management Pathways for Drinking Water Security in Kiribati

Pierre Mukheibir, Louise Boronyak-Vasco and Pelenise Alofa

Introduction

The remote Pacific Island nation of Kiribati is comprised of a group of 33 low lying coral atolls that are vulnerable to the impacts of climate change from variations in precipitation and rising sea levels (Tisdell 2008). Kiribati is considered one of the least developed Pacific Islands with few natural resources. Economic development is hampered by a shortage of skilled workers, poor infrastructure and isolation from international markets.

Kiribati has an average elevation of less than two metres above sea level exposing the local water resources of communities across Kiribati to saltwater intrusion into groundwater resulting from frequent coastal inundation and accelerated coastal erosion caused by sea level rise and the increased frequency of storms and tropical cyclones (Tisdell 2008). This makes the water unfit for people to drink and bathe, increases the risk of epidemics, and reduces potential agricultural yields. Such effects place additional strain on people's livelihoods and wellbeing (economic security, health, infrastructure etc.) (Kuruppu 2009; de Freitas et al. 2013). Illnesses such as diarrhoea and pneumonia are the leading causes of death in

---

The original version of this chapter was revised: The spelling of the second author's name was corrected. The erratum to this chapter is available at 10.1007/978-3-319-50094-2_28

P. Mukheibir (✉) · L. Boronyak-Vasco
The Institute for Sustainable Futures (ISF), University of Technology Sydney, Sydney, Australia
e-mail: pierre.mukheibir@uts.edu.au

L. Boronyak-Vasco
e-mail: louise.boronyak@uts.edu.au

P. Alofa
Climate Action Network of Kiribati (KiriCAN), Tarawa, Kiribati
e-mail: pelealofa13@gmail.com

W. Leal Filho (ed.), *Climate Change Adaptation in Pacific Countries*,
Climate Change Management, DOI 10.1007/978-3-319-50094-2_17

Kiribati and are strongly related to poor hygiene from limited access to potable water and a lack of sanitation facilities (White et al. 2008a, b).

These impacts are likely to be compounded by the projected impacts of climate change that will likely have a negative affect on both the quantity and quality of groundwater resources through variations in precipitation and rising sea level (White et al. 2007; Kuruppu and Liverman 2011). Recently the Government of Kiribati operationalised various national adaptation programmes in which the water sector has been identified as a vulnerable sector to climate change impacts (PCCSP 2011b).

Aim of this Project

Adaptation is a continuous, ever-changing process involving cycles of decision making, planning, action, observation, and above all, social learning and continuous adjustment (Biggs et al. 2011; Moser and Ekstrom 2010; Wild et al. 2015). Central to adaptation planning and implementation is the enhancement of adaptive capacity of systems. Adaptive capacity is defined by "the set of resources (natural, financial, institutional or human, access to ecosystems, information, expertise, and social networks) available for adaptation, as well as the ability of that system to use these resources effectively in the pursuit of adaptation," (Brooks and Adger 2004, p. 168). However, there are various limits and thresholds to adaptive capacity under certain dynamic climate and socio-economic scenarios (Adger et al. 2008). It is vital to identify these limits so that existing adaptation pathways can be revised and inform new adaptation strategies that are locally driven and owned.

In 2015, USAID through its Pacific-American Climate Fund (PACAM), funded the project *Supporting community adaptation to water shortages in Kiribati*. The research project aims to build the capacity of communities on two outer islands of Kiribati to identify and develop adaptation strategies that will support the diversification of water resources to ensure sufficient clean water is provided under a changing climate.

The Capacity Building Approach

The initial focus of the project was to train facilitators in the DAMP approach. Facilitators from the KiriCAN network were trained by the project team from the Institute for Sustainable Futures (University of Technology Sydney) over two days in October 2015. In March 2016, five of these facilitators delivered training on two of the outer islands of Kiribati, Tabiteuea North and Abaiang. A total of 65 community members participated in the two day workshops run on each island.

A key output of the project is a facilitators handbook to guide and support the facilitators on their future community engagements regards potable water supplies and climate change. This has been produced in both English and i-Kiribati.

Kiribati Outer Island Domestic Water Supplies

The World Health Organisation (Howard and Bartram 2003) suggests the following volumes to meet the basic water needs per person per day:

- 5 l for drinking (potable)
- 15 l for handwashing and basic food preparation (i.e. potable water)
- 20–30 l for laundry and bathing

However, many rural communities on the outer islands lack access to adequate secure and safe sources of water (National Statistics Office 2012). More than two thirds of the households on the outer islands do not have access to protected sources of drinking water viz., protected wells or rainwater tanks. None have piped water from centralised systems and rely on open wells, rainwater catchments or village scale solar water pumped systems (NSO 2012). Open wells are those without lids and thus left open all the time while closed wells comprise those that have lids or have been closed off. With the introduction of pumping and piping systems, wells are closed once pumps and piping systems have been installed.

Thus, there are distinct challenges and deprivations related to water security that curtail livelihood choices such as small-scale agricultural practices or small business opportunities (Kuruppu 2009). They also widen existing gender inequalities as women and girls may have to invest greater time accessing clean potable water, limiting time spent on livelihood and educational activities; and drive processes such as migration to urban centres to access reliable infrastructure services (Connell 2011; Kuruppu and Willie 2014).

Climate Analysis

Many of the impacts of climate change will be experienced by communities as changes in the hydrological cycle; and will exacerbate existing stresses on water resources (pollution, population growth) (Pahl-Wostl and Knieper 2014; Mukheibir 2010).

In the Gilbert Islands, the average recorded ***temperatures have increased*** by 1 °C from 1950 to 2009, while maximum temperatures have increased by 0.18 °C per decade over the past 60 years (PCCSP 2011b) (Fig. 17.1).

While annual rainfall in the Gilbert Islands can be described as highly variable, due mainly to El Nino and La Nina events, a ***gradual increase in rainfall*** can be observed for the traditionally wet seasons (PCCSP 2011b). The average rainfall is approximately 2100 mm with just over 900 mm received between May and October (Fig. 17.2).

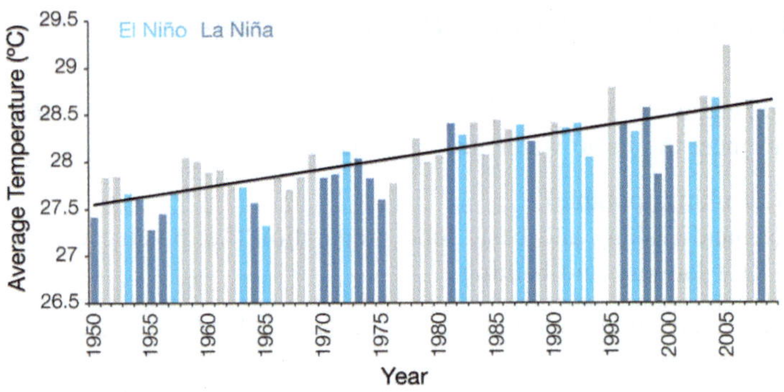

Fig. 17.1 Average annual temperatures (PCCSP 2011b)

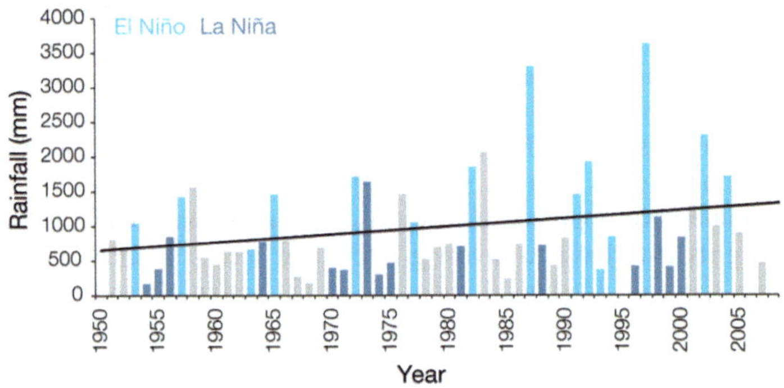

Fig. 17.2 Average annual rainfall (PCCSP 2011b)

Melting ice in the polar regions and snow-capped mountains due to the increase in temperature has led to a global increase in the sea level. At the Gilbert Islands the ***sea level has risen*** by 1–4 mm per year since 1993—10–40 mm per decade (PCCSP 2011b).

Future climate change scenarios for Kiribati are based on projections undertaken by the Pacific Climate Change Science Program (PCCSP 2011b), using 18 different models that best represent the climate of the western tropical Pacific region, specifically the Gilbert Islands.

The projections for Kiribati are based on three of the IPCC emissions scenarios for the periods around 230 and 2055 (IPCC 2007):

- **Low impact on GHG emissions** (B1)—due to a rapid change in economic structures toward a service and information economy, with reductions in material intensity, and the introduction of clean and resource-efficient technologies.
- **Medium impact on GHG emissions** (A1B)—due to very rapid economic growth, global population that peaks in mid-century and declines thereafter, and

the rapid introduction of new and more efficient technologies, with a balance of between fossil and non-fossil energy sources.
- **High impact on GHG emissions** (A2)—due to continuously increasing global population and per capita economic growth and technological change are more fragmented and slower than in other scenarios.

Under these scenarios, the following changes can be expected:

Temperature Increase

The medium and high scenarios are similar in the early to midterm years, with the higher scenario ending up with a higher average temperature increase by 2090, accompanied by a rise in the number of hot days and warm nights and a decline in cooler weather (Table 17.1).

Rainfall Increase

The projected average annual rainfall is projected to increase (and across the wet and dry seasons), with droughts projected to become less frequent (PCCSP 2011b). However, extreme rainfall days are expected to occur more often (Table 17.2).

Sea Level Rise

The sea level is expected to continue to rise in Kiribati, with seasonal fluctuations, resulting in increases in storm surges and coastal flooding (Table 17.3).

Table 17.1 Projected annual average temperature increase (in °C) for the Gilbert Islands (PCCSP 2011a)

Scenarios:	1990 (baseline)	2015 (calculated)	2030 (projected)	2045 (calculated)	2055 (projected)	2090 (projected)
Low (B1)	0	0.4	0.7	1.1	1.3	1.7
Medium (A1B)	0	0.4	0.8	1.3	1.6	2.6
High (A2)	0	0.4	0.8	1.3	1.6	3.0

Table 17.2 Projected average rainfall increase (as percentage change) for the Gilbert Islands (PCCSP 2011a)

Scenarios:	1990 (baseline)	2015 (calculated)	2030 (projected)	2045 (calculated)	2055 (projected)	2090 (projected)
Annual:						
Low (B1)	0	9	14	18	20	25
Medium (A1B)	0	8	12	19	23	37
High (A2)	0	4	7	17	23	42
Wet season:						
Low (B1)	0	6	10	12	14	19
Medium (A1B)	0	6	10	15	18	30
High (A2)	0	3	5	13	18	30
Dry season:						
Low (B1)	0	11	18	24	28	34
Medium (A1B)	0	9	15	24	30	50
High (A2)	0	8	12	23	31	57

Table 17.3 Projected annual average sea level increase (in m) for the Gilbert Islands (PCCSP 2011a)

Scenarios:	1990 (baseline)	2015 (calculated)	2030 (projected)	2045 (calculated)	2055 (projected)	2090 (projected)
Low (B1)	0	0.06	0.09	0.16	0.17	0.31
Medium (A1B)	0	0.06	0.09	0.17	0.20	0.38
High (A2)	0	0.06	0.09	0.17	0.19	0.39

Impact of Climate Change on Water Resources

The following impact map shows how the projected changes may impact the supply and quality of water resources and demand on the Gilbert Islands in the manner described in Fig. 17.3.

Increase Water Supply

With the projected increase in mean annual rainfall and the decline in drought events, both surface water and groundwater will be frequently replenished. This will in turn provide a positive impact on domestic water availability through both

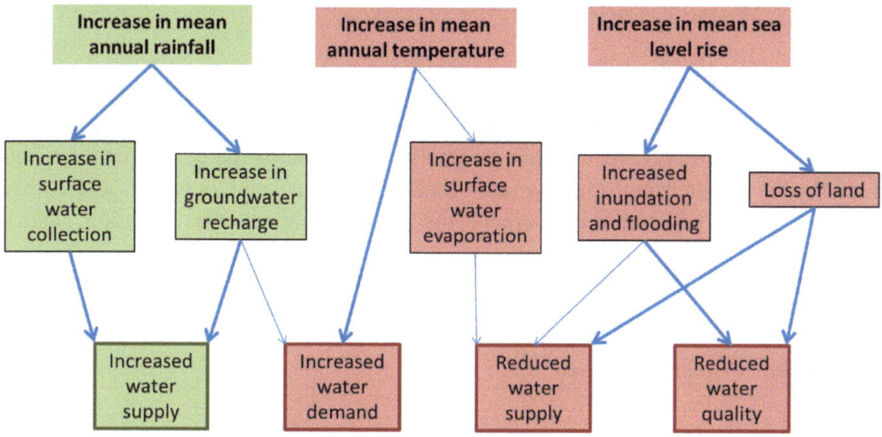

Fig. 17.3 Climate change impacts on water resources

rainwater tanks and groundwater abstraction. If average rainfall increased by 7%, the groundwater recharge is estimated to increase by 5.5% (in contrast, a 10% decline in rainfall may cause a 14% reduction in groundwater recharge) (World Bank 2000).

Reduced Water Supply

Climate change is expected to reduce water supplies in the Pacific to the point where supplies may be insufficient to meet demand during low rainfall periods (IPCC 2007). Under a warming climate it is likely that evapotranspiration would increase, but its effects on groundwater recharge would be much milder than the effect of changes in rainfall. The relatively large projected increases in rainfall that are projected for the islands near to the equator, exceed the smaller changes in potential evapotranspiration. A theoretical 10% increase in annual evapotranspiration could result in as much as a 6% decline in groundwater recharge (World Bank 2000). Higher projected temperatures will result in increased evaporation of surface water, but since freshwater lakes and dams are not a feature of water resource management in these islands, this should not be a concern.

A rise in sea level of 0.4 m (the worst-case scenario in 2090) would have little effect on the groundwater supply and could possibly even raise its volume, as the groundwater table (the top of the freshwater lens) would tend to rise while its base remained relatively unaffected. However, if the width of the islands were reduced by inundation and erosion, which is likely, the thickness of the groundwater could decline by as much as 29% (World Bank 2000, p. 24).

Reduced Water Quality

The effects of extreme events on the water supply could be significant. Currently, high sea levels during El Nino years can lead to seawater contamination of freshwater lenses. However, recovery is generally rapid due to the accompanying high rainfall. The higher overtopping and inundation that may occur in the future, however, could considerably increase the risks of saline contamination (World Bank 2000).

Increased Water Demand

The increase in average and extreme temperatures will likely increase the demand for water for drinking and cooling, although this is not considered to be significant. The increased availability of water due to the increased average annual rainfall, could result in an increase in the demand for water due to it's perceived abundance, with competing demands over potable water for domestic uses versus agricultural initiatives.

Responding Using the Dynamic Adaptive Management Pathways (DAMP) Approach

In order to respond to these potential changes in the supply and demand (of which their timing and magnitude are uncertain), accessible tools and processes are needed at the community and local government level.

The dynamic adaptive management pathway approach was designed to enable planners, as well as community leaders and members to understand the triggers that indicate the need for a new adaptive response under a range of possible future climate and risk scenarios (Mukheibir 2007). These future scenarios are not static, but are continuously changing (dynamic) and hence need to be constantly redescribed, and the effectiveness of the response options reassessed (Haasnoot et al. 2013). Therefore, strategic indicators or triggers play a key role in evaluating whether a response option/s is not performing as was originally planned, necessitating a switch to a different option/s pathway to avoid water shortages.

The approach and tools have been designed in such a way that they provide practitioners with the knowledge to undertake dynamic planning under uncertainty. Special care was taken to ensure that they are simple enough for community leaders and members to understand and follow.

The impacts of climate change will be felt at the local level, therefore it is important that any adaptation plan be managed at a local level by those directly affected by the outcomes, since the changes in the initial assumptions are best assessed locally (Pahl-wostl et al. 2007). The responses should be needs-driven from the community level, taking cultural values and gender issues into consideration (Kuruppu 2009). The integration of indigenous knowledge and practices in the information and thinking processes is equally important. In Kiribati, Village Water Committees (VWC) have been established to provide local input and guidance on water issues. The VWC attempt to have gender equity to ensure that women are directly engaged in the process.

The DAMP approach has been broken down into three stages:

(1) **Creating a hierarchy of responses and options:** The first step in process is to identify a set of viable adaptation responses that satisfy both the water demand needs as well as the socio-economic considerations of the local community – this is best done using multi-criteria analysis (MCA). The process for assessing the adaptation options against multi-criteria involves: confirming the water policy objectives, defining the problem due to climate change, assessing other uncertainties and risks, and collaboratively agreeing upon a meaningful set of criteria for assessment of the proposed options.

The assessment criteria can be drawn from social, technical, environmental, economic and political (STEEP) considerations, and should be closely linked to the community values and water policy objectives. The set of criteria should be a combination of qualitative and quantitative and could include public acceptability, impact on water quality, complexity of the option, energy intensity, and employment generation potential (Mukheibir 2007).

The number of criteria should be kept to a minimum, but should be sufficiently comprehensive i.e. the set of criteria should comprise the number of specific criteria that is just good enough to distinguish between the options. The criteria should be selected with consideration of the available data. The criteria should not require too many assumptions about the future or lead to second-guessing. Double counting of two similar or related benefits should also be avoided, such as accounting for both energy efficiency as well as GHG reductions for example (Mukheibir and Mitchell 2011).

Assigning a relative preference weighting to the criteria is a key component of the MCA process. If we consider all the criteria as having equal importance, less important criteria will have undue influence over which of the options is judged the best. Therefore, it is useful to prioritise the criteria from most preferred to least preferred, and to assign a weighting based on the number of votes they receive from the participants. Weighting a criterion means making a value-based decision as to how important it is in relation to each of the other criteria.

The second step in this process is to generate a list of options and arrange them in a hierarchy of more preferred to least preferred. Both supply and demand side options should be considered that account for both technical and social factors. However, reducing demand to below the basic minimum daily volume should not be encouraged. Options that don't meet the supply volume or water quality objectives should not be considered as viable supply options.

In order to keep the process as simple as possible, a relative ranking of all the options' performance against each criterion should be agreed by all the participants. The ranking of options against the criteria is best done through group consensus and discussion. The collective score for each option will be calculated using a weighted summation of the criteria ratings using simple arithmetic.

The options are then arranged in a hierarchy from most favourable to least favourable. The most favourable scoring option will be considered first to close the current or future gap between supply and demand. Followed by the next option, until the objective is met.

Potential responses to climate impacts could include:

- Rainwater collection could be promoted by fitting new buildings with underground cisterns and encouraging all new houses to be fitted with rainwater storage.
- Protected wells to avoid contaminating potentially good clean water.
- Consider sanitation options that do not contaminate the already limited clean potable water.
- Groundwater recharge by building swales to capture large downpours of water and give it time to seep into the ground before running off to ocean.
- Desalination should be considered only when rainwater or groundwater sources are insufficient, as the cost remains high. Future technological breakthroughs may help make desalination more affordable.
- Water importation is not considered a viable alternative due to the high costs and shipping risks.

(2) **Identifying triggers and indicators:** The outcome of this stage of the tool is a range of indicators to provide an early warning of when the objectives are not being met – for example these may include low groundwater levels, high salinity levels in the aquifer, calcification of cooking pots, unreliable rainwater supplies, some health related indicators, high cost of providing water due to high fuel prices for pumping etc. Due to future uncertainties even the best laid plans may be confronted by a new set of risks that require a new adaptive response. The type and timing of the options to be deployed will depend on the climate and risk scenario that unfolds at that time. Indicators provide the community (and local government officials) with signs that will trigger deployment of the next best option to ensure that the water supply objectives are being met.

The impact diagram (shown in Fig. 17.4) allows participants to map the connections between impacts and influences, and to identify critical indicators of change that would facilitate early warning of a supply-demand imbalance. It is ideal for understanding how climate variability and extreme events (such as big storms) will affect water resources and the subsequent cascade of impacts resulting in poor community health. Figure 17.4 that illustrates the impacts due to decreasing rainfall and the associated indicators (written in red).

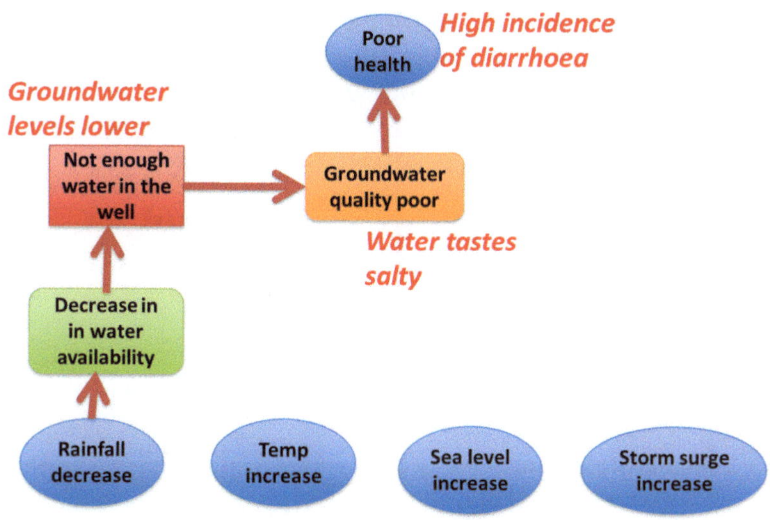

Fig. 17.4 Example of an impact map with indicators for rainfall decrease

For each trigger, it is important to agree on the threshold triggers for each impacts that would indicate that an option is not meeting its supply objective (volume), cost threshold, or other social requirements (e.g. salinity level of the water), and who would be in a position to observe and report on these changes. It may be the nurse at the local clinic who keeps a record of the cases of diarrhoea, and when this reaches a threshold of 10 per week, he or she is required to report that to the Department of Health. Using technical assessments in remote communities is often not feasible due to the lack of technical capability and equipment. Therefore the inclusion of local knowledge in this step is important, since it is the community members themselves who will be monitoring the effectiveness of the options under changing conditions.

(3) **Testing against climate and risk scenarios:** To identify what sequence of the options that would best suit the local situation under potential future risks, two possible climate scenario were used - a drying scenario and a wetting scenario. This was to ensure that the participants recognised that different adaptation management pathways would be needed to provide a resilient water supply system under different futures.

When an indicator exceeds the acceptable level, this acts as a trigger for planning for the next-best response option (in this case a drinking water supply option). If at a later date an indicator for the new supply option exceeds an acceptable level, the process is repeated to identify the next best water supply option under the prevailing conditions.

The process is best illustrated by the example shown in Fig. 17.5. Starting with the option of a hand dug well at the home, participants consider the likely impacts

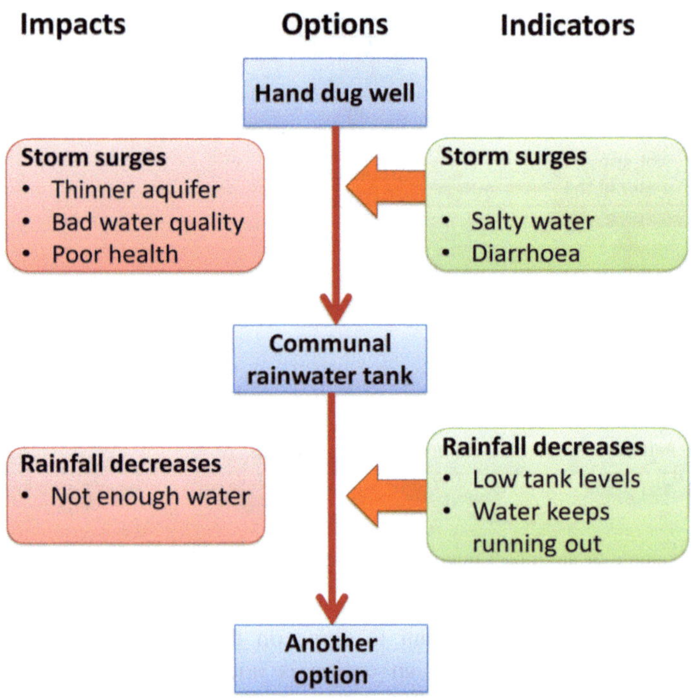

Fig. 17.5 Mapping the new responses to the impacts and indicators

(shown in the pink square) due to an increase in storm surges (with inundation and erosion) and their associated indicators (the green square). The indicators provide a signal for the need to select a new drinking water supply option. They then choose a new option from the prioritised list of options and the process is then repeated under further impacts due to rainfall decrease, for example.

Outcomes

The participants from the two workshops generated a very similar comprehensive list of future water supply options, which also included options which would not satisfy the volume and quality objectives - these have been indicated with an X, in Table 17.4. These options were not considered further.

Similarly, the two workshops generated quite similar sets of criteria for ranking the options. Criteria such as enough water and good quality water were not considered as good criteria, since they were the objectives, and hence were not discerning enough, because all the options would need to meet such an objective. The criteria that consistently scored highly were Affordability, Reliability, Ease of maintenance, Safe for the environment, and Accessibility.

Table 17.4 Master list of viable water supply options identified for Kiribati

No.	Water supply option	
1	Communal rainwater tank/s	
2	Desalination	
3	Household rainwater tank	
4	Inland groundwater well	
5	Solar water pump (inland)	
6	Catch rainwater with palm leaves	X
7	Bring water to parts of the island by boat	X
8	Collecting water from the swamp	X
9	Handpump on an inland well	
10	Moimoto (coconut water)	X
11	Protected household well	
12	Rainwater harvesting using a bucket and guttering	X

The process for mapping the impacts and associated indicators under the two scenarios revealed insights for the participants. Specifically, that if the rainfall was decreasing over time, it made little sense to augment the communal raintanks that were often empty with household raintanks. But rather an option that was not rain dependent should be sought. Similarly, if the groundwater was salty due to sea level rise, then sealing the well and placing a handpump on it was also not the answer, but to perhaps look inland or consider a rainwater tank.

The sequence of moving from a simple well, to a protected well to address storm surges, and to then to move to rainwater tanks if the groundwater became too salty was novel for the participants, but by the end of the workshop clearly understood.

All the participants agreed that the use of desalination was a last resort, and would depend on the availability of parts and technical skills in Kiribati.

Conclusion

This process of constantly reviewing the impact of climate change on water supplies and monitoring the indicators and thresholds, and then deciding if a new option is needed can be called a Dynamic Adaptive Management Pathway approach. By regularly reviewing how their supply options are functioning under changing conditions, communities will be able to plan adaptive responses to any adverse changes in a timely manner.

The challenge will then be to leverage sufficient capital to build the necessary infrastructure, and to ensure the necessary technical skills are provided and retained on the islands where it will be needed to avoid water shortages into the future.

References

Adger, N., Dessai, S., Goulden, M., Hulme, M., Lorenzoni, I., Nelson, D. R., et al. (2008). Are there social limits to adaptation to climate change? *Climatic Change, 93*(3–4), 335–354.

Biggs, C., Edwards, T., Rickards, L., & Wiseman, J. (2011). *Scenario planning for climate adaptation*. Victorian Centre for Climate Change Adaptation Research.

Brooks, N., & Adger, N. (2004). Technical paper 7: Assessing and enhancing adaptive capacity. In B. Lim & E. Spanger-Siegfried (Eds.), *Adaptation Policy Framework for Climate Change* (pp. 165–182). Cambridge, UK: Cambridge University Press.

Connell, J. (2011). Elephants in the Pacific? Pacific urbanisation and its discontents. *Asia Pacific Viewpoint, 52*(2), 121–135.

de Freitas, C. R., Helbig, M., & Matzarakis, A. (2013). Hydroclimatic assessment of water resources of low Pacific islands: Evaluating sensitivity to climatic change and variability. *International Journal of Climatology 34(3), 881–892.*

Haasnoot, M., Kwakkel, J. H., Walker, W. E., & ter Maat, J. (2013). Dynamic adaptive policy pathways: A method for crafting robust decisions for a deeply uncertain world. *Global Environmental Change, 23*(2), 485–498.

Howard, G., & Bartram, J. (2003). *Domestic Water Quantity*. Executive summary: Service Level and Health.

IPCC. (2007). *Climate Change 2007: Impacts, Adaptation and Vulnerability. Working Group II Contribution to the Intergovernmental Panel on Climate Change Fourth Assessment Report*, Geneva: IPCC.

Kuruppu, N. (2009). Adapting water resources to climate change in Kiribati: The importance of cultural values and meanings. *Environmental Science & Policy, 12*(7), 799–809.

Kuruppu, N., & Liverman, D. (2011). Mental preparation for climate adaptation: The role of cognition and culture in enhancing adaptive capacity of water management in Kiribati. *Global Environmental Change, 21*(2), 657–669.

Kuruppu, N., & Willie, R. (2014, in Press). Barriers to reducing climate enhanced disaster risks in Least Developed Country-Small Islands through anticipatory adaptation. *Weather and Climate Extremes*, pp. 1–12.

Moser, S., & Ekstrom, J. (2010). A framework to diagnose barriers to climate change adaptation. *Proceedings of the National Academy of Sciences of the United States of America, 107*(51), 1–6.

Mukheibir, P. (2007a). Qualitative assessment of municipal water resource management strategies under climate impacts: The case of the Northern Cape. *South Africa. WaterSA, 33*(4), 575–582.

Mukheibir, P. (2007b). Qualitative assessment of municipal water resource management strategies under climate impacts: The case of the Northern Cape, South Africa. *WaterSA, 33*(4), 575–582.

Mukheibir, P. (2010). Water access, water scarcity, and climate change. *Environmental Management, 45*(5), 1027–1039.

Mukheibir, P., & Mitchell, C. (2011). *Planning for resilient water systems—a water supply and demand investment options assessment framework [prepared for the Smart Water Fund]*. Sydney: University of Technology, Institute for Sustainable Futures.

National Statistics Office. (2012). *Kiribati 2010 Census of population and Housing: Volume 1 basic Information and Tables*, Ministry of Finance, Bairiki, Tarawa.

NSO. (2012). *Report on the Kiribati 2010 Census of Population and Housing*. Kiribati: National Statistics Office, Ministry of Finance and Economic Planning.

Pahl-Wostl, C., & Knieper, C. (2014). The capacity of water governance to deal with the climate change adaptation challenge: Using fuzzy set qualitative comparative analysis to distinguish between polycentric, fragmented and centralized regimes. *Global Environmental Change, 29*, 139–154.

Pahl-wostl, C., Sendzimir, J., Jeffrey, P., Aerts, J., Berkamp, G., & Cross, K. (2007). Managing change toward adaptive water management through social learning. *Ecology and Society, 12*(2), 30.

PCCSP. (2011a). *Climate Change in the Pacific: Scientific Assessment and New Research* (Vol. 2)*: Country Reports*. Pacific Climate Change Science Program.

PCCSP. (2011b). *Current and future climate of Kiribati Kiribati's current climate.* Pacific Climate Change Science Program.

Tisdell, C. (2008). Global warming and the future of Pacific Island countries. *International Journal of Social Economics, 35*(12), 889–903.

White, I., Falkland, T., & Metutera, T. (2008a). Development: Special issue on water for people. *Safe Water for People in Low, Small Island Pacific Nations: The Rural-Urban Dilemma*, 21(2), pp. 657–669.

White, I., Falkland, T., Metutera, T., Katatia, M., Abete-Reema, T., Overmars, M., et al. (2008b). Safe water for people in low, small Island Pacific Nations: The rural-urban dilemma. *Development, 51*, 282–287.

White, I., Falkland, T., Perez, P., Dray, A., Metutera, T., Metai, E., et al. (2007). Challenges in freshwater management in low coral atolls. *Journal of Cleaner Production, 15*(16), 1522–1526.

Wild, L., Booth, D., Cummings, C., Foresti, M., & Wales, J. (2015). *Adapting development: Improving services to the poor*. London: Overseas Development Institute.

World Bank. (2000). *Cities, Seas, and Storms: Managing change in Pacific island economies*, Volume IV, Adapting to Climate Change.

Chapter 18
Behind the Lens: First-Hand Images and Videos Collected by Communities That Document the Impact of Climate Change in Milne Bay, PNG

David K. Mitchell and George R. Aigoma

Introduction

Island peoples of Milne Bay Province, Eastern Papua New Guinea are subsistence farmers and artisanal fishers. Their livelihoods depend on their environment and this is reliant upon the seasons and patterns of wind, rain and sun. Environmental indicators and social activities are linked together; people in these communities having an in-depth local understanding of their environment and are attuned to change. Traditionally an oral culture, knowledge of the environment was passed along through stories and legends, however today, many of these stories are not told. Young people tend to fill their days with learning through a formal education system, and time after school is sometimes spent watching video clips on mobile phones or taking pictures to show friends. The challenge is how to blend traditional customs with formal learning and technology to reinvigorate local learning for local people.

One answer to this challenge is to put a small camera, the size of a mobile phone, into community members' hands and facilitate them to document their environment and the climate change impacts they see around them. This initiative set out to see at what level coastal island communities would take on this idea. This paper is a description of this process.

D.K. Mitchell (✉) · G.R. Aigoma
Conservation International, Alotau, Papua New Guinea
e-mail: antipodean.mitchell@gmail.com

W. Leal Filho (ed.), *Climate Change Adaptation in Pacific Countries*,
Climate Change Management, DOI 10.1007/978-3-319-50094-2_18

Methodology

Target Communities

Kwaraiwa (10° 37′22.11″S 151° 17′22.72″E) and Tubetube (10° 35′3.44″S 151° 11′14.68″E) islands are at either end of the Engineer Group, and their primary schools have their student catchment from these and nearby islands. Ware (10°57′ 45.83″S 151° 2′46.38″E) is a slightly larger and isolated island with a population of just over 1,000 people and a sufficient number of students to support its own primary school. It has the most dynamic environment of the three islands through being often impacted by cyclones.

Schools in Milne Bay Province have a Board of Management that is made up of selected community members. Therefore, the school is viewed by many as a community institution. Each school has a secure office where project equipment can be stored. Many teachers in these schools have some computer skills with which to utilise and manage the equipment. Being educators, they also can offer guidance to others in the taking of pictures and video.

Equipment

A small waterproof/shockproof digital camera (in our case Panasonic Lumix DMC-TS5) and a robust laptop (Apple MacBook Pro) with a flexible solar panel was given to three different community schools on the above three islands. The project provided a small camera because it was seen as less intrusive than larger camera/video recorders, whilst still capable of high definition images, with reasonable sound recording capability up to 4 m. Also, a laptop that could withstand tropical temperatures and humidity was chosen. As the islands do not have electricity, the communities received a solar panel that could charge the camera and laptop. The solar panel can be connected to the laptop and rolled out in sunlight to provide a power source.

Activities

The first activity after the schools received the equipment was a public demonstration of how to use the camera to take pictures and video. This was backed up with a locally-designed guide book with steps led by cartoon characters and images of the camera. It outlined the steps to follow for the following: (a) setting up the camera to take a picture or video, (b) taking a picture, and (c) planning and taking a video. Steps b and c including getting prior informed consent. The demonstration and guidebook also explained how to take pictures or video underwater. The aim of a visual guide was for it to be easily understood by all ages of the community, some of whom may have had limited formal education.

Next, the project team encouraged people to try out the cameras and laptops in their own time. Community members, initially under the guidance of primary school teachers, then downloaded their images and videos onto the laptop. Some innovative community members went through the steps in the guide themselves.

Various groups (e.g., school students, women's groups, youth, and elders) gathered in separate sessions to describe and then prioritise the climate change impacts they saw in their environment. Assisted by teachers, village recorders and others in the community, CI facilitated these group discussions, with a female extension staff leading the women's discussion. The climate change impacts were listed and participants used a process of placing a total of 3 marks each against the items in the list, resulting in a score of marks so that the most commonly perceived impacts were identified. This process identified a priority climate change impact that was then chosen as a 'project' to be photographed and then videoed. A recurring example was the highly visible impact of the ongoing loss of coastline, with the resultant loss of beach, coconuts/trees, wells and hamlet locations.

Other impacts included, altered and unseasonal weather patterns, increasing sun intensity, severe droughts and coral bleaching. Then, the group discussed how the 'project' chosen could be photographed and a story built around it. It was discussed, what steps were needed, what still images needed to be taken, what might go into a short video, what locations should be shot, what would be described in an interview near or on site, and who would like to be interviewed.

From this communal process, video footage was shot, from which short participatory videos can be produced and example of which is given in Fig. 18.1.

From the filming in Fig. 18.1 further stories were spontaneously generated as this elder mentioned that his mother told him that, when she was young the edge of the beach was where the waves are now breaking on the fringing reef. A middle aged man who was measuring the distance with a tape, then stood closer inshore where the beach was when he was in school in 1975, and the now students at the school standing on the beach suddenly had a historical timeline of coastal loss. The distance from the current beach to the foot of the higher ground is within 100 m. This coastal fringe is currently where the school, villages, wells, coconut and tree groves are. It became very evident to the students that most if not all of this would be potentially lost to the sea in their lifetime.

The project team supplemented the communities' progress through field visits where they conducted additional trainings. Community members over the next six months documented the climate change impacts they observed in their environments, along with any actions taken to adapt to those changes.

Fig. 18.1 Elder from Kwaraiwa Island indicating where the beachfront was, out from the school when he was a young student circa 1950. Essentially everything behind him on his left at that time was land. (Photo G. Aigoma)

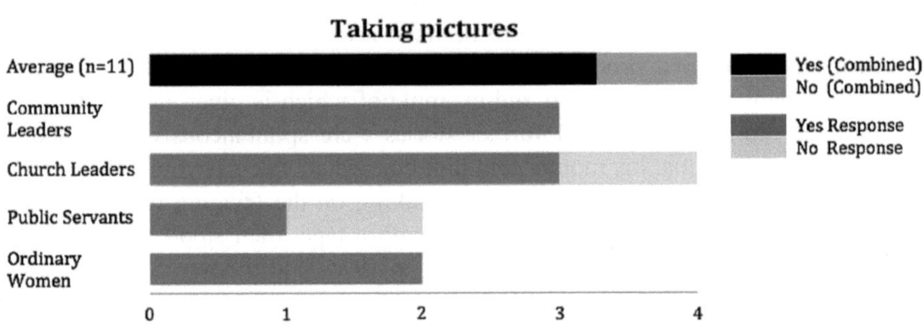

Fig. 18.2 Survey results of 11 Women at Kwaraiwa Island

Gender Evaluation

Prior to the distribution of audio-visual equipment, a survey was undertaken with women on Kwaraiwa Island to determine the potential effectiveness of the project methodology for them. The survey assessed the women's acceptance of different extension techniques and ways to express their views. In the latter series of questions, women were asked if they felt comfortable to record their views by taking pictures or video. The results are shown in Figs. 18.2 and 18.3. Across the different leadership structures of the society, overall there was acceptance of both pictures and video.

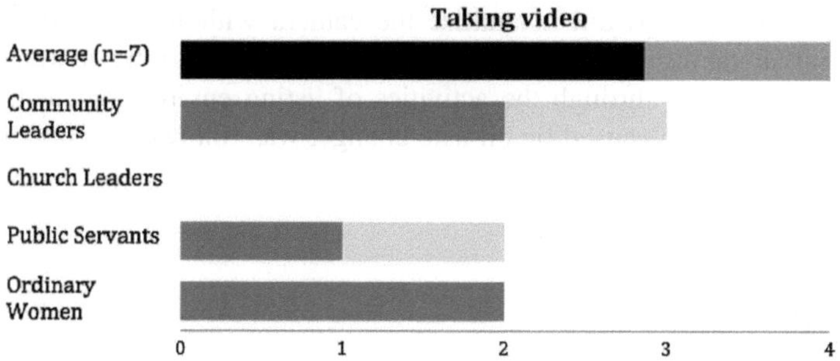

Fig. 18.3 Survey results of 7 Women at Kwaraiwa Island

In this community women were accepting of these media.

The survey was conducted through small informal group discussions with women from different sectors of the community, including 'ordinary' or village women; public servants who were teachers or health workers; church leaders; and community leaders. As the numbers of respondents were small, the results can be regarded as indicative only.

Ordinary village women tended to express that they felt at ease with both approaches. Other women gave variable responses regarding pictures and video taking, but remained positive about these as a learning technique.

Preliminary Results: Getting (or Not) Behind the Lens

Ward members on these islands used the cameras to take pictures of their food gardens and water supplies during the 2015–16 the El Niño drought and added these images into reports to government officials in the Provincial Disaster Office. Uptake of use of the camera's occurred readily in doing this and the images taken assisted in the assessment of the impact of the drought and subsequently relief supplies were dispatched to these affected communities.

Despite positive support for the project, as indicated by community members who expressed that using the camera would assist them in getting their concerns and views heard, the uptake of becoming 'videographers' did not occur spontaneously. Project staff had to directly facilitate the process within different community groups i.e. school teachers/students, clan groups and women's fellowship groups.

As the cameras were waterproof and shockproof, youth felt comfortable experimenting with the cameras to get used to collecting photos and videos underwater. Young lads on Ware Island on learning how to use the cameras took pictures underwater. In the community of Kwaraiwa, the head teacher took students to the shallows of local reefs and lagoon areas to take images underwater. The novelty of doing this helped overcome any reservations of using the camera to make recordings.

The Women's groups did not utilise the camera without facilitation and their involvement in taking pictures and video required support by a female village extension staff. Going through the activities of listing environmental impacts and discussing which were caused by climate change, was followed by taking video of these impacts.

Next Steps

At the time of writing this paper, the raw footage has not yet been edited into short videos. The intention is to initially make at least four short videos to share clear, concise messages about climate change impact and ecosystem-based climate change adaptation, within and with other communities and with a wider provincial audience. It is hoped that these videos will lead to enthusiasm from within the communities to create more video shorts. Further facilitation by the project staff will continue, building capacity in achieving this. It is evident from the project's implementation to date that by going through this process people are gaining a wider perspective of climate change and the options for coping with it.

Acknowledgements The authors are grateful to Whitney Anderson and Vivian Raksakulthai for reviewing and editing of initial versions of the manuscript. This work was supported with funding from the USAID Pacific-American Climate Fund (Grant # PACAM-14-0004).

Chapter 19
A Case for Formal Education in the Technical, Vocational Education and Training (TVET) Sector for Climate Change Adaptation and Disaster Risk Reduction in the Pacific Islands Region

Sarah Louise Hemstock, Helene Jacot Des Combes, T. Martin, F.L. Vaike, K. Maitava, L.-A. Buliruarua, V. Satiki, Nixon Kua and T. Marawa

Introduction

The Pacific Region is among the most vulnerable to climate change, while it has done little to contribute to the cause—producing less than 0.03% of current global greenhouse gas emissions (IEA 2011).

Despite this vulnerability, Pacific-African Caribbean and Pacific (P-ACP) countries appear to have the least capacity to react and adapt to adverse impacts of climate change. One of the key barriers to improving P-ACP countries' resilience to climate change impacts has been confirmed as the lack of local and regional

---

The authors are the cross-disciplinary team involved in the implementation of the European Union Pacific Technical Vocational Education and Training in Sustainable Energy and Climate Change Adaptation Project (EU PacTVET). The purpose of the project is to enhance and/or create Pacific Small Island Developing States (P-SIDS) regional and national capacity and technical expertise to respond to climate change adaptation (CCA) and sustainable energy (SE) challenges.

S.L. Hemstock (✉) · N. Kua · T. Marawa
Economic Development Division, The Pacific Community,
3rd Floor, Lotus Building, SPC – Private Mail Bag, Nabua, Fiji
e-mail: sarahh@spc.int

H. Jacot Des Combes · T. Martin · F.L. Vaike · K. Maitava · L.-A. Buliruarua
The Pacific Centre of Environment and Sustainable Development,
The University of the South Pacific, Laudala Bay, Private Mail Bag, Suva, Fiji

V. Satiki
Fiji National University, Kings Road, P.O. Box 7222, Nasinu, Fiji

W. Leal Filho (ed.), *Climate Change Adaptation in Pacific Countries*,
Climate Change Management, DOI 10.1007/978-3-319-50094-2_19

capacity and expertise resulting from the absence of sustainable accredited and quality assured formal training programmes in the climate change adaptation (CCA) and disaster risk reduction/management (DRR/DRM) sectors at the TVET level. This situation is compounded by the absence of accredited teachers/trainers and well-resourced and equipped institutions to deliver the required training programmes. For example, in the Federated States of Micronesia according to new regulations more than 60% of currently employed teachers do not meet minimum qualification requirements (Buliruarua et al. 2015).

This research, and indeed the EU PacTVET project itself, is in response to the numerous calls throughout CCA/DRR related policy, at national, regional and global levels, for "capacity building". This research attempts to put those policy calls into perspective and integrate findings from training needs and gap analyses in 15 P-ACP countries to outline key approaches that should be assimilated into a strategy for delivery of "capacity building" in the area of CCA and DRR that is both responsive to and reflective of identified P-ACP needs. This paper also examines why capacity building for CCA and DRR should be considered together.

The definitions of formal and non-formal learning adopted by this paper are: Formal learning takes place in education and training institutions, is recognized by relevant national authorities, and leads to qualifications. Non-formal learning is learning that has been acquired in addition or alternatively to formal learning. It usually takes place in community-based settings, the workplace and through the activities of civil society organizations (UNESCO Institute for Lifelong Learning 2015).

Methodology

Although this paper refers to specifically to climate change adaptation and disaster risk reduction, the remit for the EU PacTVET project training needs and gap analysis also included sustainable energy.

In order to assess national training needs in climate change adaptation and disaster risk reduction and map existing informal and formal TVET training courses and identify training and education providers (formal and non-formal); a needs and gap analysis was performed in all 15 Pacific—African, Caribbean and Pacific (P-ACP) nations (Cook Islands, Federated States of Micronesia (FSM), Fiji, Kiribati, Nauru, Niue, Palau, Papua New Guinea, Republic of Marshall Islands (RMI), Samoa, Solomon Islands, Timor-Leste, Tonga, Tuvalu and Vanuatu) (Buliruarua et al. 2015).

In trying to maximize the effectiveness of the needs and gap analysis, and to ensure it captured opinions of many stakeholders and relevant existing information, a mixed methods approach was employed including literature survey (including policy analysis) and in-country missions for all 15 P-ACPs. Prior to in-country missions, questionnaires were forwarded to government ministries for energy, climate change and education, energy utilities and TVET training institutes. This ensured that stakeholders and existing national formal qualifications and non-formal trainings were identified. With regard to stakeholder consultation, determining which methodology to apply depended on various factors ranging from availability, geographical location and sector. The following are some of the methods employed to capture the required data during the in-country mission: Consultative workshop —a consultation with stakeholders in a venue where the agendum for discussion or consultation is done for all at once. This was the methodology employed for the first two days during most in-country missions; One-to-one consultations (interviews)—where key stakeholders could not attend consultative gatherings, due to other commitments, one-to-one interviews were carried out. This was also employed where for a number of stakeholders who could not attend the consultative workshop. Workshops and interviews were carried out from February to September 2015.

The purpose of the EU PacTVET in-country-missions was to: Identify present and future market demand for training in CCA and DRR; map existing training supply in each P-ACP; list priorities for future donor project activities in CCA and DRR (since in many P-ACPs, donor aid accounts for around 50% of Gross Domestic Product so is an important employment sector); select partner TVET institutions for the EU PacTVET project.

Limitations

It was not possible to include all stakeholders in consultation workshops or visit outer islands. However, University of the South Pacific Global Climate Change Alliance project In-Country Coordinators, who work directly with at least 3 vulnerable communities in each country, were instrumental in the consultations.

Literacy and education levels amongst the stakeholders were high, but some language barriers were apparent. Group consultations can be dominated by a few confident people. Additionally, although a "standard" consultation-workshop format was devised, the consultations took on a national flavour with some areas receiving more attention than others—depending on the focus of the stakeholders or the national priorities.

The Pacific Policy Context—CCA and DRM/DRR Linkages

The significance of capacity building on climate change adaptation to the sustainable development of the P-ACP countries can be seen by the endorsement by the Forum Leaders of the Pacific Islands Framework for Action on Climate Change (PIFACC). This framework has themes on Capacity Building, Education, Training and Awareness with outcomes of:

- Increased awareness and understanding of sustainable energy and climate change issues among communities and other stakeholders
- Strengthened capacity to monitor and assess impacts of sustainable energy and climate change interventions
- Strengthened capacity to identify, design and implement effective sustainable energy and climate change measures

In recognition of the fact that climate change is a slow-acting disaster for the region amongst other factors, during the lifetimes of the Pacific Islands Framework for Action on Climate Change 2006–2015 and the Pacific Disaster Risk Reduction and Disaster Management Framework for Action 2005–2015, efforts were made across the region to integrate climate change adaptation and disaster risk reduction policy. These efforts are still ongoing. The linkages between CCA, DRM & DRR are outlined in Fig. 19.1.

To a large extent, the process by which integration of CCA and DRR/DRM has occurred regionally is via "mainstreaming" climate change into regional and national policy development and via the development and support of the regional Strategy for Climate and Disaster Resilient Development in the Pacific (SRDP). However, the SRDP has yet to be adopted regionally to replace the Pacific Islands Framework for Action on Climate Change 2006–2015 and the Pacific Disaster Risk Reduction and Disaster Management Framework for Action 2005–2015.

The overall aim of integrating CCA and DRM/DRR is to support the resilience of communities across the region to climate change and disaster impacts. This can

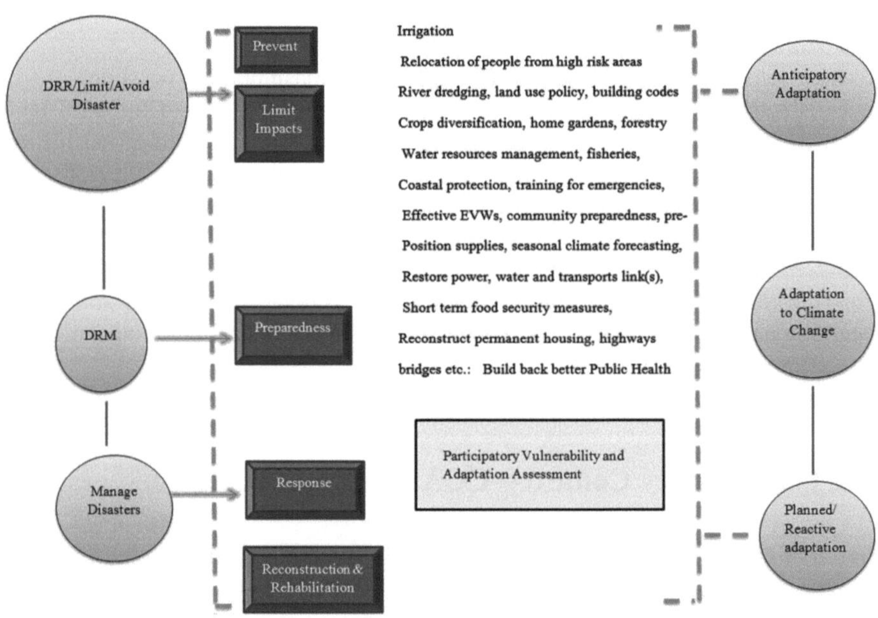

Fig. 19.1 Linkages between disaster risk management and climate change adaptation *Source* Adapted from Pacific Disaster (2015)

be achieved by developing effective, appropriate and integrated methods of risk and vulnerability assessment, planning, adaptation activities and monitoring. To be a successful integrated strategy it will need to provide input into human capacity building via training for key stakeholder groups that have a crucial role in addressing the impacts of climate change and disasters in the Pacific region. Key stakeholders have been identified as Governments (national and local), the private sector, civil society organisations, rural communities, regional organisations and development partners. The SRDP also states that "Such needs-based capacity building can provide a significant return on the investment." It has also been recognized that regional resilience starts at local level with awareness (on climate change, disaster risk and emissions reduction), training, education and action. Training and capacity development has also been highlighted as a requirement for NGOs, for national disaster agencies and other key stakeholders (such as lands, meteorological and hydrological services, health, education, tourism, planning, etc.). The SRDP process had also identified that "The education sector, in particular, has a key role to play in conducting education, training and awareness-raising in relation to climate and disaster resilience. Opportunities for their active involvement need to be identified."

The Global Policy Context—A Call for Capacity Building

At the international level, the Sendai Framework (UNISDR 2015) is a 15-year non-binding agreement which "recognizes that the State has the primary role to reduce disaster risk but that responsibility should be shared with other stakeholders including local government, the private sector and other stakeholders. It aims for the following outcome:

The substantial reduction of disaster risk and losses in lives, livelihoods and health and in the economic, physical, social, cultural and environmental assets of persons, businesses, communities and countries."

Education is seen as a key tool to achieving this outcome and the United Nations International Strategy for Disaster Reduction (UNISDR) has been tasked to support the implementation, follow-up and review of the Sendai Framework.

The Sendai Framework (UNISDR 2015) succeeds the Hyogo Framework for Action (HFA) 2005–2015 (2005) which highlighted the "use knowledge, innovation and education to build a culture of safety and resilience at all levels" as "Priority Action 3" from a total of 5 "Priority Actions".

The Sustainable Development Goals (September 2015; SDGs), and the Paris Agreement (December 2015; COP21), are two other recent landmark agreement that call for "capacity building" in disaster risk reduction, sustainable development, and climate-change adaptation and mitigation. In the case of the Paris Agreement and the Sendai Framework, this is with particular reference to science and technology transfer. The Major Group For Science and Technology (STAG) was established to ensure science and technology were integrated into all priority areas of the Sendai Framework and placed an emphasis on "Capacity development to ensure that all countries can produce, access, and effectively use scientific information" (Aitsi-Selmi 2015).

The Sendai Framework recognizes the role of capacity development in achieving its goal: "prevent new and reduce existing disaster risk…" (UNISDR 2015). Capacity building has been identified as key to achieving all its priorities for action. The key statement on accredited formal qualifications for capacity development is:

> 27 j) to promote the development of quality standards, such as certifications and awards for disaster risk management with the participation of the private sector, civil society, professional associations, scientific organisations and the United Nations.

This description also appears to be an indicator for certification at the TVET level since the TVET sector is unique in drawing upon the experiences of the private sector and other stakeholders to formulate nationally recognized and accredited qualifications.

The need for applicable capacity development is augmented by the UNISDR S&T Road Map as an outcome of the 2016 STAG conference recommendation: to "support capacity building and ensure that capacity development for disaster risk management is interdisciplinary, shared across international boundaries, and demand-driven" (UNISDR 2016a, b). In this regard, a training needs and gap analysis was conducted across all 15 Pacific—African Caribbean and Pacific (P-ACP) countries (Buliruarua et al. 2015). The needs and gap analysis enabled a "demand-driven" approach to training provision for the developing EU-PacTVET project activities, with all 15 countries requesting formal training be developed for disaster risk management in order to "professionalise" the sector. Findings from the needs and gap analysis revealed that many of those working in the government and civil society sectors on DRR and climate change had no formal qualifications in these areas; this led to a lack of confidence in their work and a perceived lack of competence. Despite over a decade of non-formal training in disaster risk management (DRM), this lack of regional "professional" capacity is also reflected in the fact that in post disaster situations it is down to external experts to carry out post-disaster assessments. Participants felt that this situation would be addressed by the introduction of accredited formal qualifications in relevant skill sets and levels. (Buliruarua 2015).

Education and capacity building have roles to play in achieving the alignment and delivering the outputs of the three recent global initiatives (the Sendai Framework, SDGs and the Paris Agreement). Findings from the needs and gap analyses of all 15 P-ACPs indicate that formal qualifications which account for local contexts are required to build national capacity to: accurately monitor and assess impacts of climate change and natural hazards; identify solutions to reduce these risks; and plan, manage and implement risk reduction projects to reduce damage and losses (Buliruarua et al. 2015). So, essentially—formal education is needed to embed the national capacity to reduce vulnerability & exposure, anticipate, respond to and recover from disasters that result from interdependent and evolving risks to sustainable development (Table 19.1).

Table 19.1 P-ACP national CCA and DRM/DRR policies (based on Buliruarua et al. 2015)

Country	Key policy
Cook Islands	Cook Islands Joint National Action Plan for DRM & CCA Climate & Disaster Compatible Development Policy 2013–2016
Fiji	National DRM Plan 1995 National Disaster Management Act 1998 Fiji National CC Policy 2012 Draft Energy Policy
Federated States of Micronesia	Joint State Action Plan for CC & DRM Draft National Policy
Republic of Kiribati	Kiribati Joint Implementation Plan for CC and DRM 2014–2023
Republic of Nauru	Nauru has not yet established a specific environmental policy, a no regrets approach has been adopted to adaption accommodating climate and sea level change considerations and implementation of the National Environmental Action Plan and the Rehabilitation Master Land use Plan
Niue	Niue's Joint Action Plan for DRM & CCA
Republic of Palau	Palau Climate Change Policy For Climate and Disaster Resilient Low Emissions Development, 2015
Papua New Guinea	The National Development Strategic Plan (2011–2030)
Republic of the Marshall Islands	RMI Joint Action Plan for CCA & DRM Vision 2018 (2003–2018) National Climate Change Policy Framework 2011 Ministry of Education Strategic Plan (2013–2016)
Independent State of Samoa	National Policy of Combating Climate Change 2007 Greenhouse Gas Abatement Strategy 2008 Strategic Action Plan 2008 Samoa National Action Plan for DRM 2011–2016
Solomon Islands	Solomon Islands National Disaster Risk Reduction Policy (2010) National Development Strategy 2011–2020 Solomon Islands Climate Change Policy (2012)
Democratic Republic of Timor-Leste	National Strategic Development Plan (2011) National Disaster Risk Management Policy/Plan (PNJRD/NDRMP) National Adaptation Programme of Action on Climate Change (NAPA)
Kingdom of Tonga	Tonga National Climate Change Policy and Joint National Action Plan for CCA & DRM 2010–2015
Tuvalu	Tuvalu National Strategic Action Plan for CCA & DRM 2012–2016
Republic of Vanuatu	Vanuatu Disaster Risk Reduction and Disaster Management Plan 2006–2016 Republic of Vanuatu National CCA Strategy (2012–2022)

The National Policy Context

A reliance on capacity building to achieve policy objectives is reflected in all the national policies listed above. For example, Tuvalu National Strategic Action Plan for CCA & DRM 2012–2016 asks specifically for capacity building support at government level to assist with "Improving Understanding and Application of Climate Change Data, Information and Site Specific Impacts Assessment to Inform Adaptation and Disaster Risk Reduction Programmes"; while Palau's Climate Change Policy For Climate and Disaster Resilient Low Emissions Development, 2015 requests that by 2020, Palau's educational system will include coordinated climate change and disaster risk information in its curriculum and will offer professional development for adults. Priority policy initiatives for Palau are outlined as: integration of climate change and disaster management into education policies and action plans; school curriculum to incorporate climate change and disaster management, and develop teachers training modules/materials; scholarship and education opportunities in climate/disaster will be prioritised; implement professional training in climate/disaster related studies including teachers; and access to information on climate change and disasters will be improved.

Like most developing countries, currently in the PIR many young people acquire workplace skills by informal means and non-formal training. Alarmingly, this is still the case in spite of the multitude of policy dialogue and unanimous agreement on the requirement of education per se to improve resilience to climate change and disasters. Additionally, although the role of "local level" action is recognised in policy dialogue (e.g. the SRDP), there are no relevant formal qualifications accessible at this level. The vast majority of offerings in this sector are at postgraduate level, which is not appropriate for the majority of stakeholders—including communities. In that regard, regional TVET qualifications aligned with the PQF levels 1 to 4 would be most appropriate. Since most of the current regional training delivery is carried out on a project basis, usually by consultants who jet in then jet out a few days later, it is unsustainable in terms of both delivery and national capacity to deliver. Ad hoc training and lack of national capacity to deliver training sustainably also means that many projects fail if those trained on project activities leave the community (Woods et al. 2006). These issues lead to the conclusion that national capacity for the delivery of quality assured regional qualifications in CCA and DRM/DRR should be a more sustainable approach.

Formal education in DRR/DRM is only offered in the region at the Postgraduate level (level 8 from 10 on the PQF). This is because most adaptation efforts to date have largely been 'top-down' in their process and approach, so limited attention has been given to integrating community experiences of climate change into adaptation actions, including the knowledge and views of community members on how to cope and adapt to localised changing environmental conditions (McNamara et al. 2012; Reid et al. 2009).

All P-ACPs essentially face the same disaster and climate change risks. A regional needs-based "interdisciplinary, shared across international boundaries" approach to DRR and CCA capacity development as outlined by STAG is essential since countries such as Tuvalu, Narau, Niue, Cook Islands and Palau have populations ranging from 1,500 to 20,000 and therefore do not have the capacity to establish credible national quality assurance systems.

Regional Accreditation of Qualifications—The Current Situation

Globally there are many "qualifications frameworks" from which to hang formal qualifications. In the Pacific Islands Region (PIR), the Pacific Qualifications Framework (PQF) allows translation of educational benchmarks for standardising national qualifications. It has 10 levels, where level 1 is basically equivalent to a school leaver's certificate (e.g. Tonga School Certificate) and level 10 is equivalent to a Doctorate (PhD). Once national qualifications have been mapped using the PQF they can then be registered on the Pacific Register of Qualifications and Standards (PRQS). This process is sanctioned by the Educational Quality Assessment Program of the Pacific Community (SPC) (Sanerivi et al. 2016). In the European Union, the Bologna Process aims to provide systems for mutual recognition of qualifications across the EU, although it recognises the importance of meeting labour market needs, it focuses on graduate and postgraduate level education (degree, masters and doctorate), not the workplace oriented TVET level (post-school and pre-degree). The P-ACP needs analysis identified the TVET level as the most appropriate for qualifications in DRR and CCA.

A key barrier to accreditation and quality assurance regionally is the lack of a clear and truly regional quality assurance structure. The need for national and regional qualifications frameworks was raised at the regional level more than three decades ago and was highlighted by Bartam (2004). A Pacific Qualifications Framework (PQF) and a Pacific Register of Qualifications and Standards (PQRS) are now in place—but qualifications on the register are owned "nationally", so there are no "regional" qualifications. The main purpose of the PRQS is to facilitate the benchmarking of nationally offered Pacific qualifications against international standards. These national qualifications are aligned to "international" qualifications using the PQF. This allows mutual recognition of nationally offered qualifications with a degree or quality assurance. Only 3 (Fiji, Samoa and Tonga) out of 15 P-ACP countries have national frameworks fully aligned to the PQF (PNG and Vanuatu also have national frameworks and Tuvalu is in process). In many instances, for example where the nation's population is below 20,000 people (Nauru, Niue, Tuvalu and Palau being 4 P-ACP examples), a national approach to quality assurance could be administratively too burdensome.

As part of the EU PacTVET project (a 6.1 million Euro European Union project funded under the 10th European Development Fund (EDF 10) Pacific regional envelope), work has already began in collaboration with the Fiji Higher Education

Commission (FHEC), the Education Quality Assessment Programme (EQAP) and the German aid agency (GIZ), with regard to achieving regional qualifications in the areas of climate change adaptation (including disaster risk reduction/management) and sustainable energy. Regional Industry Standards Advisory Committees have been established and enguaged to develop the regional qualifications at levels 1–4. At the EU-PacTVET inception meeting in October 2015 each of the 15 P-ACP countries endorsed national priority areas for TVET qualification development. This meeting also highlighted that TVET infrastructure is a common concern and priority need. In particular eleven of the fifteen countries identified some component of quality assessment (accreditation needs) as one of the key areas for assistance (Buliruarua et al. 2015).

The EU-PacTVET project is also mandated to build national capacity via a training of trainers' model. Additionally, training a cohort of nationally-based accreditors who can validate training institutions, ad hoc training and RPL is also going to be discussed at a regional level. However, given the lack of capacity at the national level for P-ACPs a regional "one-stop-shop" approach to accreditation of qualifications and quality assurance of training delivery would circumnavigate many of the barriers currently being faced. In that regard, the EU-PacTVET project is proposing to pilot an organisation for DRR and CCA (resilience) practitioners. The intention of this organization would be to develop and administer a certification scheme for practitioners based on a quality assured accreditation scheme for DRR/CCA qualifications (and RPL) and training providers, which the pilot organisation would also develop and administer. If this approach proves to be successful, it is anticipated that this scheme could be applied globally.

An Illustration of the Capacity Problems—A Case Study of the Federated States of Micronesia

According to the FSM Strategic Development Plan (2004–2023): "The FSM Education sector is at a significant stage in its development. There is wide agreement that the level of student learning and achievement is low and needs to be raised. It is also beginning to be better understood that an educated population is essential for improving economic growth and social development and that education should be considered a productive sector. The primary resources of a small island developing nation such as the FSM are its human resources." Education is also recognised as a "Critical Issue for Implementation" in all sectors covered by the plan.

The National Infrastructure Development Plan earmarked US$135.4 million for education infrastructure spending across the 20 years between 2004 and 2023.

An FSM TVET conference was held in 2011 by the National Department of Education However, there is no specific TVET policy. National curriculum benchmarks and standards were defined by the Department of Education Federated

States of Micronesia Government, Standards Development Working Group (2006–2008). Although there is no mention of climate change in the curriculum, environmental science and CC benchmarks have been developed recently. Additionally, they have been approved by the National Department of Education under the Pacific Climate Education Partnership (PCEP). Benchmarks and lesson plans/materials were developed through collaborative efforts between PREL, Micronesian Conservation Trust, and the National Department of Education and are being adopted for use in the schools.

As outlined in the Joint State Action Plans for CC & DRM (2014), governance arrangements and the cross-cutting nature of disaster and climate risk management, implementation of the draft National Policy on CC and DRM is a shared responsibility between government, private sector, civil society and communities. The draft National Policy on CC & DRM notes that national and state governments will lead the promotion, coordination and monitoring the implementation of the policy. The constitutional arrangements in FSM require the state governments to be responsible for implementing the policy. The Climate Change Act 2013 deals with governance issues relating to CC & DRM policy and introduces legal obligations for certain national government departments and agencies of FSM.

For FSM, the policy pieces appear to be in place, so what is preventing the achievement of policy objectives? Buliruarua (2015) identified the following constraints on the "Education Sector":

- low student achievement and expectation and lack of parental support (only 37% of children progress from primary to secondary school—and only 27% of that 37% go on to tertiary education)
- low student achievement and expectations may be fueled by low employment prospects—however, this is a vicious circle as an "inadequately educated workforce" has been flagged by the World Bank Enterprise Survey (2009) as the biggest constraint to enterprise development in FSM.
- a lack of well-resourced learning facilities and geographical constraints mean lack of resources on outer islands
- low numbers of people in tertiary education resulted in a lack of qualified students to take up scholarships
- teacher quality and qualifications were identified as a major constraint
- language was flagged as having a wide ranging impact on learning—it is also feared that some local languages may be lost
- a major impact on the education the sector was identified as the amended Compact Agreement—changes requiring specific subject teaching qualifications come into effect in 2015 and mean that more than 60% of currently employed teachers do not meet minimum qualification requirements and will not be funded. National restrictions (e.g. some states only employ teachers who are from that state in order to promote local language and culture) compound this issue. 17% of teachers have no formal credentials.

Constraints specific to TVET were identified as the following:

- educational pathways—e.g. colleges, and training institutions—a 2 year diploma course is a limitation as there are no established educational pathways
- national tertiary funding structures generally mean accredited courses and programmes have minimum entry requirements (completion of High School) so recognition of prior learning for TVET skills sets is not an option in FSM
- national standards—minimum benchmarks at national level versus state level and no uniformity between states (e.g. electrical program—local certified graduate versus international engineer)
- lack of infrastructure—FSM has one college and a vocationally oriented high school
- language barriers—many languages are used in the country and delivery to remote outer islands is problematic as there is no internet and transportation costly
- political will—many of the issues identified at the plenary session were previously highlighted more than a decade ago in the FSM Strategic Development Plan (2004–2023)—little has been achieved re encouraging students into secondary and tertiary education
- technology changes are reactive so it is difficult to predict future markets for skills
- limited capacity for certified people to find gainful employment compounded by a current government hiring freeze and a small private sector—this could be linked to outmigration and scholarship students not returning to FSM, despite bonding requirements (which are not enforced)
- attrition—limited on the job training is available and this is a self-limiting situation as people who are trained by one organization leave for better paying one—less than 5% of enterprises offer formal training programs to their permanent workers. This compares badly with Samoa (80%), Fiji (60%) and Timor Leste (50%) (Vandenberg et al. 2015).

Constraints faced by the College of Micronesia (FSM) the only dedicated national tertiary education provider:

- expertise and training is not in the relevant areas—resources and training are made available, but not necessarily applicable to other organisations, e.g. Red Cross, need to ensure that trainings meet their specific needs
- courses offered do not meet development needs
- poor retention of qualified instructors, the current system attracts external short term staff
- salaries should be higher and equate to expertise and performance
- priority areas identified by governments may not reflect the CC needs—over 25% of students at College of Micronesia are enrolled in education
- Programs offered, with a few exceptions, are Associate Degree level or shorter

- Only one bachelor degree program is offered, which is run jointly with the University of Guam, it is also in education
- FSM students are increasingly using distance education to access education from external universities—this could be beneficial for CCA and DRR.

The role of education in the process of job creation needs to be emphasised. The education system in FSM is modelled on the U.S. system which reflects the priorities and needs of a large developed country. It was evident from focus group discussions and interviews that this educational model was not fulfilling the needs of FSM as a small island developing nation. Many of FSM's priorities are not covered in the current system since courses offered and educational pathways do not reflect FSM's priorities (Buliruarua et al. 2015).

A Community Focused Approach

The need to integrate traditional and local knowledge on CCA and DRR into formal qualifications that were applicable at community level (i.e. levels 1 to 4 on the PQF) was highlighted by all 15 P-ACPs (Buliruarua 2015). Additionally, since a great deal of valuable non-formal training has been delivered in a community context, any new regional qualification structure needs to be able to account for recognition of prior learning (RPL) and provide pathways from further to higher education.

Job opportunities in many P-ACPs are scarce. Previous non-formal training has done little too improve employment prospects of those undertaking the training. However, in some instances, non-formal training has led to productive activities within communities, successful adaptation project outcomes and some improvements in livelihoods. Formal education at the TVET level does everything that non-formal community training does, with the added bonus of improved employment prospects. Mutual recognition of qualifications via a regional quality assurance mechanism should further improve prospects by increasing opportunities for labour force mobility. Also, if the worst predictions of climate change impacts come to fruition, and low lying atolls become uninhabitable, then having accredited qualifications will allow people to "migrate with dignity" and participate in the economies of the territories that they migrate to.

Linkages with Economic Demands

Findings from the EU PacTVET Training Needs and Gap Assessment (Buliruarua et al. 2015) indicate that formal educational pathways and the professionalization of CCA and DRM/DRR sectors need to be established as a matter of urgency since the region currently has little capacity to absorb the funding for climate change related activities entering the region. For example, the UNFCCC COP21 Paris Agreement provisions a floor of $100bn a year in climate finance for all developing countries by 2020, with a commitment to further finance in the future. Given this, adapting to climate change is at the core of major development efforts For P-ACPs.

An analysis of the situation in Tuvalu provides a good illustration of why local capacity in CCA and DRM/DRR is desperately needed to take full advantage of aid flows into the region. In 2008, Tuvalu's GDP was US$32 million, 50% of this was in the form of development aid—approximately US$4 million was spent on external technical assistance (Smith and Hemstock 2011). Due to a lack of in-country capacity, a staggering 12% of Tuvalu's GDP is spent on external experts from consulting companies and multilateral organisations. In 2013, 2.1 billion USD in overseas development aid was received by the PIR, with 48% being donated by Australia (OECD 2014).

Concluding Remarks

Across all 15 P-ACPs there appears to be strong national support of a regional approach to quality assurance (accreditation) for regionally developed qualifications in CCA and DRR/DRM. The development of qualifications in CCA and DRR/DRM is groundbreaking, as is the development of a regional qualification.

The EU PacTVET training needs and gap analysis indicated that all 15 P-ACP countries prioritised training in DRM/DRR, with some countries requesting formalised training in disaster response, including training on post-disaster assessment. For example, Tuvalu, who had been hit by TC PAM just four months before their national needs and gap analysis took place, suggested that DRM and disaster response training should be built into competencies and qualifications at certificate levels 1–4 on the PQF. "If people in communities were equipped with recognised post disaster assessment skills already we wouldn't have to wait for assessors to visit communities post disaster and disaster responses could be faster" was the finding from one group of participants. Tuvalu participants in the needs and gap analysis also concluded that recognised qualifications in disaster response would provide a professional aspect to the training offered. It was concluded that all training should be aligned toward the overall "professionalization" of disaster response and management, including identifiable career paths with sequential learning stages. (The "professionalization" aspect of this finding is in agreement with the findings of Analysis of Disaster Response Training in the Pacific Island Region Provisional Version September 2012, United Nations Office for the Coordination of Humanitarian Affairs, Regional Office for the Pacific, September 2012) (Buliruarua 2015).

Regionally devised and accredited qualifications should ensure that adaptation measures limit the impacts of climate change and natural hazards; empower locals to become actors in their own development; and limiting maladaptation and the generation of new risks.

A "one-stop-shop" for resolving accreditation issues in a trans-boundary fashion could be the implementation of a CCA/DRR-specific regional accreditation agency "The Pacific Regional Federation for Resilience" whose purpose would be to:

Professionalise climate change adaptation practice (including disaster risk management/disaster risk reduction).

This will be achieved by:

Providing a Certification Scheme through formal training and recognition of informal and non-formal learning for climate change adaptation and disaster risk reduction practitioners and organisations and an Accreditation Scheme for climate change adaptation (including disaster risk reduction) qualifications and training providers.

References

Aitsi-Selmi, A., Blanchard, K., Al-Khudhairy, D., Ammann, W., Basabe, P., Johnston, D., et al. (2015). UNISDR Science and Technical Advisory Group report 2015: Science is used for disaster risk reduction. Retrieved March 25, 2016, from http://preventionweb.net/go/42848.

Bartram, J. (2004, July). A review of Technical and Vocational Education and Training (TVET) Policy and Planning in the Pacific Islands. Pacific Association of Technical and Vocational Education and Training. Retrieved February 1, 2016, from http://www.unevoc.net/fileadmin/user_upload/pubs/TVETPolicyReviewPacific.pdf.

Buliruarua, L.-A., Hemstock, S. H., Jacot Des Combes, H., Kua, N., Martin, T., Satiki, V., et al. (2015). P-ACP Training needs and gap analysis. Reports for Fiji, Cook Islands, the Federated States of Micronesia, Kiribati, Marshall Islands, Nauru, Niue, Palau, Papua New Guinea, Samoa, Solomon Islands, Timor Leste, Tonga, Tuvalu and Vanuatu. The Pacific Community and The University of The South Pacific, Suva, Fiji. (EU PacTVET—FED/2014/347-438).

Hyogo Framework for Action (HFA) 2005–2015. (2005). Building the Resilience of Nations and Communities to Disasters. United Nations International Strategy for Disaster Reduction. Retrieved February 1, 2016, from http://www.unisdr.org/we/coordinate/hfa.

IEA International Energy Agency (2011). World Energy Outlook 2011. SBN: 978–92–64–12413–4.

McNamara, K., Hemstock, S. L., & Holland, E. (2012). Practices of Climate Change Adaptation in the Pacific: Survey of Implementing Agencies (Phase II). Pacific Centre for Environment and Sustainable Development (PaCE-SD). The University of the South Pacific, Suva, Fiji. (USP EU-GCCA: FED/2010/258-661).

OECD. (2014). Overseas Development Aid, Oceania Region Statistics. Retrieved February 1, 2016, from www.oecd.org.

Pacific Disaster. (2015). Strategy for Climate and Disaster Resilient Development in the Pacific. Retrieved February 1, 2016, from http://www.pacificdisaster.net/dox/SRDP.pdf.

Reid, H., Cannon, T., Berger, R., Alam, M., Ashley, H., Kenton, N., et al. (Eds.). (2009). *Participatory Learning and Action 60—Community-based Adaptation to Climate Change*. London: International Institute for Environment and Development.

Sanerivi, L., Prasad, R., Martin, T., Jacot Des Combes, H., & Hemstock, S. (2016, in press). Changing climate for quality assured regional qualifications in the pacific: An innovative collaboration (EU-PacTVET & EQAP). In *Symposium on Climate Change Adaptation in the Pacific Region*, Lautoka, Fiji.

Smith, R., & Hemstock, S. L. (2011). An analysis of the effectiveness of funding for climate change adaptation using Tuvalu as a case study. *International Journal of Climate Change: Impacts and Responses, 3*(1), 67–78.

UNESCO Institute for Lifelong Learning. (2015). Global Inventory of Qualifications Frameworks. UNESCO Institute for Lifelong Learning, European Training Foundation, European Centre for the Development of Vocational Training, Hamburg.

UNISDR. (2015). Sendai Framework for Disaster Risk Reduction 2015–2030. Geneva. United Nations International Strategy for Disaster Reduction. Retrieved March 3, 2016, from http://www.unisdr.org/we/coordinate/sendai-framework.

UNISDR. (2016a). Launching UNISDR science and technology partnership and the science and technology road map to 2030. Short concept note: Work stream 4, working group 2. Retrieved April 1, 2016, from http://www.preventionweb.net/files/45270_unisdrcnws4wg2capacity development.pdf.

UNISDR. (2016b). The science and technology roadmap to support the implementation of the Sendai framework for disaster risk reduction 2015–2030. Retrieved April 1, 2016, from http://www.preventionweb.net/files/45270_unisdrscienceandtechnologyroadmap.pdf.

Vandenberg, P., Miyamoto, J., & Hieda, A. (2015). Education and Workforce Skills in the Federated States of Micronesia. An assessment by the Asian Development Bank Institute, Tokyo, Japan for the 2015 Pacific Update Conference: Pacific Development in a Changing Global Economy, The University of the South Pacific, Suva, Fiji, 15–16 July 2015.

Woods, J., Hemstock, S. L., & Bunyeat, J. (2006). Bio-energy systems at the community level in the south Pacific: Impacts & monitoring. Greenhouse gas emissions and abrupt climate change: positive options and robust policy. *Journal of Mitigation and Adaptation Strategies for Global Change, 4,* 473–499.

Author Biographies

Sarah Louise Hemstock holds a PhD in bioenergy systems modelling and is an author and adviser to the Alofa Tuvalu "Small Is Beautiful" project—recognised by UNESCO as one of its "Decade of Achievement Projects". Currently she leading the European Union Pacific Technical Vocational Education and Training on Sustainable Energy and Climate Change Adaptation Project (EU PacTVET) at the Secretariat of the Pacific Community, is a Adjunct Fellow of the University of the South Pacific and Visiting Fellow at Nottingham Trent University. Formerly lecturing on "Energy and the Natural Environment" at King's College London & as a consultant for Imperial College, she researched biomass resources in several Pacific SIDS. The majority of her recent work experience relates to practical resource management and sustainable development issues relevant to some of the world's most disadvantaged communities and vulnerable ecosystems. Sarah has also worked on energy policy development (UK, Uganda, Poland, Tuvalu, EU) and has more than 30 publications. Her research now examines various aspects of community initiated sustainable development, including aid effectiveness. In 2010 Sarah was made Government of Tuvalu Honorary Ambassador—Officer for Environmental Science.

Dr. Helene Jacot des Combesis the Senior Lecturer in Climate Change Adaptation at the University of the South Pacific. She is currently developing a training needs and gap analysis on TVET in 15 Pacific island countries for the 6 million Euro European Union Pacific Technical Vocational Education and Training on Sustainable Energy and Climate Change Adaptation Project (EU PacTVET). She is Course Leader for the PGDip course on disaster risk reduction and also delivers guest lectures on the ocean response to climate change. Helene is a member of the following international Pacific regional committees: Regional Working Group on Education, Climate Change and Disaster Risk Management; the Pacific Climate Change Roundtable; and the Technical Working Group for the development of the Regional Integrated Strategy on Climate Change and Disaster Risk Management, called the Strategy for Climate and Resilient Development in the Pacific (SRDP) and a member of IAEM. Under the EU-PacTVET project, her goal is to develop competencies and competency-based TEVT trainings for DRR and reconstruction and to support the accreditation of these competency-based trainings both nationally and regionally.

Chapter 20
Piecing Together the Adaptation Puzzle for Small Island States

Tabea K. Lissner, Carl-Friedrich Schleussner, Olivia Serdeczny, Florent Baarsch, Michiel Schaeffer and Bill Hare

Introduction

Small island states in the Pacific region are exceptionally vulnerable to the impacts of climate change, due to a combination of their exposure, socio-economic conditions as well as their remoteness (Hernández-Delgado 2015). Climate change awareness is high and substantial efforts have been going into adaptation planning and implementation in the region in the last years and decades (Nurse et al. 2014). A lot of knowledge has been generated, both targeted to the specific needs of Pacific islands as well as generally applicable, and knowledge hubs to access this information have been created.[1] However, as yet efforts often lack a sufficiently robust scientific underpinning and many knowledge gaps remain (see e.g. Swart et al. (2014) for a perspective on adaptation science). Science-based adaptation requires information along the full assessment chain from regional climate change projections and associated biophysical impacts, linked to information on socio-economic sectors and their sensitivities to climate impacts to understand specific vulnerabilities as well as adaptation requirements, options and costs (see e.g. Raiser 2014). At present, this assessment chain remains fragmented and insufficiently integrated,

---

[1]See e.g. http://www.pacificclimatechange.net/; http://www.apan-gan.net/ in the Pacific. Similar resource platform are available in other regions, e.g. the Caribbean: http://ccoral.caribbeanclimate.bz/about/ as well as globally, e.g. adaptationcommunity.org or ci-grasp.org.

T.K. Lissner (✉) · C.-F. Schleussner · O. Serdeczny · F. Baarsch · M. Schaeffer · B. Hare
Climate Analytics gGmbh, Ritterstr. 3, 10969 Berlin, Germany
e-mail: tabea.lissner@climateanalytics.org

T.K. Lissner · C.-F. Schleussner · F. Baarsch · B. Hare
Potsdam Institute for Climate Impact Research, Potsdam, Germany

M. Schaeffer
Environmental System Analysis Group, Wageningen University and Research Center, Wageningen, The Netherlands

W. Leal Filho (ed.), *Climate Change Adaptation in Pacific Countries*,
Climate Change Management, DOI 10.1007/978-3-319-50094-2_20

especially with regard to the needs of small islands. Methods and tools to assess the different aspects of the analysis chain exist, but often they are not applicable to the challenges that island states face (Kelman and West 2009) and climate as well as impact models usually to not provide output targeted to the pacific region (Australian Bureau of Meteorology and CSIRO 2014). Understanding the current knowledge landscape to strategically identify gaps and to systematically link information along the full analysis chain from the climate change signal to hot-spots of vulnerability is thus key to provide a robust basis for science-based adaptation.

This paper introduces the concept of a knowledge platform and database that aims to provide the conceptual framing that is required to understand the full impact chain in order to support science-based adaptation planning. While the Pacific Small Island Developing States face some unique challenges, other aspects relevant to the adaptation process are transferable across regions. A targeted approach to understand similarities and to facilitate exchange and learning between regions is a central goal of the proposed database. The paper first provides an overview of the special challenges that small islands—in general and specifically in the Pacific—face with regard to climate change, both in terms of physical exposure as well as in terms of their socio-economic circumstances and vulnerability (section "Special Challenges for Small Islands"). This overview aims at providing insights into the most important characteristics, but is by no means a comprehensive review of impacts in small islands. The paper proceeds in providing an introduction to the concept of the proposed analysis and database framework in section "Main Climate Impacts for Island States" and summarizes the main findings and next steps in a brief conclusion (section "Important Data and Research Gaps for Robust Impact Analysis").

Special Challenges for Small Islands

Small islands are amongst those most vulnerable to the negative effects of climate change: they are uniquely exposed, limited in terms of their choices of alternative livelihood options as well as remote and hard to reach in times of crisis. This brief review of some of the impacts of climate change on small islands in the Pacific highlights some of the many interlinkages between biophysical and socio-economic processes, which govern impacts for islands, underlining the need for an integrated and scientific knowledge base to ensure targeted and efficient adaptation.

Main Climate Impacts for Island States

Sea-Level Rise

Due to their geographical setting, small islands are especially exposed to the coastal and oceanic impacts of climate change, and sea-level rise (SLR) is a direct threat to low-lying islands. Rates of SLR are not uniform across the globe and higher-than-average rates have been reported for parts the Pacific region (Meyssignac et al. 2012). Many factors affect SLR rates, some of which are driven by climate change. Global projections of SLR until 2100 are within a range of 40 to 60 cm (mean values for different RCPs) (IPCC 2013). At the same time, extreme regional SLR variability is projected to increase both in frequency and intensity under climate change (Widlansky et al. 2015). A central consequence of SLR is coast-line erosion with consequent inundation of previously inhabitable land. Research on details of coastal dynamics and the effects of climate change in this context in the Pacific remains limited. A recent study conducted in the Solomon Islands highlights the potentially disastrous effects of SLR, showing that at least 5 vegetated islands have vanished in the last 60 years, with an additional 6 islands experiencing severe shoreline recession (Albert et al. 2016). Other island states, such as Tuvalu or Vanuatu, report increasingly regular inundation of previously dry land areas over the past decades (IPCC 2012). It must be noted, however, that coastal erosion processes are determined by a variety of factors, which also include human alterations of the coastline, for example, and cannot be solely attributed to SLR as a consequence of climate change (Yamano et al. 2007). There are also indications, that coral island may be capable of some autonomous adaptation to rising sea-levels (Kench et al. 2014), though these processes are not at all well understood, especially when considering the projected accelerated SLR rates (Carson et al. 2016).

Ocean Acidification, Tropical Storms and Coral Reefs

Coral reefs are of central importance for islands, as they perform a variety of functions, including economic, ecological, recreational as well as buffering functions against storm impacts and coastal erosion. There are multiple impact pathways of how climate change affects reefs: ocean acidification as a consequence of increased uptake of CO_2 presents a major threat to the calcification of corals (Gagnon 2013). Increasing water temperatures can lead to coral bleaching events, further reducing the capacity of reefs to withstand stresses; tropical storms may damage reef ecosystems, especially if these are destabilized through other factors. Especially in conjunction, these processes can have detrimental effects and can lead to a large-scale die-back of these fragile and unique ecosystems (Lough and Cantin 2014).

Recent studies suggest that under a global temperature increase of 1.5 °C relative to pre-industrial levels, 90% (median projections) of coral reefs are at risk of regular bleaching events by 2050. At 2 °C, this increases to 100% of global reefs at risk of being affected by regular bleaching events (Schleussner et al. 2016), significantly reducing the potential for recovery. 2016 is seeing unusually intense coral bleaching events across the reefs of the world, as higher water temperatures are amplified by a strong El Niño event.[2] Large scale bleaching has been reported for the Great Barrier Reef, where 93% of the area is affected, with severe bleaching in about 30%[3] of the surveyed reef area. Similar observations have been made in reefs across the world.

Additional threats to coastal areas result from a potential increase in the intensity of tropical cyclones, though projections of likely changes in frequency and intensity remain uncertain (Walsh et al. 2012). In conjunction with reduced buffering functions from coral reefs as well as higher baseline sea-levels, the impacts of storms are likely to be additionally magnified. Recent examples of the potentially disastrous effects for small islands include tropical cyclone Pam, which hit Vanuatu in March 2015 and caused estimated economic damages of US$449.4 million, approximately 64% of Vanuatu's GDP, displaced 65,000 people, damaged 17,000 homes and caused 11 confirmed fatalities (Esler 2015). In addition to these direct impacts, negative effects on overall economic and living conditions are likely to extend over the next years.

Tropical reefs are amongst the most abundant biodiversity hot-spots in the world and small islands depend on healthy reef ecosystems for various economic activities, including food supply. A decline in species abundance has been recorded as a consequence of reef degradation (Jones et al. 2004). Declines in biodiversity are further exacerbated also as a direct consequence of ocean acidification and warming (Nagelkerken and Connell 2015) and pose a direct threat to reef based livelihoods, especially fisheries and tourism. Chen et al. (2015) estimate that by 2050, a total of between 5.85 and 11.47 billion US$ of reef value could be lost to different economic reef activities globally, with the majority of losses in the tourism and recreation sectors. In addition, pollution and other socio-economic drivers are a threat to oceanic biodiversity and reef resilience (Fenner 2012).

Water Availability and Food Production

In addition to the coastal and oceanic impacts, small islands are also highly vulnerable to changes in precipitation patterns and consequent changes in water

[2]http://www.noaanews.noaa.gov/stories2015/100815-noaa-declares-third-ever-global-coral-bleaching-event.html.

[3]https://www.coralcoe.org.au/media-releases/only-7-of-the-great-barrier-reef-has-avoided-coral-bleaching.

availability. Low-lying atoll islands, such as Kiribati or Tuvalu in the Pacific, depend on fragile groundwater lenses and the availability of fresh water is mostly determined by the ratio of consumptive use to recharge through precipitation (White and Falkland 2009). Both, rising sea-levels as well as inundation and wash-overs caused by storms can lead to salinization and render water resources unfit for human consumption. Simulations of different levels of SLR indicate that at an increase of 40 cm, the thickness of the fresh water lenses may be reduced by 50% permanently. In addition, recovery times from salinity intrusions after cyclone-induced inundation increase to over one year, leading to potential long-term fresh-water deprivation in many Pacific islands (Terry and Chui 2012). Population growth and development-related pollution further increase the pressure on scarce water resources (White and Falkland 2009) and many islands have already been facing serious water shortages in recent years.[4]

Important Data and Research Gaps for Robust Impact Analysis

In addition to their exceptional exposure to climatic hazards, small islands face additional challenges: climate science is so far insufficiently able to represent small islands in global circulation and impact models. Robust data for science-based adaptation strategies in the region remains an urgent need.

The current generation of Atmosphere Ocean General Circulation Models (AOGCM) used in the climate science community to project climate change usually provide a resolution of between 0.5 and 4°, depending on model and model components (Meehl and Bony 2011). Many island nations consist of multiple islands, smaller than the resolution of such models, therefore their signal within the oceanic environment is not adequately reflected within existing AOGCM. Regionalised climate data is so far limited, though down-scaling approaches to better represent island topography have recently been applied (Australian Bureau of Meteorology and CSIRO 2014). A recent study of aridity changes downscaled to small islands revealed the relevance of such approaches projecting mean drying of the order of observed inter-annual variability (Karnauskas et al. 2016).

Similarly, biophysical impact models to understand changes in agricultural yields or water availability, for example, are currently not able to fully represent island dynamics and regional priorities. Improved representation of Pacific Islands in biophysical models is further restricted through the limited availability of long-term data series of past observations, which are critical to calibrate models and

[4]In 2011 in Tuvalu: http://www.theguardian.com/world/2011/oct/03/pacific-nation-state-emergency-water; in 2016 in the Marshall Islands: http://thediplomat.com/2016/02/drought-in-the-marshall-islands/ as well as Palau http://www.news.com.au/travel/travel-updates/palau-water-shortage-island-paradise-of-18000-about-to-run-out-of-water/news-story/dd9c92cc9fa06517726bedb5f60ece74.

provide robust projections of potential future developments (Australian Bureau of Meteorology and CSIRO 2014).

Many islands in the Pacific only rise a few meters above sea-level. The current vertical precision and accuracy of digital elevation models that are used to understand coastal exposure to rising sea-levels (see e.g. Hinkel and Klein 2009 on the DIVA model), is insufficient to capture the exposure that low-lying atoll islands face (Sampson et al. 2016): even a few centimetres delimit the difference between survival and constant flooding. In addition, larger waves and higher wave-driven water levels along atoll islands' shorelines are projected in conjunction with sea-level rise, which may cause twice as much land to be flooded for a given value of sea-level rise than predicted by current models (Storlazzi et al. 2015). While the scientific knowledge base on climate change in the Pacific has constantly been increasing and data as well as regionalized studies are increasingly becoming available, an important restriction hampering progress that remains is the currently limited scientific capacity in the region. As a consequence, Pacific island issues are not well represented in the scientific literature and associated internationally recognised fora, such as the IPCC (Pasgaard and Strange 2013; Pasgaard et al. 2015).

Special Case for Adaptation

As the overview of selected important impacts highlights, the effects of climate change are already being felt in Pacific islands and projections indicate a further increase of these impacts, also if pathways towards limiting global temperature increase to 1.5 °C are pursued by the international community, as agreed in Paris in December 2015 (UNFCCC 2015). In addition to their exposure, small islands face several challenges for sustainable and climate resilient development: (1) physical limitations—options for livelihood diversification or relocation of critical infrastructure are limited, due to their specific topography; (2) high economic dependency on ecosystems and little room for diversification—important sources of income include fisheries-based livelihood as well as tourism, both of which critically depend on healthy reef ecosystems; (3) social change—population growth as well as changes in lifestyles and consumption patterns pose further challenges for islands with limited resources and space.

Science-based adaptation, rooted in a robust knowledge base of the full chain of climate information from projections and biophysical impact information, to vulnerability and adaptation needs and costs is essential for anticipatory adaptation, which makes most effective use of limited resources and reduces the risk of maladaptation in the context of changing conditions.

Over the next years, available climate finance to support long-term adaptation in vulnerable regions will be provided by the international community through agencies, such as the Green Climate Fund. In order to access these financial resources, robust information on climate change projections, vulnerability hot-spots and adaptation requirements is required in order to fulfil the eligibility criteria of international funding agencies.

A Regional Database for Science-Based Adaptation

In the following section, we introduce the concept of an integrated database, which will be further developed and implemented in the context of an international collaborative research and development project, conducted jointly between partners in Germany, the Pacific region, the Caribbean and in West Africa.[5] The database focusses on integrating existing experience and knowledge across regions, actors and sectors in order understand the knowledge landscape and to strategically fill existing knowledge gaps in order to work towards a science-based adaptation processes. The interregional consortium and the collaboration between institutions from both, science and practice, aims to ensure the consideration of a broad knowledge base and the inclusion of available resources, while ensuring that regional priorities are considered. The development draws on and ties into existing regional as well as global platforms, such as for example the Pacific Climate Change Portal[6], the CCORAL tool[7] or the Climate Change Knowledge Portal.[8]

Concept

Our vision of the database is to provide a framework, which links data with integrated analysis steps, to provide assessments of the full analysis chain comprising the impact pathway from biophysical to socio-economic factors that determine regionally specific vulnerability to climate change. The set-up is therefore two-fold: on the one hand, the framework aims to provide a comprehensive knowledge repository. On the other hand, a direct link to the envisaged analysis framework allows for an integrated assessment of conditions to pave the way towards science-based adaptation.

By developing a conceptual frame and filling this with existing data across regions, the approach aims at strategically identifying the most important knowledge gaps. A central component of the concept is its co-development approach,

---

[5]Partners in Germany: Climate Analytics (CA), Potsdam Institute for Climate Impact Research (PIK); Partners in the Pacific: Secretariat of the Pacific Regional Environment Programme (SPREP); Partners in the Caribbean: Caribbean Community Climate Change Centre (CCCCC), Charles&Associates Inc. Grenada; Partners in West Africa: Climate Analytics Lomé, African Climate Policy Center (ACPC).

[6]http://www.pacificclimatechange.net/.

[7]http://ccoral.caribbeanclimate.bz/about/.

[8]http://sdwebx.worldbank.org/climateportal/.

which involves scientific experts as well as stakeholders and adaptation experts from the Pacific region. Counterparts to the Pacific database are planned in the Caribbean region as well as in West Africa. Each regional database focuses on regionally-specific needs and the individual starting situation, but will be developed with a view for integration across regions to ensure knowledge sharing and learning across the most vulnerable regions of the world. Figure 20.1 summarizes the main components to be included in the database. The final set-up of the database as well as the analysis framework will be developed, refined and adjusted in the context of regional priorities in the co-development setting jointly with regional experts. The focus lies on an integration of existing knowledge, to ensure there is additional value added through the strategic identification of urgent information needs. Information sources will include scientific publications and regional reports, existing databases as well as any other material of sufficient quality and robustness, to be assessed on a case by case basis.[9]

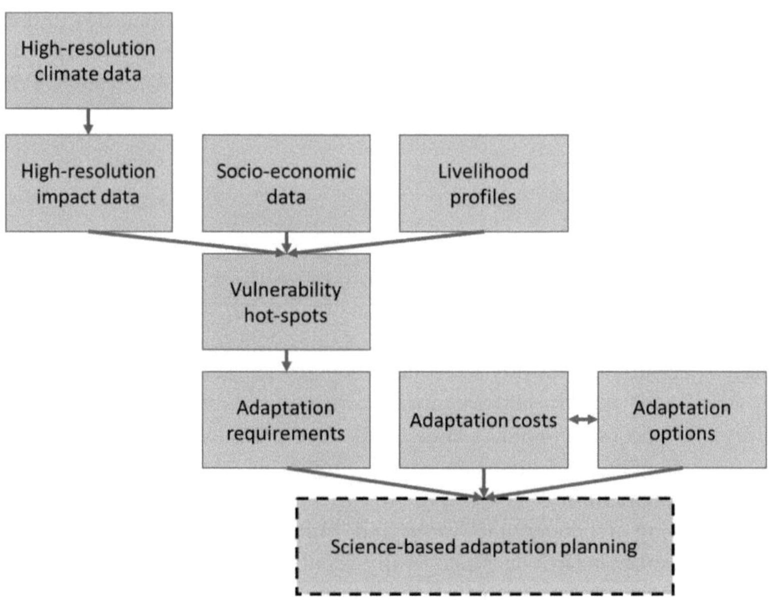

Fig. 20.1 Overview of the database concept to support science-based adaptation planning for small islands

[9]Regular updates on the progress will be available through www.climateanalytics.org. Feedback, input and commendations are welcome at any time and can be directed to the corresponding author.

Main Components of the Database Framework

High-Resolution Climate and Impact Data

A core component of the database includes a compilation of available, regionally specific biophysical climate and impact data projections. As outlined in the previous sections, especially projections of regionalised biophysical impacts, such as agricultural yields or water availability, remain limited for island regions. This database component will evaluate existing studies and modelling approaches to understand the current state of knowledge and inform the modelling community about the most urgent needs in this regard. Robust knowledge of the likely range of expected biophysical developments is essential to inform effective adaptation and as a basis to assess the vulnerability of sectors and population groups.

Socio-economic Data, Livelihood Profiles

In addition to the biophysical data, a focus of the database framework lies also on the human side of climate change impacts and the consequent needs and options for adaptation. Two main components are included to understand the landscape of vulnerability: on the one hand, available socio-economic data is compiled from a variety of sources to identify priority sectors and thematic areas in the regional context. On the other hand, based on research conducted in the context of the overarching research project, a typology of livelihood profiles will be developed jointly by all regional project partners, to understand the livelihood-specific impact pathways and consequent targeted adaptation needs.

Vulnerability Hot-Spots

Based on the information compiled on biophysical as well as socio-economic impact determinants, regional vulnerability profiles can be derived. The concept of vulnerability is a topic of continuous debate, both in the scientific and policy arenas: many different definitions and approaches for measurement have emerged (Moret 2014). One of the central challenges is the fact that different methods lead to different results and consequent differences in defining of who or what is most vulnerable (Füssel 2010; O'Brien and Wolf 2010). This challenge increasingly leads to the view that vulnerability assessments may actually hinder decision on adaptations, due to an inability to determine who is most vulnerable, rather than providing a robust information base.

The present framework explicitly addresses this challenge by providing a flexible interface to understand the different dimensions of vulnerability. It allows to transparently explore vulnerabilities and provides the opportunity for an informed integrated assessment which makes explicit the underlying assumptions that lead to a specific ranking of vulnerability. Sectoral prioritisation, choice of indicators and different methods are some examples of these assumptions and determinants, which affect results of vulnerability assessments and have made informed decision difficult in the past. The framework provides a platform to assess vulnerability in a collaborative setting, allowing for a discussion between all relevant stakeholders to come to a joint understanding of the situation.

Adaptation: Requirements, Costs and Options as well as Limits, Barriers and Potential Residual Damages

The adaptation module of the database framework contains information on existing adaptation options, showcasing best (and worst) practices for adaptation. The framing aims to provide a direct link to the outcome of the impact as well as vulnerability profiles, pointing towards adequate adaptation options. These options are also linked to information on potential costs of measures and strategies. In addition, it is also planned to include information on potential limits and barriers to successful adaptation, in order to provide information on potential residual impacts, which may require other coping or risk management mechanisms.

The integrated structure and approach of the database framework aims to link existing information and assessment approaches in order to provide a comprehensive and science-based knowledge base, which can effectively inform adaptation. The development and continuous use and refinement of the concept is critically dependant on the co-development approach, which includes a variety of stakeholders and experts from a diverse range of fields and regions. The concept it not aimed at providing a static, closed structure, but wants to provide a dynamic and flexible platform, which can be adapted to changing needs and priorities and can be updated to reflect new concepts and knowledge continuously across time.

Critically, the comprehensive and systematic overview generated through the creation of the framework will shed light on the gaps in information and knowledge along the impact chain. These insights can guide prioritization of investments into capacity building and allow for focussed efforts to fill these gaps.

Summary and Conclusions

While adaptation experience is increasing, including an evolving understanding of prerequisites and limitations to specific forms of adaptation, knowledge is fragmented, due to the mostly local nature of adaptation. The special circumstances of small islands across the world in terms of exposure, geographical setting and socio-economic options make well-planned, science-based and anticipatory adaptation an imperative. While awareness and willingness to act in the Pacific region is high, adaptation initiatives are urgently in need of a strengthened comprehensive and integrated scientific knowledge base to consolidate adaptation efforts. The database framework presented in this contribution aims at providing an open structure, which allows to gradually address knowledge gaps and contribute to an informed and effective adaptation process. Successful science-based adaptation is critically dependant on collaboration between experts, to exploit available experience, also across regions.

References

Albert, S., et al. (2016). Interactions between sea-level rise and wave exposure on reef island dynamics in the Solomon Islands. *Environmental Research Letters, 11*(5), 054011. doi:10.1088/1748-9326/11/5/054011.

Australian Bureau of Meteorology and CSIRO. (2014). *Climate variability, extremes and change in the Western Tropical Pacific: New science and updated country reports*. Melbourne, Australia: Pacific-Australia Climate Change Science and Adaptation Planning Program Technical Report, Australian Bureau of Meteorology and Commonwealth Scientific and Industrial Research Organisation.

Carson, M., et al. (2016). Coastal sea level changes, observed and projected during the 20th and 21st century. *Climatic Change, 134*(1–2), 269–281. doi:10.1007/s10584-015-1520-1.

Chen, P.-Y., Chen, C.-C., Chu, L., & McCarl, B. (2015). Evaluating the economic damage of climate change on global coral reefs. *Global Environmental Change, 30*, 12–20. doi:10.1016/j.gloenvcha.2014.10.011.

Esler, S. (2015). *Post-disaster needs assessment: Tropical cyclone pam*. Vanuatu: Port Vila.

Fenner, D. (2012). Challenges for managing fisheries on diverse coral reefs. *Diversity, 4*(1), 105–160. doi:10.3390/d4010105.

Füssel, H. M. (2010). How inequitable is the global distribution of responsibility, capability, and vulnerability to climate change: A comprehensive indicator-based assessment. *Global Environmental Change, 20*(4), 597–611. doi:10.1016/j.gloenvcha.2010.07.009.

Gagnon, A. C. (2013). Coral calcification feels the acid. *Proceedings of the National Academy of Sciences, 110*(5), 1567–1568. doi:10.1073/pnas.1221308110.

Hernández-Delgado, E. A. (2015). The emerging threats of climate change on tropical coastal ecosystem services, public health, local economies and livelihood sustainability of small islands: cumulative impacts and synergies. *Marine Pollution Bulletin, 101*(1), 5–28. doi:10.1016/j.marpolbul.2015.09.018.

Hinkel, J., & Klein, R. J. T. (2009). Integrating knowledge to assess coastal vulnerability to sea-level rise: The development of the DIVA tool. *Global Environmental Change, 19*(3), 384–395. doi:10.1016/j.gloenvcha.2009.03.002.

IPCC. (2012). Special report on managing the risks of extreme events and disasters to advance climate change adaptation (SREX). In C. B. Field, V. Barros, T. F. Stocker, D. Qin, D. J. Dokken, K. L. Ebi, M. D. Mastrandrea, K. J. Mach, G.-K. Plattner, S. K. Allen, M. Tignor, & P. M. Midgley (Eds.), *A special report of working groups I and II of the intergovernmental panel on climate change*. Geneva: Cambridge University Press. doi:10.1017/CBO9781139177245.

IPCC. (2013). Summary for policymakers. In T. F. Stocker et al. (Eds.), *Climate Change 2013: The physical science basis. Contribution of working group I to the fifth assessment report of the intergovernmental panel on climate change* (pp. 1–100). Cambridge, UK and New York, NY, USA: IPCC AR5 WGI, Cambridge University Press.

Jones, G. P., McCormick, M. I., Srinivasan, M., & Eagle, J. V. (2004). Coral decline threatens fish biodiversity in marine reserves. *Proceedings of the National Academy of Sciences of the United States of America, 101*(21), 8251–8253. doi:10.1073/pnas.0401277101.

Karnauskas, K. B., Donnelly, J. P., & Anchukaitis, K. J. (2016). Future freshwater stress for island populations. *Nature Climate Change*, (April), 1–7. doi:10.1038/NCLIMATE2987.

Kelman, I., & West, J. J. (2009). Climate change and small island developing states: A critical review. *Ecological and Environmental Anthropology, 5*(1), 1–16.

Kench, P. S., Owen, S. D., & Ford, M. R. (2014). Evidence for coral island formation during rising sea level in the central Pacific Ocean. *Geophysical Research Letters, 41*(3), 2013GL059000. doi:10.1002/2013GL059000.

Lough, J. M., & Cantin, N. E. (2014). Perspectives on massive coral growth rates in a changing ocean. *The Biological Bulletin, 226*(3), 187–202.

Meehl, G. A., & Bony, S. (2011). Introduction to CMIP5. *CLIVAR Exchanges, 16*(56), 4–5.

Meyssignac, B., Becker, M., Llovel, W., & Cazenave, A. (2012). An assessment of two-dimensional past sea level reconstructions over 1950–2009 based on tide-gauge data and different input sea level grids. *Surveys in Geophysics, 33*(5), 945–972. doi:10.1007/s10712-011-9171-x.

Moret, W. (2014). *Vulnerability assessment methodologies: A review of the literature* (pp. 1–89).

Nagelkerken, I., & Connell, S. D. (2015). Global alteration of ocean ecosystem functioning due to increasing human CO_2 emissions. *Proceedings of the National Academy of Sciences, 2015*, 201510856. doi:10.1073/pnas.1510856112.

Nurse, L. A. et al. (2014). Small islands. In: V. R. Barros, C. B. Field, D. J. Dokken, & M. D. Mastrandre (Eds.), *Climate change 2014: Impacts, adaptation, and vulnerability. Part B: Regional aspects. Contribution of working group II to the fifth assessment report of the intergovernmental panel on climate change* (pp. 1613–1654). Cambridge and New York: Cambridge Univ Press, Cambridge, United Kingdom and New York, NY, USA.

O'Brien, K. L., & Wolf, J. (2010). A values-based approach to vulnerability and adaptation to climate change. *Wiley Interdisciplinary Reviews: Climate Change, 1*(2), s. 232–242. doi:10.1002/wcc.30.

Pasgaard, M., Dalsgaard, B., Maruyama, P. K., Sandel, B., & Strange, N. (2015). Geographical imbalances and divides in the scientific production of climate change knowledge. *Global Environmental Change, 35*, 279–288. doi:10.1016/j.gloenvcha.2015.09.018.

Pasgaard, M., & Strange, N. (2013). A quantitative analysis of the causes of the global climate change research distribution. *Global Environmental Change, 23*(6), 1684–1693. doi:10.1016/j.gloenvcha.2013.08.013.

Raiser, K. (2014). Adaptation to climate change: Inciting yet another top-down/bottom-up debate. *Climate Exchange*.

Sampson, C. C., Smith, A. M., Bates, P. D., Neal, J. C., & Trigg, M. A. (2016, January). Perspectives on open access high resolution digital elevation models to produce global flood hazard layers. *Frontiers in Earth Science, 3*, 1–6. doi:10.3389/feart.2015.00085.

Schleussner, C.-F., et al. (2016). Differential climate impacts for policy relevant limits to global warming: The case of 1.5 and 2 °C. *Earth System Dynamics, 7*, 327–351. doi:10.5194/esd-7-327-2016.

Storlazzi, C. D., Elias, E. P. L., & Berkowitz, P. (2015). Many atolls may be uninhabitable within decades due to climate change. *Nature Scientific Reports, 5*(14546), 1–9. doi:10.1038/srep14546.

Swart, R., Biesbroek, R., & Lourenço, T. C. (2014). Science of adaptation to climate change and science for adaptation. *Frontiers in Environmental Science, 2*(29), 1–8. doi:10.3389/fenvs.2014.00029.

Terry, J. P., & Chui, T. F. M. (2012). Evaluating the fate of freshwater lenses on atoll islands after eustatic sea-level rise and cyclone-driven inundation: A modelling approach. *Global and Planetary Change, 88–89,* 76–84. doi:10.1016/j.gloplacha.2012.03.008.

UNFCCC. (2015). *Adoption of the Paris agreement.*

Walsh, K. J. E., McInnes, K. L., & McBride, J. L. (2012). Climate change impacts on tropical cyclones and extreme sea levels in the South Pacific—a regional assessment. *Global and Planetary Change, 80–81,* 149–164. doi:10.1016/j.gloplacha.2011.10.006.

White, I., & Falkland, T. (2009). Management of freshwater lenses on small Pacific islands. *Hydrogeology Journal, 18*(1), 227–246. doi:10.1007/s10040-009-0525-0.

Widlansky, M. J., Timmermann, A., & Cai, W. (2015, September). Future extreme sea level seesaws in the tropical Pacific. *Science Advances, 1,* 1–9. doi:10.1126/sciadv.1500560.

Yamano, H., et al. (2007). Atoll island vulnerability to flooding and inundation revealed by historical reconstruction: Fongafale Islet, Funafuti Atoll. *Tuvalu. Global and Planetary Change, 57*(3–4), 407–416. doi:10.1016/j.gloplacha.2007.02.007.

Chapter 21
Using Modelling Outputs to Inform Coastal Climate Change Adaptation Studies: Practical Tips for Adaptation Planners and Scientists

Gregory Wilford Fisk, Philip Edward Haines and Beth Frances Toki

Introduction

Understanding, analysing and spatially representing the risk of impacts from future climate change has been a key focus of coastal climate change vulnerability assessments in Australia over the past 10 years. While our ability to numerically model these impacts has improved dramatically over this time, we now have the problem of interpreting an increasing quantity of simulated results and predicted data.

The key driver for undertaking climate change impact modelling is communication of risk and supporting decision makers on the need to address potential hazards that will manifest from future climate change. But providing 'lines on maps' has been met with community and political resistance, which has hindered effective climate change adaptation and planning for extreme weather events, setting back significant progress that has been made in this field since the late 1990s. This is particularly the case in Australia where for the last few years (since 2012) conservative governments at the National, State and local level have largely wound back the need for consideration of climate change in development planning, and in some cases, removing mention of climate change from planning laws and documentation altogether.

This paper examines the evolution of how modelling outputs have been used and presented for use in climate change adaptation planning. Assessment of the

G.W. Fisk (✉) · P.E. Haines · B.F. Toki
BMT WBM, Level 8, 200 Creek Street, Brisbane, QLD, Australia
e-mail: greg.fisk@bmtwbm.com.au

P.E. Haines
e-mail: Philip.haines@bmtwbm.com.au

B.F. Toki
e-mail: beth.toki@bmtwbm.com.au

W. Leal Filho (ed.), *Climate Change Adaptation in Pacific Countries*,
Climate Change Management, DOI 10.1007/978-3-319-50094-2_21

evolving techniques has included consideration of the uncertainty in actual modelled outputs and the degree of usability of these outputs by strategic planners and other decision-makers.

Evolution in Mapping Coastal Climate Change Impacts

Climate change vulnerability studies are increasingly reliant on deterministic numerical modelling to represent future climate scenarios in order to identify areas potentially subject to a future hazard. These models generally simulate surface water hydraulics (e.g. flooding, storm-tide inundation), shoreline response (e.g. coastal erosion and recession), groundwater hydraulics and geotechnical stability.

More than 10 years ago it was common for coastal erosion modelling to produce just three scenarios: present day, 2050 and 2100 conditions (see Fig. 21.1). While simple to interpret, the significant number of assumptions that were intrinsic to the modelling meant that actual shoreline response could feasibly be vastly different to the modelling conditions. Yet the significant uncertainty in the modelling was rarely discussed or challenged.

As engineers became more aware of some of these limitations, such as potential slope adjustment after coastal erosion events, the response was simply 'more lines on the map' (see Fig. 21.2 for example). While this was convenient for the

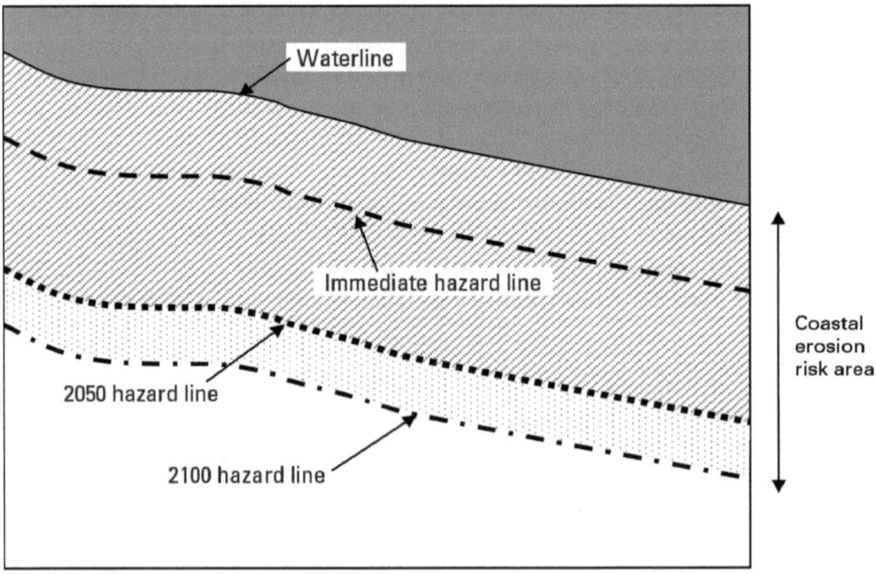

Fig. 21.1 Typical definition of coastal climate change erosion hazards (*Source* ESC 2016)

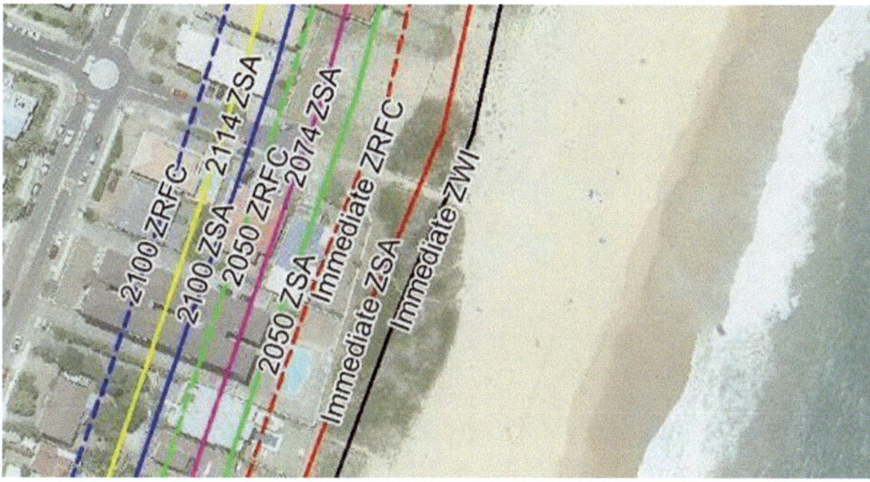

Fig. 21.2 Nine hazard lines reflecting different modelling assumptions and different timeframes for consideration (*Source* Horton and Britton 2015)

modellers, it still did not address the underlying uncertainty in the outputs. In fact, the additional lines implied a false sense of greater certainty.

Over the past 5 years or so there has been a paradigm shift towards risk-based management when considering coastal climate change impacts (Rollason et al. 2010; Rollason and Haines 2011). A risk-based approach allows for greater consideration of uncertainty within modelled outputs. Thus, instead of just one line representing a predicted shoreline position at 2100 (as per shown in Fig. 21.1), the risk-based approach enabled the use of multiple lines, each with a different 'likelihood' or chance of occurrence (typically ranging from an almost certain chance to a very rare chance). Figure 21.3 shows three likelihood lines for shoreline recession at Coffs Harbour, NSW for the current timeframe (BMT WBM 2011). Lines were also produced for future timeframes, including 2050 and 2100.

Probabilistic approaches are common for floodplain mapping, and this has been extended to coastal tsunami hazard mapping as well, as shown in Fig. 21.4 for an assessment in the Choiseul Province of the Solomon Islands (BMT WBM 2014a). Again, mapping can be presented at varying timeframes based on underlying assumptions associated with future sea level conditions.

Indeed, most probabilistic inundation mapping of coastal floodplains (see Fig. 21.5 for example) requires basic assumptions on future sea level conditions, as well as other factors such as coincident timing of sea and rain events, tide conditions, and barometric set-up levels.

A risk-based approach requires consideration of the 'likelihood' of impact within a particular area, as well as the 'consequence' of impact if it is to occur. NGIS (2013) applied this approach, in a simple manner, when evaluating the impacts of sea level rise in Nuku'alofa, Tonga (Fig. 21.6). The plot highlights the vulnerability

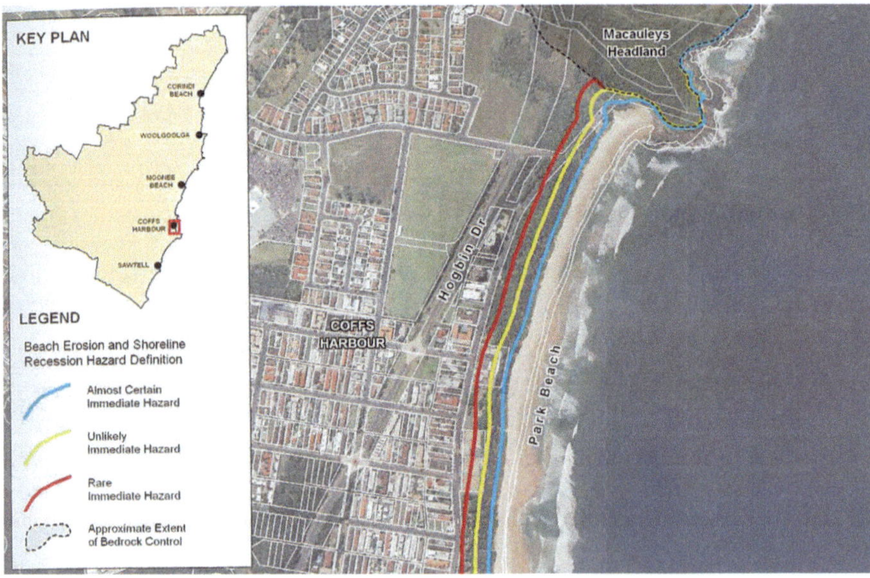

Fig. 21.3 Example of shoreline mapping showing different likelihoods of occurrence (for current timeframe) (*Source* BMT WBM 2011)

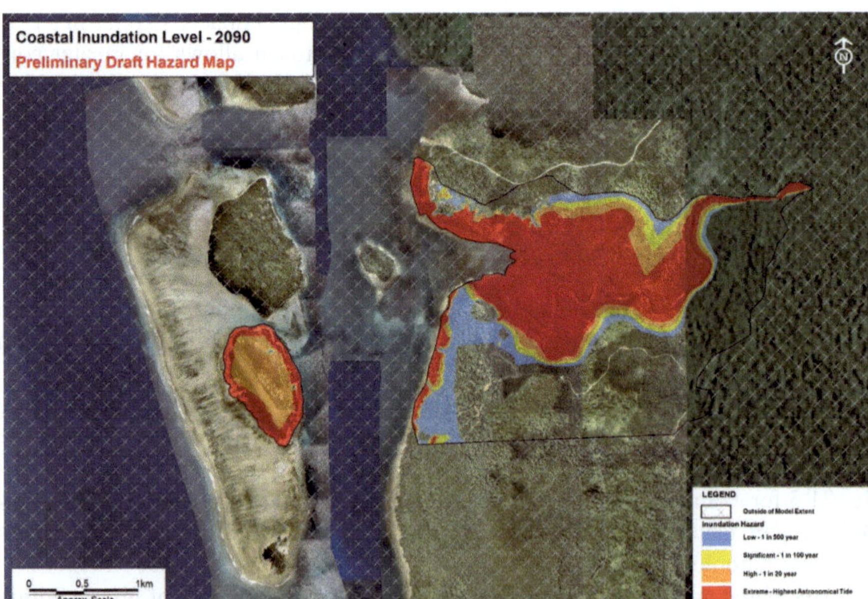

Fig. 21.4 Example coastal inundation mapping based on probability of occurrence (future timeframe) (*Source* BMT WBM 2014a)

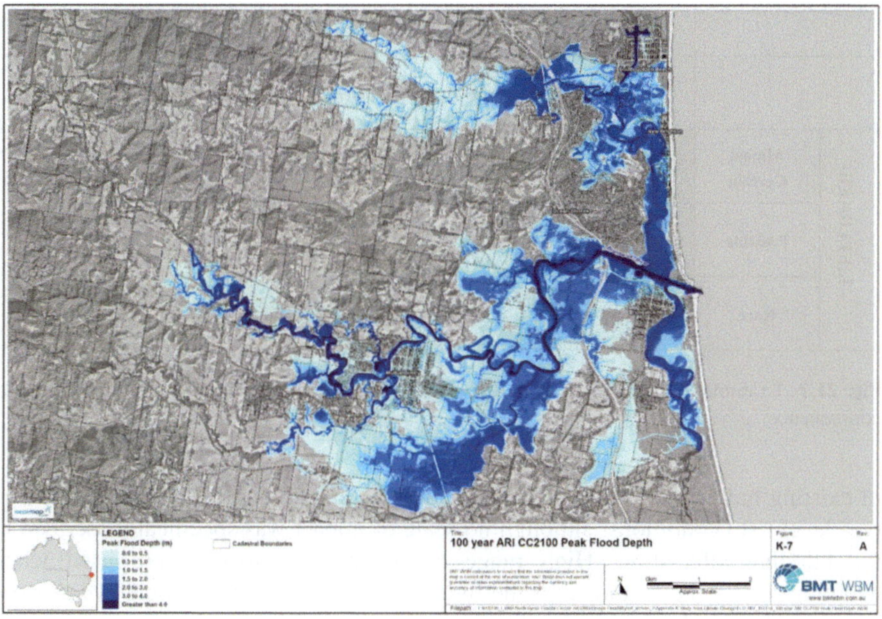

Fig. 21.5 Example of flood inundation depths for fixed probability (for future timeframe) (*Source* BMT WBM 2016)

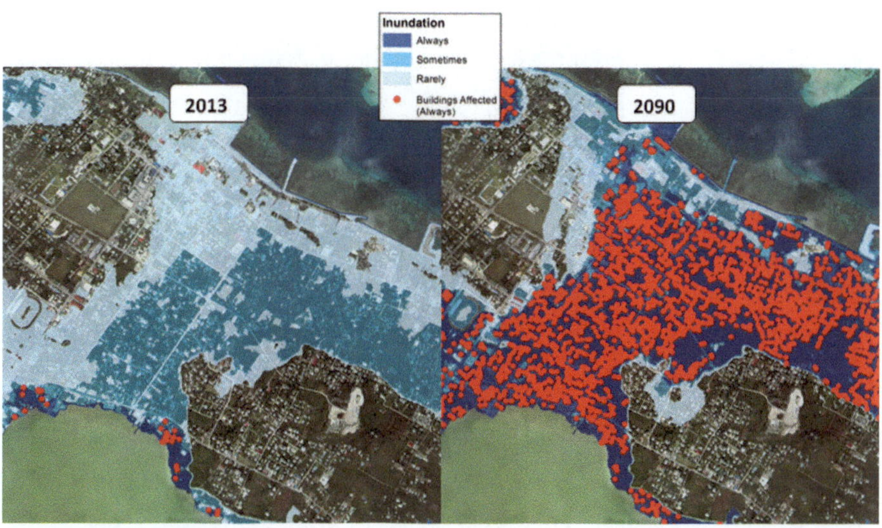

Fig. 21.6 Example of asset inundation mapping for variable probability (for current and future timeframe) (*Source* NGIS 2013)

		CONSEQUENCE				
		Insignificant	*Minor*	*Moderate*	*Major*	*Catastrophic*
LIKELIHOOD	Almost Certain	Low	Medium	High	Extreme	Extreme
	Possible	Low	Low	Medium	High	Extreme
	Rare	Low	Low	Low	Medium	High

Fig. 21.7 Example risk matrix that defines the level of risk based on both likelihood and consequence

of existing buildings to future sea inundation when adopting future climate change (sea level rise) conditions, although no scale of impact was specifically ascribed to the building affectation. High range conservative estimates were used when adopting sea level conditions for the future timeframe, while historical events were used as the basis for inundation likelihood.

The most contemporary and sophisticated risk maps involve integration of spatially represented 'likelihood' and 'consequence'. BMT WBM (2014b) applied

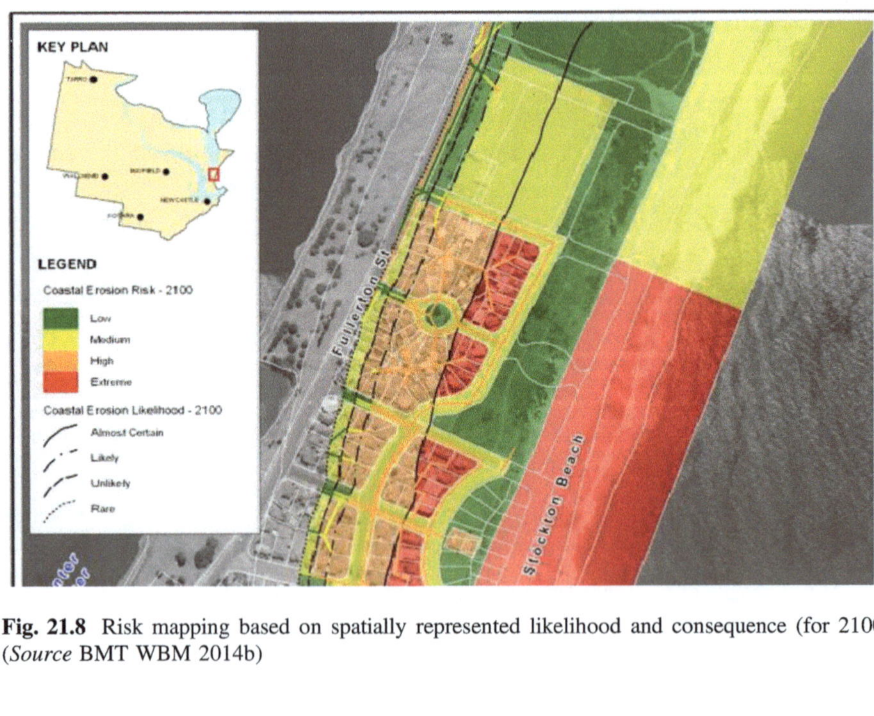

Fig. 21.8 Risk mapping based on spatially represented likelihood and consequence (for 2100) (*Source* BMT WBM 2014b)

this approach at Stockton Beach, NSW, wherein an agreed risk matrix (Fig. 21.7 for example) of likelihood and consequence was used to derive overall risk levels for areas subject to future hazard (Fig. 21.8). In this example, levels of consequence were related to land use zoning as a proxy for value of the land and assets at risk. Similar risk maps were generated for different timeframes demonstrating a dynamic risk profile for the coastal fringe with time.

Challenges in the Interpretation of Model Outputs

Model outputs are typically presented as GIS-based maps or lines as part of hazard vulnerability studies. The resultant hazard lines and hazard areas can be an effective way to communicate current and future risk of impact to land owners and occupiers in vulnerable areas. The generation and publication of hazard maps can also satisfy a planning authority's duty of care to notifying occupants of known risks, thus limiting future liability for the authority.

Our experience indicates though that hazard mapping on its own, without a complementary plan for future adaptation, can be counterproductive and in fact can become a barrier to effective decision making. We highlight below five common challenges associated with interpreting modelled outputs for climate change and extreme weather hazard studies.

Not Enough Information—Oversimplification

Through a lack of awareness or simply a lack of appreciation, humans tend to simplify our understanding of complex systems down to a relatively basic level. This includes a lack of recognition and consideration of natural variability and the inherent uncertainty and unpredictability of weather and climatic systems. Not recognising this uncertainty becomes a barrier to effective adaptive management in its own right but also precludes developing customised responses to climate change and extreme weather events that can be differentiated based on different risk and probability levels.

While presenting a 'single' line that represents future conditions (see Fig. 21.1 for example) is alluring from a planning perspective, the high uncertainty attached to a single line is potentially lost on end users and the community that will ultimately rely on this information for future decision making. In many instances though, decisions need to be made on this simple representation of risk. The result is inevitably overly conservative planning and development decisions.

Too Much Information

There are a number of variables that need to be considered when assessing future hazard vulnerability. This includes for example the likelihood of occurrence of extreme events and the future timeframe for consideration. The introduction of ISO 31000:2009 standard risk management approaches to coastal management (Rollason et al. 2010; Rollason and Haines 2011) allows a mechanism for considering a greater number of possible scenarios. Conceivably, hazard lines or areas can be generated for any adopted likelihood and for any timeframe. Commonly up to nine hazard lines are produced representing a range of likelihoods for a range of climate conditions (see Fig. 21.9 for example).

This presents a significant challenge for end users as they attempt to interpret the relevance of each line, which is made more difficult when there are vastly different consequences as a result of adopting different lines. Multiple lines are also very hard for the community to interpret as they generally have a poor understanding of how the lines are derived, how the numerical models work, and their inherent uncertainties. Unfortunately, the complexity involved in interpreting the meaning of multiple lines is lost on many end-users, who simply want to default hazards to a single line.

Having a lot of information may be advantageous but only if it is used to demonstrate that choices made for future land use and asset management are based on the most reasonable or appropriate estimate (amongst a range of alternatives). This therefore allows for a considered level of risk tolerance noting that some

Fig. 21.9 Interpretive descriptors for hazard lines to aid in end use for planning (*Source* BMT WBM 2013)

communities will seek to implement a more risk-averse approach (asserting they are 'leading the way' and 'best practice') while other communities will seek to implement a more modest or passive approach ('following others' and adopting minimum standards of compliance only).

Where there is an appropriate level of differentiation between scenarios (but not too many in the context of the argument above) decision makers have been able to develop innovative, transparent and adaptive planning responses. An example, using the three hazard conditions presented in Fig. 21.8, would be to use the hazard study to set planning controls as follows:

- prohibit all new development seaward of the 'Minimum' hazard line;
- permit limited development seaward of the 'Best Estimate' hazard line that can accommodate coastal erosion impacts or are otherwise considered sacrificial; and
- impose no practical constraints/preferentially place development landward of 'Maximum' hazard line.

This graded approach to risk would not be possible with a single line definition of hazard; but noting any approach still has to be adaptive and reviewed as the climatic system is neither static or spatially homogenous and the lines will need to be reviewed and amended over time in response to emerging science.

Defaulting to the Worst Case

Faced with uncertainty there is a predilection to show both a 'most likely' case but also a 'worst' case based on the need for conservatism, concern about future liability and the application of the widely used 'precautionary principle'. Most climate change policies adopted in Australia and the Pacific use a sea level rise value of 0.8 to 0.9 m by 2100, which is clearly at the high end of projections issued by IPCC AR5 (see Fig. 21.10).

While there is a sound mantra of 'planning for the worst, but hoping for the best', adopting a worst case may be a significant barrier to effective climate change adaptation, as it may lead to a level of mal-adaption response where measures are put in place that are pre-emptory to impacts that may never be realized, resulting in key economic and social benefits that are potentially sterilised.

Ultimately, community and other decision makers will respond more favourably to adopting climate change adaptation measures where they can see a clear trajectory of impact—particularly at a local scale. This underlies the need to collect data and information over the few next decades to capture this detail as well as to build our knowledge of the effects of major storms and weather events to benchmark their frequency and severity against natural variability.

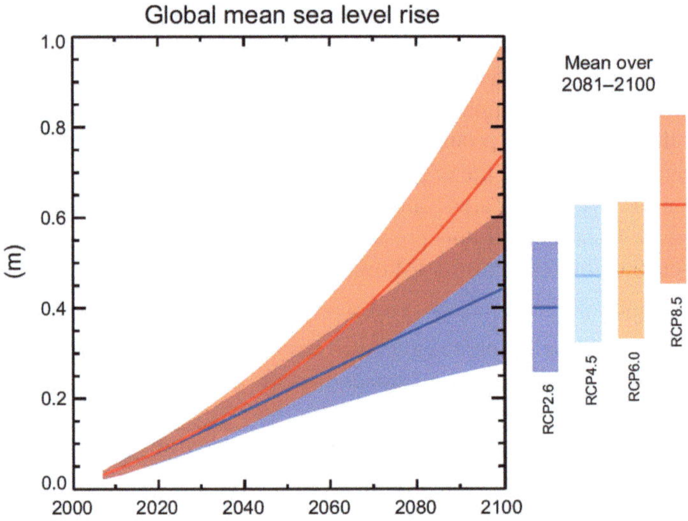

Fig. 21.10 Global projections for sea level under a range of climate scenarios relative to 1986–2005 (*Source* IPCC 2013)

1.1 *Understanding the consequences of being in a hazard zone—is it the end of the world?*

As stated above, the move towards ISO 31000:2009 risk-based approaches allows consideration of variability in likelihood or probability of an impact occurring, and the spatial extent of the impact. Although this is critical to understanding vulnerability, it is in fact only half the risk equation noting a level of analysis must also be undertaken to ascertain the consequence of the impact on affected land and assets.

Land and assets may have a high degree of resilience to impact, meaning that actual risk to future climate change may be relatively low even though likelihood is high. For example:

- an important social value or asset affected by extreme temperature may be able to be done elsewhere or undertaken at a different time of day with little impact on the community;
- a built asset may be able to withstand periodic inundation without permanent damage; and
- an environmental asset may be able to readily and naturally recover from storm damage, such as beach dune vegetation.

All of these examples demonstrate that although these assets are located within an area of exposure from climate change and extreme weather event impacts, the consequence of this exposure may not be significant or can easily be ameliorated. This is why it is critical to consider both likelihood and consequence as part of adaptation planning.

1.2 *Understanding the confidence level and dealing with uncertainty*

Decision makers prefer to make decisions based on reliable and certain information. Unless uncertainty is explicitly presented, the outputs from hazard models could potentially be treated 'as gospel' without consideration of reliability and defensibility.

For the scientists and engineers that have derived the lines, there is generally a much broader understanding of associated uncertainty due to the range of variables and assumptions made in the modelling, including:

- the quality and reliability of input data,
- the level of calibration or 'goodness of fit' to real data,
- the inherent uncertainty around the adopted climate change projections and timeframes of impacts, and
- how climate change impacts will vary between global projections and the local context.

For highly conservative decision makers, these variables and the uncertainty they produce can be a significant barrier to ANY planning response purely on the basis that the justification for adaptation action cannot be unequivocally proved.

Discussion and Suggested Approach to Coastal Adaptation Planning

To overcome the challenges in decision-making that manifest from the uncertainties in timing and extent of future climate change impacts, a 'continuum' concept model has been developed by Fisk and Kay (2010) that employs the use of shorter term warning triggers to overcome the inherent uncertainties associated with the timing and severity of longer term impacts (Fig. 21.11). The model aims to facilitate a pragmatic adaptive approach to future uncertain outcomes. The use of triggers to define when future actions will take effect enables clear intent to be given, but without the need to actually do works and actions until certainty of potential impact has been significantly improved.

Advantages to setting trigger levels for adaptation actions are as follows:

- Assuming the trigger if set effectively should give communities adequate time to act prior to realising the undesirable impact. This could include for example, development of a response plan, time to run an adequate public consultation programme, and sufficient time to more quantitatively assess the various adaptation options such as the feasibility of constructing physical works (or other expensive and/or controversial adaptation response);

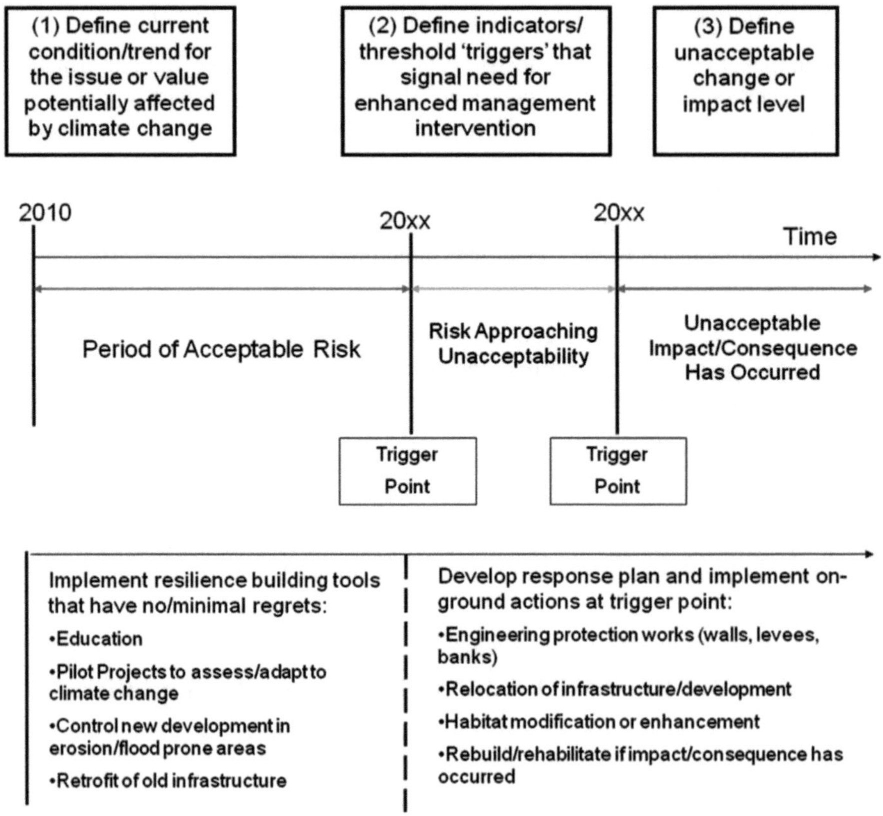

Fig. 21.11 Continuum model for climate change adaptation action (*Source* Fisk and Kay 2010)

- It prevents decision makers from acting too early and mal-adapting. Premature action in response to climate change can have significant cost and political implications such as exposing decision makers to appeals against development decisions and compensation claims. Acting too early also does not take advantage of any research and technological advancements that continue to occur during the period of acceptable risk (e.g. the period leading up to the trigger point);
- The approach can help to 'lock' decision makers into an action plan over a longer period of time and facilitate evaluation of plans and strategies; and
- The approach demonstrates to the broader community a level of preparedness and a willingness to act as climate change impacts are realised and as trigger points are approached in the future; but inversely noting that if climate change impacts are not realised to the extent predicted, the community is not unnecessarily burdened or worse off.

Limitations of Assessment

Advances in coastal science and engineering allows us to better understand complex coastal processes and the interactions between man-made structures and development and these processes. The assessment and recommendations provided in this paper are limited to consideration of coastal management in the Australasia region. Coastal processes differ around the world, and as such, impacts of future climate change are likely to be different, necessitating different management and planning responses.

Conclusions

Climate change vulnerability and related hazard studies are evolving, making use of ever improving data sets and modelling platforms to illustrate and predict future impacts. However, it is increasingly important that such studies have the end user in mind both in terms of explaining how hazard areas and lines have been derived but also how they should be interpreted in the context of response plans that will have potentially significant financial and social cost.

Challenges described above are in our view some of the most common traps and pitfalls of interpreting numerical model results by end users. The examples provided start to demonstrate the need to develop and implement more adaptive planning approaches that can take into account and reconcile the uncertainties present in the both climate change science and our ability to predict it through numerical models.

References

BMT WBM. (2011). *Coffs harbour coastal processes and hazards definitions study*. Report prepared for Coffs Harbour City Council by BMT WBM Pty Ltd, Broadmeadow NSW.

BMT WBM. (2013). *Byron shire coastline hazards assessment update*. Report prepared for Byron Shire Council by BMT WBM Pty Ltd, Brisbane.

BMT WBM. (2014a). *Integrated climate change risk and adaptation assessment to inform settlement planning in Choiseul Bay, Solomon Islands*. Report prepared for Australian Department of the Environment by BMT WBM Pty Ltd, Brisbane.

BMT WBM. (2014b). *Newcastle coastal zone hazards study*. Report prepared for Newcastle City Council by BMT WBM Pty Ltd, Broadmeadow.

BMT WBM. (2016). *North byron shire flood study report*. Report prepared for Byron Shire Council by BMT WBM Pty Ltd, Brisbane.

ESC. (2016). Development assessment and sea level rise. Retrieved April 28, 2016, from http://www.esc.nsw.gov.au/development-and-planning/considerations/coastal-and-flooding-considerations/coastal-and-flood-management/interim-coastal-hazards-adaptation-code/development-assessment-and-sea-level-rise. Eurobodalla Shire Council.

Fisk, G. W., & Kay, R. (2010). Dealing with uncertainty in climate change adaptation planning and developing triggers for future action. In *Proceedings of Practical Responses to Climate Change National Conference 2010, Melbourne, Australia.*

Horton, P., & Britton, G. (2015). Defining beachfront setbacks based on 'Acceptable risk'—is it the new approach? In *Proceedings of Australasian Coasts and Ports Conference, Auckland, New Zealand.*

IPCC. (2013). Climate Change 2013: The Physical Science Basis. Intergovernmental Panel on Climate Change, Cambridge University Press, Cambridge.

NGIS. (2013). CRC014 Pacific Islands Capacity Building: Coastal Risk Assessment Report—Tonga. Report prepared for Australian Department of the Environment by NGIS Pty Ltd, Perth.

Rollason, V., Fisk, G., & Haines, P. (2010). Applying the ISO 31000 risk assessment framework to coastal zone management. In *Proceedings of 19th NSW Coastal Conference, Batemans Bay, Australia.*

Rollason, V., & Haines, P. (2011). Outcomes from the application of ISO 31000:2009 risk management principles to coastal zone management. In *Proceedings of 20th NSW Coastal Conference, Tweed Heads, Australia.*

Chapter 22
Lifelong Learning (LLL) for Energy Practitioners in Small Island Developing States (SIDS): The Pivotal Role of Education in Energy Efficiency and Demand Side Management

Pravesh Raghoo, Pratima Jeetah and Dinesh Surroop

Introduction

Small Island Developing States (SIDS) is a group of heterogeneous islands that share common geographical features, economic traits, social situation and environmental problems as illustrated in Table 22.1.

Compared to continental areas, SIDS are highly susceptible to the impacts of climate change and global warming. These threats include sea–level rise, droughts, earthquakes, floods and inundations, coral reef deterioration and extreme storms events among others. Sea–level rise and global warming are direct threats to these island nations as rising seawater level leads to erosion of beaches affecting tourism and virtually all other sectors like financial services, agriculture, fisheries, water and sanitation (UNEP 2014). Natural disasters affecting islands have adverse economic impacts as in the case of Haiti for which, from the period 2000–2014 the total damage accrued from natural disasters was 118% of their Gross Domestic Product (GDP) (following January earthquake) and in the case of Samoa where total damage from natural calamities amounted to 41.8% of their GDP.[1] Climate change impacts in island nations were rather neglected until its first consequences started to appear. In Anjouan Island, Comoros and in Mauritius, it was found that beach aggregate mining was a major influencer of beach erosion (Sinane et al. 2010). In Papua New Guinea and in the Solomon Islands, 63, 000 people were relocated

[1]Statement based on authors' calculations from World Bank, World Development Indicators (WDI) and International Development Database, Centre for the Epidemiology of Disasters (CRED).

P. Raghoo (✉) · P. Jeetah · D. Surroop
Department of Chemical and Environmental Engineering,
University of Mauritius, Réduit, Mauritius
e-mail: praghoo@gmail.com

W. Leal Filho (ed.), *Climate Change Adaptation in Pacific Countries*,
Climate Change Management, DOI 10.1007/978-3-319-50094-2_22

Table 22.1 Some common features of Small Island Developing States (SIDS). [*Sources* Bass and Dalal–Clayton (1995), Betzold (2015), Blancard and Hoarau (2013)]

Geographical features	Economic traits	Social situation	Environment
– Small physical scale – Distance from international markets – Islands are either low–lying atolls or mountainous	– Diseconomies of scale – Not competitiveness on global market – Thin financial base	– Lack of employment – Small but growing population	– Land degradation – Loss of biodiversity – Waste management – Vulnerability to climate change

following an inundation event in 2008 (Hoeke et al. 2013). Russell (2009) showed the link between diseases like dengue and malaria and climate change by taking three Pacific Islands namely Samoa, Tonga and Kiribati as examples. In addition, severe droughts and floods and freshwater unavailability can deteriorate the level of sanitation and hygiene in some Caribbean, Pacific and Africa islands as showed in McMichael and Lindgren (2011). Tourism which is a major pillar for most SIDSs' small open economies, is a weather dependent and climate–sensitive sector which can surely affect a visitors' perception and choice for a vacation (Nurse 2014). In the foremost, it should be noted that islands contribute less than 1% of greenhouse gas emissions (GHG) which is the main contributor to climate change and global warming but are paradoxically the first victims of climate change (Gomes 2014). In Mauritius, for instance, carbon dioxide emissions amounts to 0.0096% of global emissions (CSO 2015).

Climate change adaptation and mitigation measures in islands are very cost intensive as these measures involve much infrastructural works with high upfront costs. The fact that islands are distant from international markets increases freight costs for the imports of raw materials and industrial supplies (Nurse 2014). SIDS economy are basically constricted to a thin base of economic sectors with a high dependence on imports for food and energy and relatively low profits from product exports. Hence, SIDS face significant issues for funds allocation in climate change mitigation and adaptation projects due to thin financial base and many priority sectors requiring government financial support to sustain.

As the two most feasible options to mitigate climate change impacts in island nations, widespread adoption of sustainable energy sources and energy efficiency (EE) measures are widely sought. The huge potential of renewable energy (RE) in SIDS is indeed a plus point for climate change abatement and to curb down any rise in anthropogenic carbon dioxide and other GHGs emissions. However, RE expansion in SIDS has known some financial and institutional constraints as elaborated in Dornan (2015a). EE on the other hand, is regarded by the International Energy Agency as the largest contributor of energy services as being the 'first fuel' (IEA 2012). EE is basically a compelling option, as effective as that of shifting to more sustainable energy sources in order to combat the devastating effects of climate change. To recall, EE means the limited use of energy to carry out a function that would otherwise require more energy to be completed, without any loss of quality in the final product from the function. It can be achieved by substituting less efficient technologies by more efficient ones, through laws and legislations, through energy policies and through a change in energy consumers' lifestyle and behaviour via awareness schemes and education. EE can be applied at a household, commercial, transport and at industrial levels—basically everywhere energy is used. A concept derived from energy efficiency is demand side management (DSM) which deals with planning, implementing and monitoring activities that alters consumers' energy demand.

The uniqueness of this paper lies here, where a focus is made on educating energy practitioners[2] in SIDS. There are two related issues gaining widespread attention in SIDS. Firstly, SIDS show less resilience against climate change impacts; have a high exposure to energy security and in some regions have low electrification rates. Outcomes of some energy policies to expand RE and EE practices in SIDS in order to boost the energy sector have emerged less praiseworthy in some cases (e.g. Summer Time in Mauritius). On the other hand, studies (see Singh et al. 2015) concluded that there was a lack of knowledge among energy practitioners in some Pacific and African SIDS based on a study they conducted under Project L^3EAP (Lifelong Learning for Energy Security, Access and Efficiency in African and Pacific Small Island Developing States). Ironically in SIDS, on one side there is a poor base of energy experts and on the other a less performing energy sector. This paper therefore attempts to address this gap in literature. The direct and immediate response to this issue is therefore a course on Energy for SIDS to provide further training to energy practitioners and increase knowledge base in islands. However, as simple as it may seem, this will undoubtedly not be possible without an intermediate step: curriculum planning. In the development of a course, it is important to identify the need for energy education, key matters that has to be addressed through education and the technicalities to ensure that the objectives of the propose course are met. The paper also highlights challenges for EE education implementation and a potential solution through global cooperation and private sector participation.

While most emphasis is made on SIDS in this paper, the information and claims presented is not limited to island nations only. They can presumably be extended to other non-island countries which share similar sustainable and energy problems like SIDS. For example, sub-Saharan African countries where energy courses are rare. The paper is arranged as follows: section "SIDS and Energy Issues" reviews the major energy issues in SIDS. Section "Conclusion from Previous Assessment" describes past research and the need to design EE courses; sections "Importance of EE and DSM Education for Energy Practitioners" and "Desirable Features of an EE

[2]In this text, the terms *energy practitioner*, *energy professional* and *learners* are used interchangeably.

and DSM LLL Course for SIDS" outline the importance of EE and DSM education and some desirable features of an effective EE course. Section "EE as an Online Lifelong Learning Course" deals with the mode of delivery of such course such that it attains a high participation rate and sections "Issues and Challenges in EE Education" and "Overcoming Barriers Through Global Cooperation" describes some barriers and solutions that course developers, teachers as well as learners can faced.

SIDS and Energy Issues

This section lists two of the major issues in SIDS which are energy access and energy security. It also describes EE and DSM concepts and implications. Table 22.2 gives a summary of other energy issues in SIDS which can potentially be addressed by energy efficiency.

Energy Access: A Major Island Issue

Different authors have defined energy access based on their understanding of the term. Brew–Hammond (2010) defined 'energy access' as the 'ability to use energy' whether it is electricity, natural gas, LPG or charcoal or in any other form. According to Modi et al. (2005), energy services can be described as those facilities

Table 22.2 Summary of major energy issues in SIDS. [*Source* Adapted from UNEP (2014)]

	High	Low	Description
Fossil fuel dependence	✓		Most electricity is produced from diesel generators and transport is entirely dependent on fossil fuels
Access to electricity	✓		Some remote and rural areas in SIDS are still without electricity
Energy prices	✓		Electricity costs are relatively higher. Un–electrified households spent on batteries and kerosene for energy needs
Energy supplied by RE		✓	RE in sparingly tapped. Exceptions are Fiji with 50% power from hydro; Cape Verde with 30% from wind energy
Transmission losses	✓		Aging infrastructure and poor maintenance are poor resulting in high transmission and distribution losses
RE potential	✓		Wind, solar, tidal and geothermal energy have huge potential. Efforts are towards expanding biomass energy
Investment in RE	✓		SIDS are seeking investment from donor agencies to achieve established policies and RE targets and diversify energy mix

that electricity, fuels and mechanical power provide such as lighting, space heating, transportation, water pumping for agriculture and heat for cooking among others. A widely accepted definition of energy access was given by IEA (2014) as englobing these four elements: (1) minimum electricity access in households, (2) access to safer and sustainable cooking fuels and stoves, (3) access to modern energy sources for productive uses and (4) access to sustainable energy sources for public services. Often undermined in energy discussions, the issue of energy access in SIDS is real, especially in rural and remote areas. Energy poverty in the Pacific Islands Countries (PICs) is mostly concentrated in Papua New Guinea (PNG), Solomon Islands and Vanuatu with 10, 14 and 17% electrification rate respectively (Dornan 2014). In the Caribbean SIDS, Haiti has the lowest electrification rate (38.5%) and among African SIDS, Guinea–Bissau and the Comoros have electrification rates with 20% and 45% respectively (IEA 2014; Martin et al. 2013).

Access to electricity is essential and a key player for a comfortable and productive lifestyle. Electricity is critical for industry–based economies, in ensuring adequate medical services and in education where students can study till late at night. Other benefits of having access to energy is thoroughly dealt in Karekezi et al. (2012). In SIDS, electrification issues arise as small and isolated families living in poor rural areas fail to create an optimum demand for electricity that motivates investment from power utilities. This is because of low income and low ownership of electrical appliances in rural areas. Hence, grid expansion to supply these families with electricity is often not lucrative for power utilities. Off–grid electrification options such as implementing stand–alone diesel–powered generators and sustainable energy technologies like solar photovoltaics or wind turbines can eventually achieve higher electrification in rural and isolated areas but the lack of trained personnel to carry out maintenance and reparation works on these power systems make the latter dormant for months and affecting electricity supply Dornan (2015b).

There are however, a number of energy policies and electrification programmes that were launched to widen electrification in SIDS. While some failed, others gathered much success. Dornan (2014) highlighted some policy recommendations to widen energy access in SIDS but nonetheless, claimed that politically motivated decisions is often a significant barrier to widen electrification.

Energy Security

While the issue of inadequate electrification is concentrated in a group of SIDS, energy security is widespread in almost all SIDS due to their petroleum–intensive energy sector and their high dependence on fossil fuels imports. Oil is basically the fuel of choice for most SIDS for transport, power generation and other industrial and commercial activities. In Antigua and Barbuda, Tonga, Kiribati, Seychelles and Niue almost 100% of their energy sector is dominated by oil. High dependence on imported oil is basically, the primary cause of energy security.

So far there is no consensus on a general definition of energy security. Winzer (2012) gathered 37 definitions, Sovacool (2011) gathered 45 definitions and Ang et al. (2015) compiled a plethora of 83 definitions from an intensive literature review from scientific publications and other think–tank reports. A number of authors have also attempted to quantify energy security by proposing indexes while others have attempted to determine dimensions and indicators of energy security. A well acclaimed definition of energy security was given by the Asian Pacific Energy Research Centre as the capacity of an economy to ensure adequate supply of energy resources at an energy price that it affordable and reasonable to a level that will not compromise the proper economic functioning of a country APERC (2007). Most oil is sourced from the Middle East but since they are politically unstable countries, they affect considerably the price of oil on the international market. Volatile oil prices is a real game–changer for SIDSs' small economy. SIDS lack preparedness and efficient strategies to deal with these prices fluctuations and subsequent economic shocks. Volatile prices can cause inflation, high import bills and trade deficits. Consequently, high oil prices can disproportionately affect competitive businesses as the latter will have to increase input costs with less aptitude to change final costs. On a microeconomic level, rising oil prices can surcharge household budget affecting those who have less flexibility to cope with price shocks Davies and Sugden (2010).

Among the most feasible strategy to counteract energy security issues is import diversification, that is, exporting oil from different countries but as argued by Vivoda (2009), import diversification depends on several external factors such as geography, political ties with main oil exporter and financial resources which makes diversification less plausible in the case of SIDS. Stockpiling oil and diversifying the energy mix with more sustainable energy sources are innovative and cost–effective options to dampen inconsistencies in oil prices.

Energy Efficiency and DSM Concepts

Energy efficiency and DSM are two concepts that can be grouped under the general heading of Energy Management. The latter is defined as the efficient and effective consumption of energy (Capehart et al. 2008) and involves implementing various types of energy policies to achieve cost savings. Hence, through reduced per capita energy consumption, the dependence on fossil fuels is lessened and the amount of GHGs emitted along with most direct and indirect negative impacts of climate change and global warming can eventually be reduced Yacout (2014). A number of DSM techniques exists but their application in SIDS and successful outcomes demand a lot of environment, demographic and territorial data analysis Sanseverino et al. (2014). A comprehensive review of DSM practices, benefits and challenges can be found elsewhere (see Strbac 2008). Figure 22.1 illustrates these techniques diagrammatically.

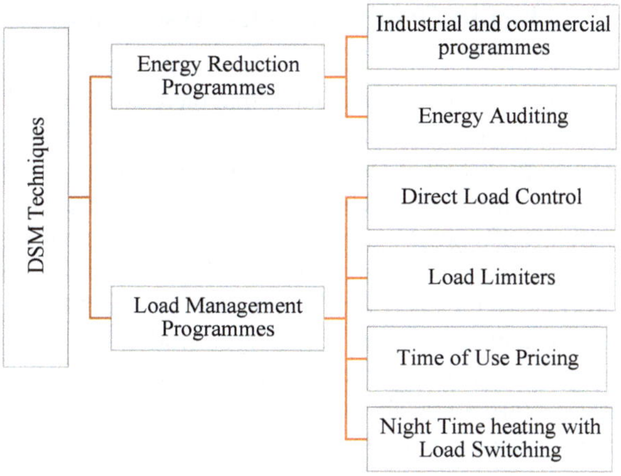

Fig. 22.1 DSM techniques (authors' illustration based on Strbac 2008)

An energy professional is thus someone who is well versed with all these techniques, illustrated in Fig. 22.1, their limitations, prospects and mode of application. EE and DSM holds many benefits to SIDS and can potentially improve their energy sector. There is however, no clear and understandable link between energy efficiency and improving electricity access but definitely, energy security issues can be tackled through energy management. Most probably, EE and DSM applied in power utilities can decrease load shedding and frequent outages which will enhance electric supply reliability. There are also some SIDS where urban populations consume excessive energy while rural population do not have adequate access to same. In this context, energy efficiency in electrified urban regions is essential to achieve cost savings. These savings can be invested in electrification programmes in poor rural and isolated regions. With regard to energy security, EE and DSM will subsequently decrease the amount of fossil fuels consumption in all energy intensive sectors and thus decrease dependence on oil imports and exposure to volatile oil prices. Intermittency of renewable energy sources can be addressed by DSM. There are constant criticisms on the flexibility, variability and non–controllability of renewable energy sources. DSM techniques can therefore be applied to align renewable energy sources with thermal units to strike the right balance between electricity demand and supply in the most sustainable way (Strbac 2008). For example, in an energy system consisting of a wind turbine to supply to households, legislations and awareness schemes can demotivate households to consume more electricity during low wind regimes.

There is a number of DSM and EE programmes that were launched in SIDS in domestic, industrial and commercial sectors. Some of the most widely applied are: the replacement of low efficient Compact Fluorescent Lamps (CFL) with high efficient Light Emitting Diodes (LED) lamps; awareness campaigns; standards and labelling of electrical appliances and incentives to conduct industrial energy audits. In Mauritius in 2008–2009 the replacement of CFL with LED lamps yielded an average energy saving of 2 GW and a reduction in peak demand by 14° MW Elahee (2011).

Conclusion from Previous Assessment

Singh et al. (2015) concluded 'that there is a demand for training in Energy Access, Security and Efficiency (EASE) in the African and Pacific Small Islands Developing States (SIDS)' based on a survey in small island communities. The survey–based approach identified training needs for energy practitioners in island nations. The study surveyed two groups of people—energy practitioners (NGOs, energy analysts, engineers in utility companies) and university academic staff working in the energy field in the PICs and in Mauritius. The first group received 61 respondents who were asked open–ended questions related to energy. The questionnaire consisted of seven sections. The first section (Section A) sought basic bio–data information about the respondents. In section B and onwards, respondents were asked general questions on energy issues to gauge their personal knowledge on specific areas of EASE. In the last section, respondents proposed recommendations to be included in a course on energy. From the second group of respondents comprising of university staff from Higher Education Institution (HEIs) their knowledge on the topics on EASE where also judged through open–ended questions. Even though the level of participation of academics in the survey was low, the results were quite compelling and reliable. The aim of surveying academics was to assess whether university staff had the required knowledge and expertise or if they needed more training in energy issues which energy practitioners claimed to have 'low knowledge'. The survey revealed that almost 50% of the sample of energy practitioners have 'low' knowledge of specific energy issues in the PICs. For Mauritius, similar results were obtained. These energy issues were mostly legislation and energy policies, energy management, energy poverty and security. For university academic staff, the results showed that academics had good knowledge of EASE but little knowledge of energy policies and legislations.

Singh et al. (2015) did not clearly elaborated on the reasons for the lack of knowledge among energy practitioners. The lack of knowledge on the topics of EASE and other such energy related themes can be attributed to few courses being dispensed in these fields at tertiary and post–secondary level. In SIDS, there are practically no technical courses or lifelong learning courses (LLL) on EASE which has probably resulted in a poor base of energy experts. Past studies have highlighted the need to incorporate energy programmes in almost all levels of education, be it primary, secondary and tertiary. A study by DeWaters and Powers (2011) concluded that energy literacy among US students was discouragingly low. Lee et al. (2015) showed that energy literacy among Taiwanese students varied based on gender, socio–economic status and grade. Other authors (Gelegenis and Harris 2014) compared undergraduate courses in energy education between Greece and UK universities to identify some peculiarities in Greek energy programmes. Mälkki and Paatero (2015) research was based on curriculum planning in energy engineering education.

In this paper, we built up on the work of Singh et al. (2015). Through an intensive literature search from peer-reviewed papers and reports from development banks, intergovernmental societies and research institutes, we attempted to identify the aspects that are required for successful adoption of an EE educational programme for energy professionals.

Importance of EE and DSM Education for Energy Practitioners

UN (2005) famously stated that education, apart from being a human right, is a 'prerequisite for achieving sustainable development and an essential tool for good governance, informed decision–making and promotion of democracy. Education for sustainable development develops and strengthens the capacity of individuals, groups, communities, organizations and countries to make judgements and choices in favour of sustainable development'. Education is a vital tool to inculcate the habit of energy saving and an energy efficient lifestyle and attitudes in society Zografakis et al. (2007). Newborough and Probert (1994) stated that rather than technology, education and legislation are two essential mechanisms to remediate to energy squandering. In addition, adequate knowledge about EE and DSM for energy practitioners will enable this group of people to understand an energy system, to know how much energy a particular process or machine consumes and where this energy comes from. Thus energy practitioners can positively lead more informed decisions which can improve the energy security condition, promote economic development and reduce environmental risks in SIDS. Additionally, knowledge of EE and DSM will help energy practitioners to develop and implement policies at his/her workplace that will allow his/her organisation to save energy and achieve savings in monetary terms. Being more energy literate will allow energy practitioners to understand the link and interconnectedness among energy issues and how these issues can be solved.

Desirable Features of an EE and DSM LLL Course for SIDS

Desirable features of an energy management programme has been rarely discussed in past literature. Some suggestions of desirable features of an educational LLL programme on EE and DSM for energy practitioners may include (Kandpal and Broman 2014; KEEP 2003):

(a) Knowledge of the different forms of energy (potential energy and kinetic energy), the founding concepts of energy (units, the concept of 'Power') and the natural laws that govern energy (e.g. First and Second Laws of Thermodynamics);

(b) Knowledge on the development and use of energy resources for a comfortable lifestyle. As such the distinction between primary and secondary energy resources, the way industrial and commercial societies obtain and use their energy, the demand and supply process from a power plant to requiring sectors;
(c) Concepts on how energy flows through systems, its value as a resource and the effects, both positive and negative, they have on human well–being and the environment;
(d) Courses should cover all the aspects of DSM as illustrated in Fig. 22.1, that is, (i) the different DSM management techniques, (ii) their applications and limitations, (iii) their working principle, (iv) installations troubleshooting, maintenance and performance monitoring and (v) any financial assessment;
(e) Project works to motivate energy practitioners to identify energy inefficiency at their workplace and to propose for solutions;
(f) Techniques to perform industrial energy audits and to write compelling energy audits reports to trigger the attention of senior management against the economic and environmental consequences of energy squandering and inefficient technologies;
(g) The relationship between energy and the environment, on how excessive electricity consumption derived from fossil fuels is detrimental to the environment;
(h) Courses should include both theoretical work and practical aspects. Therefore, curriculum should include laboratory and demonstration experiments, tutorials, assignments and seminars;
(i) Course design must be in line with global efforts so as to enhance experience and knowledge sharing between other HEIs and organisations;
(j) Knowledge about designing energy efficiency policies at workplace and to effectively monitor them should be included in curriculum;
(k) A thorough knowledge of the Energy Management System (EnMS), ISO 50001, which is a systematic approach that allows an organisation to achieve continual improvement of energy performance;
(l) Teaching language for such course should be as per students request to facilitate learning and understanding

EE as an Online Lifelong Learning Course

The Need for Lifelong Learning (LLL)

DSM and EE fall in most of the undergraduate or postgraduate modules or subjects on energy engineering or energy management. However, energy engineering and/or management courses are a subset or component of another huge discipline. For example, energy courses is a subset of the bachelor programme Chemical

Engineering at the University of Mauritius (called BEng Chemical Engineering (Minor: Energy Engineering)); similarly a subset of the postgraduate programme in Electrical Engineering at the University of Birmingham (called MEng Electrical and Energy Engineering). Undergraduate courses are likely to be of a duration of 3 to 4 years full–time and postgraduate programmes either 1 year full–time or 2 to 3 years part–time. Thus this offers learners who are already in service, less flexibility to enrol for these courses and gain more knowledge on the required topics. Therefore, there was a need to come up with an alternative that can potentially provide more flexibility to learners and a lifelong learning course fits definitely this position. It should not be ignored that energy practitioners come from different professional background having worked in different energy fields. While some power plant engineers are specialised in electricity generation and transmission systems, those energy experts working in the Department of Energy and/or the Ministry have more knowledge on energy policies and legislations. However, there is no need for defining course prerequisites for LLL online courses which will enable a larger number of learners, irrespective of past academic qualifications, to participate.

Mode of Information Delivery

In curriculum planning, there is also a need to identify the mode of delivery of lectures. When it comes to flexibility of learners, it is obvious that *in-class* (face-to-face interaction) type learning will be less tempting. An alternative is mixed-mode learning that is, both a combination of face-to-face interaction and online study. Maintaining contact between trainers and learners is important as it allows students to ask questions on areas that seemed complicated and unclear to him/her. However, in recent years, the coming up of the online system has motivated a number of HEIs to exploit online study. Online programmes allows students to study at home or anywhere they want at any time of the day. They can therefore better coordinate professional commitments with family time, leisure and learning. They also have the freedom to work at their own pace (Kandpal and Broman 2014).

Issues and Challenges in EE Education

Contextualisation

Mitigating and adapting with the impacts of climate change and global warming are two main motivations to shift towards sustainable energy resources and EE in SIDS. But in an organisation, the senior management is concerned mostly with profits and savings. Programmes on energy efficiency should explicitly demonstrate how adopting EE on an industrial level can ultimately help to mitigate climate change and energy issues. This will allow learners who take up an EE course to have a better idea of the seriousness of energy issues in SIDS and their implications on the environment. Learners will have more incentives to adopt energy efficient behaviour and devise energy efficiency policies at its workplace. In addition, EE programmes has to be tailored in a way that students can, in general determine savings that can be made to an organisation. In a nutshell, EE education should cover two fundamental aspects namely, environmental sustainability and economics. These two aspects require the inclusion of numerical workings, basically to use some mathematics and science which can be very demanding for some learners especially if they achieved a certain age and was disconnected from academics for a long time or are poor in mathematical computations.

Course Structure

A course on EE tailor made for energy professionals needs to disseminate both practical and theoretical knowledge. Nonetheless, teachers have more available expertise and knowledge on theoretical work and therefore, curriculum developed by teachers is focussed more on their available expertise. Hence, this ends up with a mismatch between the knowledge disseminated by teachers and the skills required by learners. Consequently, students are not provided with the knowledge they need and the areas of expertise of students are given undue importance. Therefore, the aims and objectives of EE education to energy practitioners are not basically met. Students lose interest in the subject which returns us back to square one.

Unavailability of Books and Learning Materials

A cursory search on the internet revealed that learning materials on EE is restricted mostly to theoretical and mundane matters. For example, the benefits of adopting energy efficiency in household or industrial scale, the barriers for widespread EE adoption and success–stories of EE policies implemented—among other findings on the internet. However, such sources do not actually share the technicalities, mathematical and economics level calculations that is required for learners to learn and apply. Besides, a lack of good quality textbooks and other learning materials is a significant bottleneck to disseminate knowledge and to fulfil the training needs of learners.

The Need for Competent Teachers/Trainers

Teachers should effectively be well trained and prepared to disseminate EE knowledge. In this context, regular short courses should be organised for teachers to be involved in EE education. Incentives should be given to teachers of the tertiary level to conduct research on innovative EE techniques. This can help to motivate teachers on the topics of EE and DSM where he can bring more up-to-date information in the course. Teachers should have the opportunities to participate in local and international seminars on EE and to bring home a broader experience.

Overcoming Barriers Through Global Cooperation

Barriers and hurdles to implement a high quality EE programme can be addressed through global corporation and resources from other development partners. As such, the collaboration of the private sector and university administration is critical. For the development of such course, the private sector has to share its vision and goal toward sustainability and how through training of their employees these goals can be met. The private sector can highlight areas where they believe energy efficient processes and knowledge are essential and course developers can take all this into consideration while developing the course. This will allow the course to fulfil the needs of a wide range of industries when suggestions come from different private sector sources. Funds from the private sector can also be of great help in designing such course. Additionally, collaboration with local and international networks (such as in Mauritius, there is the Energy Management Office) and NGOs is an important step to strengthen EE education.

Conclusion

It is clear that EE is a vital tool to mitigate the disastrous effects of climate change and global warming and as highlighted by past studies EE knowledge disseminated through education is very effective. In this paper, challenges and opportunities for successful EE education in SIDS was studied in an attempt to devise an educational online LLL course for energy practitioners in SIDS. In the foremost, it was seen that a LLL course on EE education especially designed for energy practitioners in SIDS should not be a 'basic' or 'elementary' course that provide information in EE practices. Rather it should consider the economics, mathematics and scientific knowledge behind energy efficiency and form a comprehensive educational programme that can be very impactful for small island communities. It is agreed upon that energy needs in developing islands differ from developed continental countries. Thus, EE education designed for energy professionals in islands which make better illustration of day to day energy problems in island communities are more likely to train energy professional in every aspect that allow them to combine their newfound acquired knowledge to address energy issues in SIDS. In addition, it was seen that barriers to implement such programme are a lack of commitment from teachers/trainers, lack of educational resources, poorly designed and out-of-context curriculum It can however, be addressed through collaboration with other networks. This study is essential to course developers as it provides some ingredients for a successful EE programme designed for energy practitioners. In the quest to form a more dynamic energy sector in SIDS, EE education in mandatory as it represent a viable option for SIDS to uplift their energy sector.

Acknowledgements The authors of this article would like to express their sincere gratitude to the anonymous reviewers of this article. The paper was produced in the frame of an EDULINK II funded project (with the financial assistance of the European Union) entitled L^3EAP (Lifelong Learning for Energy Security, Access and Efficiency in African and Pacific Small Island Developing States—http://www.project-l3eap.eu/). It is also pointed out that the arguments and contents of this article are solely the responsibility of the authors and cannot, under any circumstances be regarded as the reflection of the European Union.

References

Ang, B. W., Choong, W. L., & Ng, T. S. (2015). Energy security: Definitions, dimensions and indexes. *Renewable and Sustainable Energy Reviews, 42,* 1077–1093.

APERC. (2007). *A quest for energy security in the 21st century.* Japan: Asian Pacific Energy Research Centre.

Bass, S., & Dalal–Clayton, B. (1995). Small Island states and sustainable development: Strategic issues and experience. Environmental planning issues, No. 8, International Institute for Environment and Development (IIED).

Betzold, C. (2015). Adapting to climate change in small island developing states. *Climatic Change, 133*(3), 481–489.

Blancard, S., & Hoarau, J. F. (2013). A new sustainable human development indicator for small island developing states: A reappraisal from data envelopment analysis. *Economic Modelling, 30,* 623–635.

Brew-Hammond, A. (2010). Energy access in Africa: Challenges ahead. *Energy Policy, 38,* 2291–2301.

Capehart, B. L., Kennedy, W. J., & Turner, W. C. (2008). *Guide to energy management–5th edition.* New York: CRC Press.

CSO. (2015). *Digest for energy and water.* Mauritius: Central Statistics Office.

Davies, M., & Sugden, C. (2010). *Microeconomic impacts of energy prices in the pacific.* Suva: Pacific Financial Technical Assistance Centre.

DeWaters, J. E., & Powers, S. E. (2011). Energy literacy of secondary students in New York State (USA): A measure of knowledge, affect and behavior. *Energy Policy, 39,* 1699–1710. Elsevier.

Dornan, M. (2014). Access to electricity in small Island developing states of the Pacific: Issues and challenges. *Renewable and Sustainable Energy Reviews, 31,* 726–735. Elsevier.

Dornan, M. (2015a). Renewable energy development in the small Island developing states of the Pacific. *Resources, 4,* 490–506. doi:10.3390/resources4030490.

Dornan, M. (2015b). Reforms for the expansion of electricity access and rural electrification in small island developing states. *AIMS Energy, 3*(3), 463–467.

Elahee, K. (2011). The challenges and potential options to meet the peak electricity demand in Mauritius. *Journal of Energy in Southern Africa, 22,* 8–15. Energy Research Centre, University of Cape Town.

Gelegenis, J. J., & Harris, D. J. (2014). Undergraduate studies in energy education—a comparative study of Greek and British courses. *Renewable Energy, 62,* 349–352.

Gomes, C. (2014). The case small island developing States of the Caribbean: The challenge of building resilience. Economic Commission for Latin America and the Caribbean (ECLAC).

Hoeke, R. K., McInnes, K. L., Kruger, J., McNaught, R., Hunter, J., & Smithers, S. (2013). Widespread inundation of Pacific Islands by distant-source wind-waves. *Global Environmental Change, 108,* 128–138. Elsevier.

IEA. (2012). World Energy Outlook 2012. International Energy Agency, Paris, France.

IEA. (2014). Africa Energy Outlook 2014. International Energy Agency, Paris, France.

Kandpal, T. C., & Broman, L. (2014). Renewable energy education: A global status review. *Renewable and Sustainable Energy Reviews, 34,* 300–324.

Karekezi, S., McDade, S., Boardman, B., & Kimani, J. (2012). Chapter 2—energy, poverty and development. In *Global energy assessment—towards a sustainable future*, Cambridge University Press, Cambridge, UK and New York, NY, USA and the International Institute for Applied Systems Analysis, Laxenburg, Austria, pp. 151–190.

KEEP (2003). Wisconsin K–12 energy education program. Wisconsin Centre for Environmental Education, University of Wisconsin–Stevens Point.

Lee, L. S., Lee, Y. F., Altschuld, J. W., & Pan, Y. J. (2015). Energy literacy: Evaluating knowledge, affect and behavior in Taiwan. *Energy Policy, 76,* 96–106. Elsevier.

Mälkki, H., & Paatero, J. V. (2015). Curriculum planning in energy engineering education. *Journal of Cleaner Production, 106,* 292–299. Elsevier.

Martin, R., Gomes, C., Alleyne, D., & Philips, W. (2013). An assessment of the economic and social impacts of climate change on the energy sector in the Caribbean. United Nations Economic Commission for Latin America and the Caribbean.

McMichael, A. J., & Lindgren, E. (2011). Climate change: present and future risks to health and necessary responses. *Journal of Internal Medicine, 270,* 401–413. doi:10.1111/j.1365-2796.2011.02415.x.

Modi, V., McDade, S., Lallement, D., & Saghir, J. (2005). Energy services for the Millennium Development Goals. UN Millennium Project, UNDP. Washington, DC: The World Bank and the Energy Sector Management Assistance (ESWAP).

Newborough, M., & Probert, D. (1994). Purposeful energy education in the UK. *Applied Energy, 48,* 243–259. Elsevier.

Nurse, L. A., McLean, R. F., Agard, J., Briguglio, L. P., Duvat-Magnan, V., Pelesikoti, N., Tompkins, E., & Webb, A. (2014). Small Islands. *In Climate change 2014: impacts, adaptation, and vulnerability,* pp. 1613–1654. UK and New York, USA: Cambridge University Press.

Russell, L. (2009). Poverty, climate change and health in Pacific Island countries. Menzies Centre for Health Policy, pp. 17.

Sanseverino, E. R., Sanseverino, R. R., Favuzza, S., & Vaccaro, V. (2014). Near zero energy islands in the mediterranean: Supporting policies and local obstacles. *Energy Policy, 66,* 592–602. Elsevier.

Sinane, K., David, G., Pennober, G., & Troadec, R. (2010). Fragilisation et modification des formations littorales meubles sur l'île d'Anjouan (Comores): Quand l'érosion d'origine anthropique se conjugue au changement climatique. VertigO—la revue électronique en sciences de l'environnement. Retrieved April 16, 2016, from http://vertigo.revues.org/10528

Singh, A., Naidu, S., & Bijay, P. (2015). The training needs of energy practitioners in the African and Pacific SIDS—outcomes of Project L^{3}EAP. In *Proceedings of international conference on energy, environment and climate change* (ICEECC), 8–9 July, pp. 335–348.

Sovacool, B. (2011). *The Routledge Handbook of Energy Security.* New York: Routledge.

Strbac, G. (2008). Demand side management: Benefits and challenges. *Energy Policy, 36,* 4419–4426.

UN (2005). UNECE Strategy for education for sustainable development. United Nations Economic and Social Council.

UNEP (2014). GEO small Island developing states outlook. United Nations Environment Programme, Nairobi, Kenya.

Vivoda, V. (2009). Diversification of oil imports sources and energy security: A key strategy or an elusive objective? *Energy Policy, 37,* 4615–4623.

Winzer, C. (2012). Conceptualising energy security. *Energy Policy, 46,* 38–46. Elsevier.

Yacout, D. (2014). Concept of energy management. Retrieved October 24, 2015, from http://www.ecomena.org/energy-management/

Zografakis N., Dasenakis D., Kalitsounakis M., & Paraskaki I., (2007). Strenghtening of energy education in Crete. In *Proceedings of the International Conference on Environmental Management, Engineering, Planning and Economics*, Skiathos, 24–28 June, pp. 2933–2938.

Part IV
Trends on Climate Change Adaptation

Chapter 23
Beekeeping as Pro-forest Income Diversification in Solomon Islands

Wilko Bosma, Stephen Suti and Payton Deeks

Forests of Solomon Islands

Some 76% of Solomon Islands is covered by tropical rainforest, one of the highest in the Pacific. The forests are very rich in commercial tree species (Pauku 2009). The forest ecosystems has been recognized as one of the "Global 200" eco-regions for conservation due to its biodiversity (Olson and Dinnerstein 2002).

Nearly all (87%) forested land in Solomon Islands is customary land, meaning it is held by communal landowning units (tribes). Following a legal change in the 1980s, commercial logging was allowed on customary land (where previously it had only been allowed on government-owned land). Enticed by promises of royalty payments and infrastructure development, many tribes sold timber rights to commercial logging ventures which were owned primarily by Asia-based foreign companies (Frazer 1997; Wairu 2007).

Logging has been a major contributor to Solomon Islands' national and rural economies for several decades. Exports of forest products contributed 50–70% of the country's annual export revenue between 2000 and 2009 (CBSI 2009) and continue to be a large component of its economy. Timber stocks in the country have been severely over-harvested during the last two decades. Harvesting within this time period revealed an average yearly increase of 68,500 m^3/yr, reaching seven times the estimated sustainable level of 250,000 m^3/yr rate within the last five years. Moreover, log production still continues to increase despite earlier assertions that peak production has been reached in 2009 (Fig. 23.1) (Katovai et al. 2015).

W. Bosma (✉) · S. Suti
Natural Resources Development Foundation (NRDF), Gizo, Western Province, Solomon Islands
e-mail: nrdf@solomon.com.sb

P. Deeks
Pacific-American Climate Fund, Manila, Philippines

W. Leal Filho (ed.), *Climate Change Adaptation in Pacific Countries*,
Climate Change Management, DOI 10.1007/978-3-319-50094-2_23

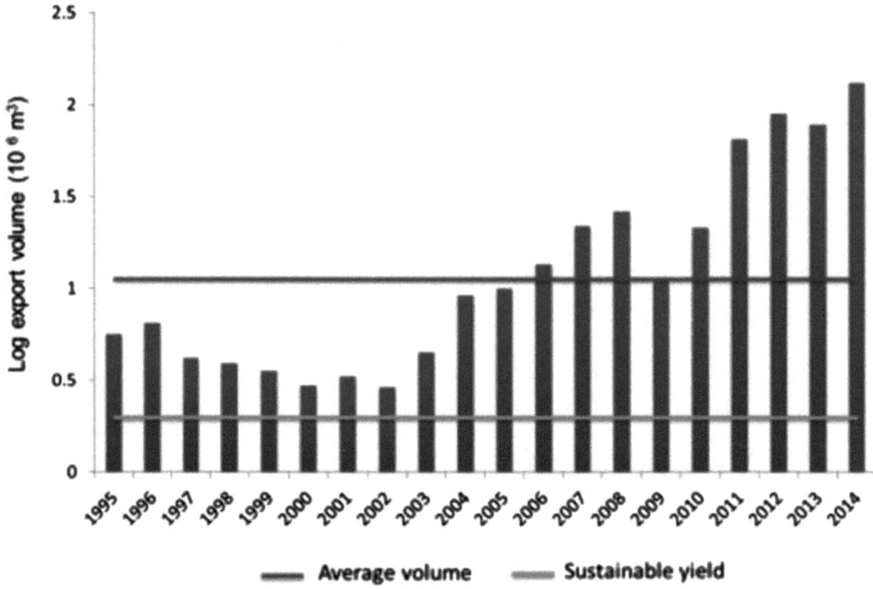

Fig. 23.1 Log export volume for Solomon Islands between 1995–2014. The rapid increase in logging licenses issued during this period has resulted in a steep increase in logging activities in the country, with harvest quadrupling the sustainable yield. (Katovai et al. 2015)

While the commercial logging ventures do provide employment for local communities and build some schools and clinics, local people are not seeing the benefits that were anticipated, with infrastructure being limited and of poor quality and royalty payments highly variable (Dauverge 1999; Wairu 2007). In a 2013 survey of Choiseul Province, even communities who had sold their forests rights rarely mentioned royalties as a major source of income (Mataki et al. 2013).

Impacts of Poor Forestry Management and Climate Change

In addition to not reaping the rewards of the logging operations, the communities are suffering from the effects of poor forestry management practices (Duncan 1994; Frazer 1997). In addition to the environmental implications discussed below, many social issues are associated with forestry in Solomon Islands including corruption, land disputes, prostitution, child abuse and corruption (Dauvergne 1999; Herbert 2007; Wairiu 2007; Pauku 2009; Katovei et al. 2015).

Unsustainable logging has had serious impacts on forests and the larger environment. Logging has changed the vegetation cover of the main islands of Solomon Islands, potentially reducing biodiversity and threatening endemic species (Pauku 2009). Further, the forests provide significant "ecosystem services" which include but are not limited to habitat for plants/animal; carbon sinks; water catchment areas; soil filtration; erosion control; food and shelter; runoff avoidance, and flood control (Lindberg et al. 1997; Turner et al. 2009; Locatelli and Pramova 2010). One study estimated the value of forest goods (e.g., for food, medicine, housing, and canoes) in Choiseul Province, Solomon Islands, to be worth SBD 49,533,251 (USD 7,076,178) (Mataki et al. 2013).

The environmental impacts of poor forest management are exacerbated by impacts of climate change. Heavy rainfall and the resulting flooding have displaced people from their homes and damaged their food sources (see Fig. 23.2). In April 2014 Solomon Islands experienced extreme rainfall, resulting in the worst floods in the history of the country, resulting in at least 21 deaths, hundreds of homes destroyed, and approximately 10,000 people displaced. Honiara, the capital of Solomon Islands, sustained serious damage to its bridges, roads, electrical services and water conduits, dealing a serious blow to its weak economy. The disaster in Honiara was linked to poor watershed management of the upper catchment areas of the Mataniko river system (Pikacha et al. 2014). Healthy forests can help buffer communities from these anticipated threats. The forests safeguard lives by storing clean water, preventing landslides in mountainous areas, and protecting villages against storms and cyclones.

In addition to the increased rainfall (and increased intensity of storms), Solomon Islands is predicted to experience more "very hot days" in the future. And while projections indicate a decreasing *number* of tropical cyclones, the storms that do come to fruition may be higher intensity (Pacific Climate Change Science Program 2011). The unpredictability of weather events is of significant concern for local communities whose very source of livelihoods are very much intertwined with their surrounding environment – the forests. Solomon Islanders are acutely aware of these threats; in a recent study, rural households identified climate-related changes and natural disasters as the greatest future threat they are facing (Schwarz et al. 2011).

Fig. 23.2 Examples of poor logging practices resulting in damage and pollution of rivers and water catchments in Choiseul Province, 2013. Photo © NRDF

NRDF's *Forest for Life* Programme: Rationale and Methods

Through its *Forest for Life* programme, NRDF supports local communities' adaptation to climate change impacts by improving forest management and conservation. At the same time, the programme works to create, develop, and encourage livelihood and income diversification (away from commercial logging) in order to increase resilience and adaptive capacity to the impacts of climate change. Secure and diverse livelihoods help protect households from exogenous shocks such as disasters but also underpins their nutrition, education, health and welfare (Ellis 1999; Cannon 2008). NRDF aims to build livelihoods that thrive in a healthy forest environment rather than livelihoods that subsist on large-scale commercial logging.

Forest for Life began in 2012, and aims to turn sustainable forest management from a conceptual model into practice. This has resulted in an approach that addresses the multiple functions of the forest, or "multifunctional forest management". NRDF's multifunctional management approach includes the following components and activities: Table. 23.1

This paper addresses the livelihood development component of *Forest for Life*. Data are based on mixed method approach using both quantitative and qualitative techniques. Quantitative data are obtained from ledger books from savings clubs as well as beekeepers' recollections. Qualitative data are sourced from key informant interviews, focus group discussions, and personal observations of NRDF's interventions since 2012.

NRDF works to create medium-term (5–10 years) partnerships with forest-owning communal landowning units (tribes) and implement sustainable forest and land-use activities in these areas. NRDF has learned from experience that interventions such as sawmilling, beekeeping, and forest conservation need at least an implementation period of two to five years to become successful. NRDF currently works with nine tribes, sometimes referred to as project areas or partners. The tribes sign a mutual agreement (usually with 5 years' duration) with NRDF in which both parties commit to sustainable use and development of the forest areas that are

Table 23.1 The components of NRDF's forest for life programme

Component	Activities
Sustainable forest management	• Small-scale timber harvesting • Forest Stewardship Council (FSC™) certification for community groups
Forest conservation	• Setting up Protected Areas under the 2010 Protected Area Act • Payment for Environmental Services (PES) • Low-impact activities such as tourism and research facilitation
Livelihood development	• Beekeeping • Women's saving clubs

owned by the tribe. This means that the tribe refrains from any large scale extractive developments such as logging and mining or larger scale forest conversion for agricultural purposes such as oil palm plantations. In 2016 all current partners have renewed their agreements with NRDF.

A typical project area features one or two settlements, with a population ranging from 80 to 400 people per area. The forest are as which are owned by tribes and involved with programme activities range from 500 to 6000 hectares per tribe. As of 2016 NRDF's project areas cover about 25,000 hectares of unlogged forest and included approximately 2600 beneficiaries.

Beekeeping in Solomon Islands

In Solomon Islands beekeeping with European honeybees (*Apismellifera)* began in the early 1960s and currently beekeeping takes place in rural areas of all the country's nine provinces (Ramoiau 1999).

Beekeeping remains a small-scale activity in Solomon Islands. Only a few commercial beekeepers owning 50 or more hives exist in Solomon Islands, with the majority of commercial enterprises having 25 or fewer hives. Most family-based farms have only 2–6 bee hives under their management.

Beekeeping provides the beekeepers an opportunity for income with relatively low startup costs, low labour input, low environmental impact, and with the bonus of increased pollination which is beneficial to crops and may increase yields. Beekeeping also takes up minimal land area and soil quality is not an issue. Hives, which are stationary, can be kept in close proximity to a household and require little daily maintenance, so women (who often have additional household chores) can be beekeepers (Hilmi et al. 2011). These factors make beekeeping a very suitable industry for Solomon Islanders.

Boxes and frames which comprise the hives are made from local timber and usually placed in the backyards of the farmers or in nearby garden areas placing hives under small leaf huts to protect them against rain and sun. Bees are purchased from other local farmers or ordered from the Solomon Islands Ministry of Agriculture. Solomon Island beekeepers sometimes have trouble getting other supplies that need to be imported, such as foundations, hive tools, smokers, veils, and extractors. As this equipment is expensive, beekeepers often share this equipment among many people.

Although some overseas export occurred in the late nineties it is thought that most of the honey produced in the Solomon Islands is consumed locally. Each hive can produce around 25–30 kg/year. Most of the honey is very dark with high moisture content (19–21%) which can lead to quicker fermentation (Bee-craft.com 2016). The majority of honey is collected during the drier season from October to April .

Beekeeping as a Pro-forest Income Diversification Option

Under the livelihood development component of its *Forests for Life* programme, NRDF supports production of honey through beekeeping.

This intervention aims to:

- Reduce vulnerability to climate change among rural communities by increasing alternative income generating options.
- Give more opportunities for women and youths in the community to engage more in NRDF's programme and become beneficiaries of the activities.

Solomon Islander women do not have the opportunities available to men to earn income or participate in decision making in forest management, a reality that is not likely to change in the near future. In this context, NRDF sought ways to bring a level of gender equity into the programme by introducing livelihood activities that were of interest to and appropriate for women. However, While NRDF initially conceived of beekeeping as a women's project, it turns out that most farms are managed by all the family members and thus has become a family run enterprise where family members work side by side and gain the benefits.

NRDF started to introduce beekeeping activities to its partners in 2012. In each partner community (which has already signed an agreement with NRDF to manage and conserve its forests), beekeeping is introduced after consultation with the community members. If community members show enough interest and commitment, a workshop or a training (3 days long) is organized for the whole community to learn more about beekeeping and obtain skills to set up and manage a honey farm/beehives. During this training, NRDF provides the community with a hive and bees along with some basic equipment so that they can start up their first hive. If an experienced farmer is already present (male or female) he/she will be appointed to take the lead in managing the newly-established hive. After 6–8 months the hive should be strong enough to split into a new hive which is given to a new farmer family who is selected based on their interest, commitment and skills and capacity to keep bees.

To set up a productive hive, a family needs to invest about USD 150 for materials and bees (USD 90 if they are making their own box). Once the hive is productive the farmer spends an average of 100 work hours a year on one hive, including harvesting and monitoring the hive.

A farmer begins with a nucleus colony of bees (also called a nuc). This nucleus colony includes small versions of everything that a full colony has. This means they have honey, pollen, baby bees in various stages of development, and a queen. They also have a comb which is a labor-intensive (for the bees) thing to produce (Keeping Backyard Bees 2015). A nuc is placed in a brood box, which is the layer of the hive where the eggs, larvae and pupae develop. This is usually placed under another container (called a super) which has removable frames inside, where the bees make the comb and then fill it with honey Fig. 23.3.

Fig. 23.3 A honey farm in Choiseul with 5 nuc boxes in front (left), single brood (starter) box at the right and two double hives boxes (productive) at the back. Photo © NRDF

Under good conditions it takes approximately 4–6 months before a hive starts producing honey. In theory, one hive can produce about 20–30 kg of honey a year and under current prices (1 kg for about USD12) this could provide a family with an income of 240–360 USD/hive/yr. Considering the annual minimum wage for an unskilled worker in Solomon Islands is about 1060 USD, beekeeping is an appealing source of income. In a village, USD 63 can cover the annual school fees and costs for a primary school child.

NRDF has found through interviews that most earnings from beekeeping is used for direct family needs such as school fees, clothes, and housing (Suti 2016). Income has also been saved in the local women's saving clubs (see below). From the current number of 29 farmer families who are beekeeping, 12 women are members of a savings club. From interviews it was also clear that money saved in clubs came partly from beekeeping (Suti 2016) (Fig. 23.4).

The beekeeping programme has been very successful and in 2015 a total of 308 kilograms of honey were produced by 29 farmers, valued at SBD 25,630 (USD 3,244). Beekeeping has produced short-term benefits while partners continue to commit and progress in longer-term activities under the *Forest for Life* programme such as forest conservation, Forest Stewardship Council certification, and carbon trade.

Fig. 23.4 MendanaTutikera of Pine village, Solomon Islands harvests honey. With income from his bees, he pays the school fees for his granddaughter (pictured far left). Photo © USAID/S. Gulick

The Katazo Honey Farmers Network: "Turning Your Backyards into a Golden Spring"

The Sirebe tribe lives along the Kolombangara River in southwest Choiseul Province. The Sirebe tribe—partnered with NRDF since 2007—has committed to protect their unlogged forest area (approximately 1100 hectares), designating it the Sirebe Rainforest and Biodiversity Conservation Area. Other parts of the area were set aside for sustainable timber harvesting and some future gardening. The area is part of the Mt. Maetambe to Kolombangara River Corridor an area of almost 87,000 hectares. It forms the largest water catchment on the island of Choiseul. The intact nature of the vegetation of the upper portion and the lower catchment area is crucial to maintain the constancy of, and uncontaminated nature of, the water source (Boseto and Pikacha 2015).

In 2012, two tribe members began beekeeping to have honey for their own consumption but also sell in order to support their community training centre. In 2013, their initial efforts were further bolstered by NRDF's interventions and the Sirebe tribe took this opportunity to build up their farms to supply the high demand for honey in Sasamungga village and the nearby Taro township. In 2014, the tribal committee decided to set up a beekeepers' network, the Katazo Honey Farmers Network, which has as its motto "turning your backyards into a golden spring," referring to both the color of honeybees and the value of the honey. Sirebe tribe members can become members of the network, as can members of the neighboring Vuri tribe with which Sirebe has a strong alliance. To become a member of the network one must:

1. Show clear interest in beekeeping (take their own initiative)
2. Have a good farm location which is free of dispute
3. Supply own timber for boxes and frames
4. Make a small leaf shelter for hive boxes (before hive is released)
5. Have at least two members in family who are interested in beekeeping
6. Be willing to learn/an open person

Once a member of the Katazo Honey Farmers Network, a farmer (who usually has 1–5 family members assisting him/her) will be given one nuc box and is provided on-the-job training on necessary beekeeping skills. This training includes splitting of bees, making hive boxes and frames, hive management, honey harvesting, and honey quality control. When a member has established five hives, he needs to split and return one nuc to the network. All members have free access to use the network's honey extractor and other tools (Fig. 23.5).

The network expects members to build their farm up to 10 hives which, as mentioned before, could potentially mean up to USD 3600 in annual income for

Fig. 23.5 Female beekeeper and member of the Katazo Honey Farmers Network, Sasamungga. Photo © NRDF

Table 23.2 Statistics on Katazo honey farmers network

	2014	2015	2016 (1st quarter)
Members	5	8	10
Double hives	6	11	14
Total honey production (kg)	43	92	149
Approximate income (USD)	344	736	1,192

those with fully producing hives. The data below show that members of the Katazo Honey Farmers Network are increasing production but are not yet at the optimized hive production level (Table 23.2).

The Katazo model has been successful and has the potential to be duplicated to other areas. Its membership, number of hives, and production have continued to grow since its inception (see Table 23.2). However it must be noted that the success of the network is largely due to the work and vision of Linford Jahjo, the founder and coordinator of the network. Other partner areas are aware of the Katazo network but have yet to find a suitable person(s) to take the lead to start a similar farmers' network model. However, for the time being farmers elsewhere will continue operating as individuals (Fig. 23.6).

Fig. 23.6 Labeling of 0.33 ml honey bottles sold locally for SBD 30 (USD 3.80). Photo © NRDF

Boeboe Village Has a More Challenging Path

Boeboe Village is located along the coast in Choiseul Province. The inhabitants, members of the Kamaboe tribe, have been partners with NRDF since 2012. The Kamaboe Tribal land is one of the larger project areas (about 6000 hectares) working with NRDF. Most of the people live in the village of Boeboe and the approximate population is about 262 of which 108 are female. Part of the Kamaboe land is the Guerre Forest Conservation area which comprises approximately 3000 hectares of mostly undisturbed primary forest (see Fig. 23.7). Besides beekeeping, the Tribeis involved in other aspects of NRDF's work as well.

Beekeeping in Boeboe village started in 2013. Following NRDF's protocol at the time, in the initial stage one central community farm with two (brood) boxes was set up. After a workshop and additional training a small committee of four persons was selected to start up the project and manage the two hives. The assumption was that the central community farm would initially harvest some honey to build up capital and then extend to individual honey farmers in the village.

However it soon became apparent that the central honey farm was not a success. The main reason for failure was that the farm lacked commitment or ownership; no one felt responsible for managing the hives. Additionally, beekeeping was new to Boeboe and so inexperience led to some mistakes being made when handling the hives. After about six months, NRDF decided to move the hives to a family of

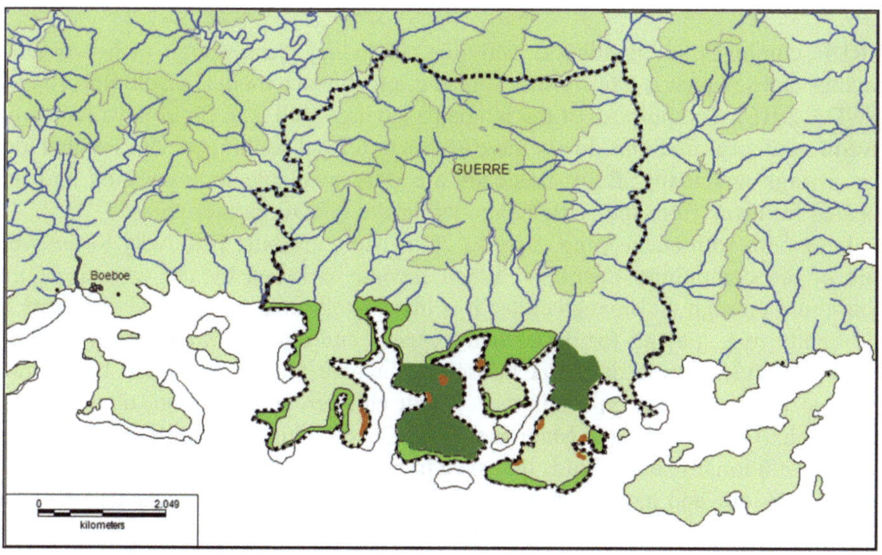

Fig. 23.7 Land use map of Guerre conservation area with the village of Boeboe on the left. *Light green* are Mangroves, *dark green* proposed timber harvesting areas and *brown* some scattered gardens. Most of the Northern part of the area is conserved forest

which one member had some experience with bee keeping. This change was for the better and within a year this family managed to split the hive in four new hives.

Although the number of farmers increased through splitting of hives, the production figures were far below expectations. In 2014 less than one liter of honey was produced by all four farmers combined. NRDF and the farmers suspected that the area around the village lacked the type of flowering vegetation bees collect nectar from. The area where the hives were located is surrounded by mangroves and some secondary growth forest and gardens. However, in 2015 the current six farmer families owning one hive each managed to increase their honey production to eight kg (six liters) and in the first quarter of 2016 already 14 kg (10 L) was harvested, suggesting that the honey production was not influenced by surrounding vegetation. An interesting observation in Boeboe (and in other areas) was the increase of flowering among trees and plants in and around the village. This increase seems to be associated with the introduction of the bees resulting in more and better pollination.

Through the lessons from Boeboe and Katazo, NRDF has decided to move away from the initial community farm model. Rather, the first step in a new village will be to identify the person(s) who has the right commitment and experience to start up the project. If those "champions" are successful and even have some harvest in six months, then new farmers families can be identified who can start a new farm.

The Role of Women's Savings Clubs

NRDF combines its livelihood activities including beekeeping with women's savings clubs, which is a fairly widespread micro-finance model in Solomon Islands (IWDA 2016). The purpose of the women's savings club is to provide women who live in rural areas and thus do not have access to formal banking services with a tool for savings and loans. Bank branches are only available in townships and are located far away from the settlements within the NRDF partner areas.

In tandem with beekeeping, savings clubs can contribute positively to the welfare of the community members. When income is saved by the women of the household through the savings clubs, it is more likely to be used for investments in the family (e.g., paying for direct family needs and children's education).The literature provides ample data that household income that is controlled by women tends to be spent more on items that benefit children such as food, clothing, and education. (Quisumbing and Maluccio 1999; Quisumbing and Smith 2007; Kenney 2008). Additionally, households that save money are less vulnerable to shocks such as disasters, etc. and have a greater capacity to cope with and recover from shocks (Davis et al. 2004; Gaillard et al. 2009). Similarly, these savings help households adapt to climate change impacts as people can relocate homes (e.g., away from rising sea levels or away from areas prone to flooding), or reinforce their homes to be more resilient to high wind from typhoons, etc. For these reasons, women's savings clubs are a component of climate change adaptation in Solomon Islands.

In the savings club model, women are encouraged to put aside small sums of money which is tracked in their Individual Savings account and secured by the group in a lockbox. Women are allowed to withdraw from their Individual Savings or obtain a loan from the General Fund (funded by member contributions or fundraisers). The loan from the General Fund must be paid back according to a schedule, and with interest, and generally the amount should not exceed 80% of what the individual has in her Individual Savings (IWDA 2016) though other models allow a loan of up to 100% of an individual's savings.

Savings clubs are established in a short 1–2 days training for the women and girls in the communities. NRDF provides the women with the most necessary stationary (e.g., passbooks, record book) and hardware for the boxes where the money is kept (usually featuring three locks with the keys kept by three different individuals to ensure its integrity). A savings club should not have more than 30 members. If more women are interested to join they are encouraged to start their own sub-group under the main village saving club. NRDF has produced a village saving club guide that it used in the short training courses and given as reference handbook to each club (Dyer 2014).

Savings clubs in the Solomon Islands can and do exist separate from beekeeping. In 2015, it was estimated that some 300 women's savings clubs existed throughout the country (Microfinance Pasifika 2015). In NRDF project sites, beekeepers are not automatically club members and club members are not always beekeepers. However, there is a nice complementarity to encourage wise use of new income that beekeepers are receiving from honey. This, in turn, lets them expand their operations or buy materials to benefit their bees and other benefits from saving money as noted above.

Fig. 23.8 Women's savings club in Voza village (Choiseul) with their locked cash box. Photo © NRDF

To date, NRDF has assisted women in setting up eight saving clubs with 154 members in total, with membership continually increasing (see Fig. 23.8). According to the last figures collected (December 2015) the women have saved more than USD 8,000 . In March 2016, the female family members and farmers in the Katazo Honey Farmers Network began the process of starting their own women's savings club. It is expected that much of the honey earnings will be saved in this club once fully functional.

Conclusion

In a country where commercial logging has promised communities much but delivered little, beekeeping is an alternative livelihood that is "pro-forest". With little or no footprint, beekeeping has a minimal negative impact on the forests of Solomon Islands yet provides income diversification that can reduce vulnerability to climate change impacts.

NRDF is working with communities who are committed to protecting their forest to enable them to make up to USD 500 per family year from beekeeping. This income generation is complemented by the women's savings clubs. As the income from honey—among other sources—is saved and managed by women, there are positive knock-on effects such as an increase in children attending school and households better protected from climate change impacts.

While many communities have had marked success with beekeeping, others have struggled. To date, the lessons learned by NRDF include:

1. Giving the right training to the right people and finding "champions" within communities;
2. Making sure that everyone understands the project (repeating the messages);
3. Conducting frequent visits to community for monitoring and coaching (at least once every two months in the first and second year) see Fig. 23.9;
4. Facilitating exchange visits/training among NRDF's partner communities; and
5. Assisting farmers through the entire process beginning with the establishment of hives up until the sales and marketing of their honey products.

Future prospects of *Forests for Life* include a refinement of approach (incorporating the abovementioned lessons learned); an increase in honey production; and an expansion of beekeeping to other communities.

Fig. 23.9 NRDF visiting with bee farmers in Boeboe (2014). Photo © NRDF

References

Bee-craft.com. (2016). *Water content of honey*. Retrieved March 18, 2016, from http://www.bee-craft.com/water-content-of-honey/.

Boseto, D., & Pikacha, P. (2015). *Choiseul biological expedition summary report*. Solomon Islands: Ecological Solutions.

Cannon, T. (2008). *Reducing people's vulnerability to natural hazards: Communities and resilience*. Helsinki, Finland. (Research paper 2008/034, UNU-WIDER).

Central Bank Solomon Islands (CBSI). (2009). *Annual report, central bank of Solomon Islands*. Retrieved March, 2016, from http://www.cbsi.com.sb/fileadmin/publications/ar/AR-2009.pdf.

Dauvergne, P. (1999). Corporate power in the forests of the Solomon Islands. *Pacific Affairs, 71* (4), 524–546.

Davis, I., Haghebeart, B., & Peppiatt, D. (2004). *Social vulnerability and capacity analysis*. Geneva: ProVention Consortium. (Discussion paper and workshop report).

Duncan, R. C. (1994). *Melanesian forestry sector study* (international development issues) (Vol. 36). Canberra: Australian International Development Assistance Bureau.

Dyer, M. (2014). *Women'svillage savings clubs: Financial empowerment of Solomon Island women* (an internal NRDF document describing village savings clubs procedures and establishment). Gizo, Solomon Islands: NRDF.

Ellis, F. (1999). Rural livelihood diversity in developing countries: Evidence and policy implications. *Natural Resource Perspectives*, 40.

Frazer, I. (1997). The struggle for control of Solomon island forests. *The Contemporary Pacific, 9* (1), 39–72. Spring.

Gaillard, J. C., Maceda, E., Stasiak, E., Le Berre, I., & Espaldon, M. V. (2009). Sustainable livelihoods and people's vulnerability in the face of coastal hazards. *Journal of Coastal Conservation, 13*(2/3), 119–129.

Herbert, T. (2007). *Sexual exploitation of children in the Solomon Islands: A report focusing on the presence of the logging industry in a remote region*. Church of Melanesia, Solomon Islands: Christian Care Centre.

Hilmi, M., Bradbear, N., & Mejia, D. (2011). *Beekeeping and sustainable livelihoods* (Diversification booklet number 1 second edition). Rome, Italy: Rural Infrastructure and Agro-Industries Division, Food and Agriculture Organization of the United Nations.

International Women's Development Agency (IWDA). (2016). *Women's savings club training guide: a guide for community facilitators and peer trainers to support set up and running of women's savings clubs in the Solomon Islands*. www.livelearn.org.

Katovai, E., Edwards, W., & Laurance, W.F. (2015). Dynamics of logging in Solomon Islands: The need for restoration and conservation alternatives. *Tropical Conservation Science, 8*(3), 718–731. www.tropicalconservationscience.org.

Keeping Backyard Bees. (2015). *Packages or nucs- which is a better start?* Retrieved March 21, 2016, from http://www.keepingbackyardbees.com/packages-nucs-better-start/.

Kenney, C. (2008). Father doesn't know best? Parents' control of money and children's food insecurity. *Journal of Marriage and Family, 70*(3), 654–669.

Lindberg, K., Furze, B., Staff, M., & Black, R. (1997). *Ecotourism and other services derived from forests in the Asia-Pacific region: Outlook to 2010*. Asia-Pacific forestry sector outlook study working paper series, working paper no: APFSOS/WP/24 food and agriculture organization of the United Nations, Regional Office for Asia and the Pacific.

Locatelli, B., & Pramova, E. (2010). *Forests and adaptation to climate change: What is at stake? World resources report 2010 expert perspectives*. Question 4, Washington, DC. www.worldresourcesreport.org/nl/responses/forests-and-adaptation-climate-change-what-stake-0.

Mataki, M., Solo, G., Donohoe, P., Alele, D., & Sikajajaka, L. (2013). *Choiseul province climate change vulnerability and adaptation assessment report, secretariat of the pacific community (SPC)*. Fiji: Suva.

Microfinance Pasifika. (2015, September 21–25). *Plenary 5: Financial inclusion in the Solomon Islands, session synopsis*. Pacific microfinance week 2015: Next mile financial services in the Pacific. Honiara, Solomon Islands.

Olson, D., & Dinnerstein, E. (2002). The global 200: priority ecoregions for global conservation. *Annals of the Missouri Botanical Garden, 89,* 199–224.

Pacific Climate Change Science Program. (2011). *Current and future climate of the Solomon Islands*. http://www.pacificclimatechangescience.org/wp-content/uploads/2013/06/13_PCCSP_Solomon_Islands_8pp.pdf.

Pauku, R. (2009). Solomon Islands forestry outlook. *Asia-Pacific forest sector outlook study II*. Working paper no. APFSOS II/WP/2009/31. Bangkok, Thailand: Food and agriculture organization of the United Nations, regional office for Asia and the Pacific.

Pikacha, P., Sirikolo, M., & Boseto, D. (2014). *Watershed Management for HoniaraCatchments* (Article in Solomon Star), May 1, 2014. Retrieved March 18, 2016, from http://ecologicalsolutions-si.com/watershed-management-for-honiara-catchments.html.

Quisumbing, A., & Maluccio, J. (1999). Intrahousehold allocation and gender relations: New empirical evidence. Policy research report on gender and development. *Working paper series no. 2*. Washington, DC: Development Research Group, The World Bank.

Quisumbing, A., & Smith, L. (2007). Intrahousehold allocation, gender relations, and food security in developing countries. Case study #4–5 of the program: *Food policy for developing countries: The role of government in the global food system*. Ithaca, NY: Cornell University.

Ramoiau, R. (1999). *How to start beekeeping in the Solomon Islands*. Honiara, Solomon Islands: Solomon Islands Honey Co-op. ISBN 0473-05993-2.

Schwarz, A.-M., Bene, C., Bennett, G., Boso, D., Hilly, Z., Paul, C., et al. (2011). Vulnerability and resilience of remote rural communities to shocks and global changes; empirical analysis from Solomon Islands. *Global Environmental Change, 21,* 1128–1140.

Suti, S. (2016). *Personal communication between Stephen Suti Agalo and beekeepers from Boeboe, Rarakisi, and Gorabara villages of Solomon Islands*, February/March.

Turner, W., Oppenheimer, M., & Wilcove, D. (2009). A force to fight global warming. *Nature, 462*, 278–279.

Wairu, M. (2007). History of the forestry industry in Solomon Islands: The case of Guadalcanal. *The Journal of Pacific History, 42*(20), 233–246.

Chapter 24
Coupling Disaster and Financial Management to Reduce Vulnerability: Challenging the Traditional Samoan Mindset, Experiences from the Community

Su'a Julia Wallwork

Introduction

The Intergovernmental Panel on Climate Change (IPCC) has defined climate change as the "adjustment in natural or human system in response to actual or expected climatic stimuli or their effects, which moderates harm or exploits beneficial opportunities" (IPPC 2007).

The increasing frequency of natural hazards due to climate change poses more vulnerability to less equipped countries (Downing 1991). Pacific Island Communities (PICs) are less equipped due to their small size and location. As smaller economies and less developed nations, the region as a whole is more vulnerable than other parts of the globe. Vulnerability, or the "state of susceptibility to harm from exposure to stresses associated with the environmental and social changes and from the absence of capacity to adapt," (Adger 2006).

This paper attempts to reconstruct the last 30 years of experience of ADRA in front-line disaster and how it has adapted these to deal with climate change responses from both the human, cultural and natural perspectives. The paper presents ADRA's ongoing challenges and then aims to draw out some lessons learned from Samoa's rich experiences in preparing to adapt to climate change. If we are to build resilience for climate change in the Pacific it is going to need to be done from the lens of Pacific Islanders; this paper is an attempt to share the experience of ADRA in taking up this challenge in Samoa.

Pacific approaches research should aim to be responsive to changing Pacific contexts and therefore should be underpinned by Pacific cultural values and beliefs (HRC 2005). The primary and secondary research for this article was conducted in

S.J. Wallwork (✉)
Adventist Development and Relief Agency, Apia, Samoa
e-mail: sjwalwork@adra.org.ws

W. Leal Filho (ed.), *Climate Change Adaptation in Pacific Countries*,
Climate Change Management, DOI 10.1007/978-3-319-50094-2_24

ADRA's main office, the villages and communities of Samoa. Mindful of the need to conduct research in a culturally responsible manner, this paper is based on a long series of conversations, focus group discussions and interviews with ADRA staff, Government officials, chiefs and local community members. Although the interviews and conversations were not conducted in a systematic academic manner, they are nevertheless the outputs of many hundreds of hours or conversations and many years of implementation of social and natural resource programs at the frontlines of climate change and disasters from the communities of Samoa over the last 30 years.

Samoa and Its Traditions

Samoa lies in the southwest Pacific within an exclusive economic zone of 120,000 km^2. It has a population of 200,000 that sprawls across a landmass of only 200 km^2. Samoa relies significantly on its location in the middle of the Pacific Ocean for its food, transport and the majority of its economy. Moreover, its limited size, proneness to natural hazards and external shocks enhance the vulnerability of islands to climate change (Mimura et al. 2007). According to the World Bank, Samoa is ranked 30th of countries most exposed to three or more hazards. Samoa ranked 51st out of 179 countries in the Global Climate Risk Index 2012 report on countries that suffer most from extreme weather events (WB 2013). Samoa ranked 51 out of 179 countries in the Global Climate Risk Index 2012 report on countries that suffer most from extreme weather events (WB 2013).

Samoans rely heavily on their culture, traditions and indigenous knowledge to guide their journey of life through the traditions of their forefathers. Rarely do they look for a scientific explanation for anything. Samoans have their unique seasonal calendar and look to the moon and the sun, and the stars for the weather forecast. The Samoan seasonal calendar is predominantly based on the observations of local environmental changes, which are in turn influenced by weather and climate. Monitoring changes in plants and animal behaviour, for example, are key indicators used by the Samoans to forecast changes in weather and climate (Lefale 2009).

We turn our faces to the wind to know what kind of threatening storm is approaching. We look to the plants and animals for signs. When a certain bird the "Atafa" is sighted, we prepare for a storm. When the "Mosooi" tree blossoms its fragrant flowers used for weaving '*leis*' (garlands of flowers usually worn around the neck) that signals certain seashells are ready to be harvested. When cockroaches fly at night, we know that it will rain in the morning. When sea levels rise, we always knew that it was just a temporary positioning of the moon that will quickly readjust.

While not directly linked to climate change, the earthquake and Tsunami of 2009 was a major wake-up call for Samoa. This "Wave of Fire" (as it later became known) was the greatest natural disaster loss of life and devastation experienced, with 189 deaths. This incident all of a sudden questioned our elders and traditions; Samoans were at a loss to understand. The event caused us to rethink our relationship with the sea, for a long time afterwards the fish were untouchable because

Samoans believed the sea was tainted with the blood of their loved ones. The question that resonated with all Samoans was how could our primary source of life bring us such death and devastation?

Samoa is a deeply religious nation in which citizens believe their country and the world are in the hands of the Creator. The majority of households perceive natural disasters, the rising sea, the increasing intensity of the storms, and flooding as some punishment from God for the sins of our nation and its leaders. They do not consider the need to adapt to changes in the climate, nor do they see any causes for change. If climate change is God's will, then so be it.

So, *Will all this talk of climate change impact the mindset of the old man rolling the strand of the dried coconut husk on his knee (to make the sinnet rope to strengthen the beams of his 'fale'), as he looks out to the ocean*?

Traditional knowledge, values, beliefs and ancestral practices underpin the ability of the Samoan communities to live successfully in their environment (Flores-Palacio 2015). The 2009 Tsunami and recent storms have begun to bring our people slowly to terms with the realization that climate change is happening, and that these events are a very real scientific phenomenon.

Climate Change in Samoa

Temperatures have increased by 0.22 C per decade since 1950; sea level has risen by about 4 mm per year, and Ocean acidification has been slowly increasing since the 18th Century. Samoas climate is also strongly influenced by its proximity to the Pacific Ocean and the strength of the South Pacific Convergence Zone that undulates with the more common El Nino and La Nina oscillations.

Projections of future climate-related risk indicate long-term systematic changes will continue across Samoa. Estimates include an increase in sea level, higher occurrences of high-intensity rainfall events, increased average temperature, rising sea levels and less frequent but more intense tropical cyclones (IPCC 2007) all of which are increasing in intensity.

Vulnerability, or the "state of susceptibility to harm from exposure, to stresses associated with the environmental and social changes, and from the absence of capacity to adapt," (Adger 2006) offers a key lens to understanding the impacts of climate change on any given region. Samoa is vulnerable from a variety of perspectives. Coastal villages remain high-risk communities for climate change effects. As a semi-subsistence society with 70% of the population and infrastructure located in low-lying coastal areas, the impacts of sea-level rise, increasingly tense storms, changing rainfall patterns and impacts on ecosystems will pose considerable risks to community livelihoods and Samoa's sustainable development. Under the Samoan Coastal Infrastructure Management Strategy, some 68% of Samoa's national coastline of 578 km was found to be either extremely vulnerable or vulnerable to coastal erosion (Fakhruddin et al. 2015).

Projected sea level is expected to exacerbate coastal erosion, loss of land and property, and dislocation of island settlements. Coastal floods are also likely to become more frequent and severe. High winds, storm surges, and heavy rains cause serious damage to agricultural plantations, infrastructure, and the country's socio-economic base (UPRS 2015). Changes in tropical cyclone systems increase the risk to life, property, and ecosystems. With 70% of the population and infrastructure in low-lying coastal areas, climate change cuts at the country's lifelines, water, food and its natural resources.

Samoa's heavy reliance on agriculture, both extensive and intensive, makes it sensitive. Crops can be destroyed or damaged by extreme climate conditions such as drought, prolonged rainfall as well as isolated extreme events like cyclones and tropical storms. The hardship to farmers and households is immeasurable given their heavy reliance on agriculture, both as a source of food supply, as well as a source of income.

Water is vital not only for hydropower generation but also basic needs of communities and agricultural development. Water, however, is heavily dependent on rainfall and land-use practices. Increased rainfall run-off will accelerate soil erosion and sedimentation in existing water supplies (Chase and Veitayaki 1992). Tropical cyclones and storm surge also affect water supplies by damaging water supply infrastructure while sea level rise increases intrusion of salt water into fresh water supplies.

Climate change includes increased burden of waterborne, foodborne and vectorborne diseases; traumatic injuries and deaths from extreme weather events; increased burden of respiratory illnesses; compromised food security; and heat-related illnesses (Flores-Palacios 2015). It is widely agreed in the literature and from our experiences that human survival, in the face of impacts of climate change, is dependent on mitigation and adaptation strategies (Costello et al. 2009).

Recent Natural Disasters and Their Impacts on Samoa

The 2009 Tsunami displaced hundreds of households from their traditional villages. The tsunami of 2009 not only claimed lives, but it left communities forced to relocate inland. The ocean which had traditionally been a source of life for Samoans had suddenly become untouchable because it had taken the lives of loved ones. Villages which were housed by the Ocean now shunned it.

According to the Samoan Post-Disaster National Assessment, "The total estimated damage and loss on Cyclone Evan which hit Samoa in December 2012 were equivalent to about 28% of the total value of goods and services produced in the country in 2011". The value of durable physical assets across all economic and social sectors destroyed by Evan (referred to as damage) is estimated at USD 103.3 million. A significant amount, given the relatively small size of Samoa's economy, almost 28% of the total value of goods and services produced in the country in 2011.

Adaptation Programs in Samoa and the Role of the Government

In 2005, Samoa prepared a National Adaptation Programme of Action (NAPA)—a document which prioritises Samoa's climate-related projects. The NAPA aims to communicate urgent and immediate adaptation needs and activities; it also implements projects which will mitigate and reduce the economic and social costs of climate change. More specifically, the NAPA provides a process by which Least Developed Countries such as Samoa can prioritise their vulnerabilities, formulate strategies and activities to build resilience.

The NAPA project builds on the national development goals, strategies and action plans implemented by the Government of Samoa. The 2005–2007 Strategy for the Development of Samoa (SDS) with its theme of 'enhancing people's choices' had six priority strategic areas that guided Samoa's development for the next 3 years continuing from past SDS periods. These include private sector development, agricultural development, tourism, community, education, and health development. The four National Environmental Management Strategies (NEMS), including the national waste management policy, national land use policy, the water resource policy and the national policy on population and sustainable development, have a common interest in promoting sustainable development (NAPA 2016).

What is notable about the NAPA is its focus on adopting an integrated approach where relevant stakeholders can work hand in hand to ensure that those whose livelihoods are most vulnerable to climate change. It also places significant emphasis on projects being country-driven, local and community-based.

The NAPA has provided a strong framework for the implementation of climate change adaptation projects in Samoa. While the majority of these focus on forestry and agriculture, there has recently been a greater focus on coastal zone management, human health, food security, infrastructure, water and policy and planning. Most, if not all, of these projects place considerable emphasis on community participation.

In implementing the PACAM project, ADRA has begun to incorporate the same community-based integrated approach to implementation that NAPA emanates. Working in collaboration with government ministries, local village authorities, and other disaster agencies, the ultimate objective of ADRA is to enhance livelihoods and improve disaster risk management at the community level. More importantly, ADRA attempts to build understanding between implementing agencies and communities regarding climate change, and what it means to those truly affected by the changes afoot.

The ADRA/PACAM Project—Community Disaster Management and Livelihood

The ADRA mandate is to alleviate suffering and reduce poverty. Throughout the world, this continues to be the focus for ADRA with its main motto being —"Changing the World, One Life at a Time". In our little spot on the world map, we cannot effect any significant change in the world. We are simply not big enough. We can, however, "Make a difference, one life at a time". The 'Development' and the *'Relief'* in our name are the essences around which we align our community development and adaptation work.

Recognising the traditional mindset and adapting our projects to encompass these beliefs while also steering communities and traditions to grasp this unfortunate new reality of unpredictability is ADRA's vision. Our approach needs to bridge traditional values, religion, and current realities and answer the question on how to prepare our families and communities to be resilient given the changes that lay ahead.

The project holds its roots in the Samoan earthquake and tsunami experience of 2009. The tsunami highlighted the incapacity and lack of preparedness of local communities to cope with such events. Villages were isolated and cut-off for extended periods of time with little access to basic needs. Equipping such communities with the skills not only to prepare for a natural disaster but also to manage the long-term effects of these disasters is vital, if not life-saving.

In recent years, ADRA Samoa has implemented projects that have focussed on sustainable economic development, with emphasis on developing good nutrition, to help reduce the non-communicable diseases rife in Samoa. In our experiences, we have found success is most likely when we focus on strengthening the core economic status of the community. Once we have momentum for socio-economic development, we can focus on other activities like planning and disaster preparation and planning.

ADRA's project, therefore, follows two core and parallel approaches. Firstly, it seeks to develop disaster risk management and preparedness activities among village communities. Secondly, it aims to increase and diversify livelihoods through improving the financial marketing literacy of low-income households.

ADRA was careful in identifying those villages which had not previously had any adaptation project. Following the 2009 tsunami, extensive rehabilitation and adaptive programs have already been conducted in many of these coastal villages. It was still difficult to prioritise which communities to focus efforts given the majority of village communities are located in coastal areas.

The success of previous projects indicated that the support of traditional village councils implementing adaptation projects was pivotal to the success of any community-based project. The Council plays a vital role in ensuring the security of catchment areas by having members of families responsible for their catchment areas. The project activities are supported by the Council of Matai (Chiefs), and women's and youth groups in each of the ADRA villages. Embraced by a culture founded on love and respect, the authority of the Council of Chiefs is well recognised and adhered to.

Building on this experience, ADRA chose to work only in villages which still followed the traditional hierarchy (Chief) system, who are in a stronger position to implement their Community Action Plans. After a rapid appraisal of suitable sites, the project was able to identify six villages on the island of Savaii, which had not previously benefitted from any climate adaptation program. These are namely the villages of Sagone, Siufaga, Vailoa Palauli, Falelima, Satufia, and Samata. The populations of these communities range between 289 to almost 800 individuals with the smallest village consisting of only forty (40) households. ADRA believes success will only be achieved with a two forked approach to its work, developing communities to be prepared (community plans) while at the same time enhancing socio-economic community resilience, both are shared below.

Developing Community Action Plan and Village Disaster and Climate Risk Management Plans (VDCRM)

This program aims to bolster households and communities disaster preparedness. Improved resilience should provide the communities with the edge to better withstand climate-related economic, social and natural shocks.

One of the lessons learned from the 2009 tsunami was that in the hours following a disaster, search and rescue, provision of immediate assistance to injured and homeless, is almost entirely carried out by family members, relatives, and neighbours. Communities were unprepared, due to an ineffective warning system and no prior planning. Escape routes in villages that were surrounded by steep cliffs were close to impossible. The extent of this tragedy underlined the need for greater community understanding of natural disaster events and community-level preparedness and planning. Top-down disaster risk reduction programmes and responses failed to address the specific vulnerabilities, needs and demands of at-risk communities. Engaging communities in direct consultation and dialogue is essential from the beginning of any program.

To prepare communities for disasters, ADRA delivered a Disaster Preparedness and Climate Risk Management Training Toolkit. The toolkit walks households through a process of understanding, planning and ultimately improving their adaptive capacity to disasters and long-term changes due to climate change. By enabling villages to identify their vulnerabilities, they are better able to understand them and identify locally suitable ways to respond to mitigate these risks using the resources available to them. More importantly, communities are organized and trained to respond themselves in times of disaster.

The training focuses on Disaster Risk Management (DRM), facilitation techniques, participatory approaches and project management. The toolkit includes a series of training modules focusing on such areas as situation analysis, prioritising and action planning, DRM structure and planning, training/resourcing response teams and simulation exercises and drills. From these training modules, villages formulate their Community Action Plans that lay out what needs to be done before, during and post episodal and long-term changes due to climate change.

All Village action plans, community action, and disaster management plans are compiled into the Village Community Disaster Reduction Management (VCDRM) Plan, overseen by the village elders. This plan is then reproduced and shared with partners, government ministries, and other initial first response agencies and shared into the national database.

Enhancing Livelihoods and Improved Financial and Marketing Literacy of the More Vulnerable Lower Income Households

From the six villages, we have engaged over 600 households with a target of both diversifying and increasing incomes for all households by 20%. We also support an education program for households to equip them with the necessary financial and management skills to maintain and sustain their income-generating activities. An underlying element of this work is the financial literacy training in partnership with the Bank of the South Pacific and the Central Bank of Samoa. After the training, villagers can record incomes and savings in ledger books before depositing in a bank account.

Diversification not only reduces sensitivity to the effects of climate stressors but more importantly it has the added co-benefit of generating income for village communities. The introduction of vegetable crops that are less susceptible to adverse weather conditions is vital (local varieties of root crops) as is having a diverse food basket that does not rely on just one or two crops. Diverse Livelihoods take an integrated approach and include the establishment of all aspects of livelihood with fish catchment areas, vegetable, herb, and seaweed farming being part of a well-balanced livelihood. ADRA is also providing market information for each village and their products and identifying possible market opportunities.

There are considerable benefits to this form of participative approach. Engaging village communities in these projects empowers them to take responsibility for the success of each activity. It not only results in Village Plans which are reflective of each village's unique situation but just as importantly, it allows communities to participate actively in undertaking activities. During the hands-on activities community members have recently begun to notice the small and sometimes subtle changes due to climate change, and have begun discussing the changes and implications on their livelihoods in the long run.

Discussion

Unfortunately, the traditional Samoan mindset, steeped in religious faith and traditional knowledge, perceives climate change differently from the 'gloom and doom' theme that seems to surround the phenomenon of climate change today. Our people have little fear of climate change. A lack of scientific knowledge coupled with the incremental way in which climate change takes place may be behind this lack of awareness.

Despite sporadic cyclones and extreme environmental effects, the majority of Samoans fail to see any major changes, and for that matter lack concern over future projections of climate change. The majority of Samoans do not directly correlate changes to the environment with climate change yet, nor do they perceive it as a future threat. They feel the sunshine is getting warmer, but they have no understanding that it is because of increasing global temperatures or the melting of the poles.

An ongoing challenge for ADRA is how to encourage and change the Knowledge, Attitudes and ultimately behaviours. Samoa's geographical isolation has not helped; rural communities are cut off from global events and most media channels.

The strong leadership of the National Government that laid out the NAPA provides an excellent framework around which to implement and align between civil society and national Government. By integrating the concepts of *faamatai (the way of the Chiefs)* and *faasamoa (the Samoan way)* into this project, ADRA can tap into the traditional model of a community decision making by consensus under the leadership of the village *matai* (chiefs). We are finding that when we have strong buy-in from traditional leaders, we are achieving success in our work.

Another opportunity we are exploring through our work is how to bridge the young and the older generations, their knowledge, and traditions. ADRA makes every effort to work respectfully with village elders to ensure the passing of information down and working with the younger generation (has access to social media and other mediums of communication to share up).

We are working with villagers to learn more about ocean tides, migration patterns and breeding patterns of fish. Many fishers already possess traditional knowledge of fish patterns, passed down through generations of fishing families. With the changes in climate and migratory patterns, ADRA hopes that combining this technical knowledge with traditional practices will better equip fish farmers to manage their fish catchment areas sustainably.

We are learning that unless the project is "holistic" and "integrated" (i.e., focus on needs (social, economic and ecological) they will not be truly successful. As part of the integrated approach, we engage communities in economic development activities. Improved economic status is part and parcel of a more resilient community. To highlight this, we are now working with communities on showcasing new commercial products. We initially began with Chillies for the highly demanded Samoan Chilli Sauce. Since introducing these small farms, we have now introduced

Turmeric, sweet potato, cassava, and yams to diversify local crops and build more options for community harvest. We are also finding that seaweed farming for Carrageenan does hold potential as an income provider and nutritious food security crop. Current harvest techniques can be improved, as can access to markets.

As these small wins continue, we are finding communities are beginning to open up to other changes and discussions around the impacts of climate change. Most certainly, the path ahead is not without challenges. Foremost is the need to try and bring about change to mindsets so that a better understanding and appreciation of impacts from climate change can be developed.

Core to that we need to understand people's perceptions about the changes occurring in the land and seas of Samoa and using this project and other work to identify the best opportunities to align likely climate changes with a future adaptation of all sectors of Samoan society.

Although this paper describes the work in progress, ADRA is finding that its role is that of a Bridge. A bridge between science and local knowledge, between young and old and traditions. The biggest challenge we face is how to turn a future plagued with problems, to one that allows Samoans to see the opportunities and minimize the risks that will come with climate change.

Conclusions and Prospects—Applying the Learning

- Programs and implementation with communities are most effective when they address an integrated approach with social, cultural, economic and ecological objectives and outputs.
- Looking for opportunities to integrate science into the overall narrative of Samoan society, ongoing change is important if we are to encompass the positive and negative implications of climate change. How to enhance the Samoan "seasonal natural calendar" with some climate change science is an important endeavour we are looking into (i.e., identifying changes that animals make that may provide early warning systems, or at least identifying the key months when threats are greatest).
- ADRA has been working with these communities such that it has developed enough trust to become a very powerful messenger to bring the scientific realities of climate change and begin to bridge the conversations and knowledge that need integration into our traditional oral culture.
- Introducing livelihood options to communities and addressing the socio-economic desires of communities first is a solid gateway to starting to build resilience and the overall concepts behind climate change. Diversifying incomes and preparing for emergencies is important. ADRA is preparing itself for these challenges.

References

Adger, W. (2006). Vulnerability. *Global Environmental Change, 16,* 268–281.

Chase, R., & Veitayaki, J. (1992). *Climate change: Challenges for Samoa—pacificdisaster.net.* Retrieved April 29, 2016, from http://www.pacificdisaster.net/pdnadmin/data/documents/2350.html.

Costello, A., Abbas, M., Allen, A., Ball, S., Bell, S., Bellamy, R., et al. (2009). Managing the health effects of climate change. *The Lancet, 373,* 1693–1733. London, UK. Retrieved May 16, 2009 from http://www.thelancet.com/climate-change.

Downing, T. E. (1991). *Assessing socioeconomic vulnerability to famine: Frameworks, concepts, and applications.* Research Report 91-1. Providence, RI: World Hunger Program, Brown University.

Fakhruddin, S. H. M., Babel, M. S., & Kawasaki, A. (2015). *NHESS—Abstract—Assessing the vulnerability of infrastructure to climate change on the Islands of Samoa.* Retrieved April 29, 2016, from http://www.nat-hazards-earth-syst-sci.net/15/1343/2015/nhess-15-1343-2015.html.

Flores-Palacios, X. (2015). *Samoa—Exploring the linkages between climate change and population movements.* Retrieved April 29, 2016 from http://samoanstudies.ws/wp-content/uploads/2015/03/Ximena-Flores-Palacios.pdf.

Health Research Council of New Zealand (HRC). (2005). *Guidelines on Pacific health research.* Auckland: Health Research Council of New Zealand. Retrieved April 29, 2016, from http://www.hrc.govt.nz/assets/pdfs/publications/May%202005%20Pac%20Guidelines.

IPPC. (2007). *Climate change 2007: Working group II: Impacts, adaptation, and vulnerability, natural hazards center||from the editor.* Retrieved April 17, 2016, from https://hazards.colorado.edu/article/from-the-editor-5.

Lefale, P. (2009). Ua 'afa le Aso Stormy weather today: Traditional ecological knowledge of weather and climate, the Samoa experience. *Climate Change, 100*(2), 317–335. doi:10.1007/s10584-009-9722-z.

Mimura, N., et al. (Eds.). *Climate change 2007: Impacts, adaptation, and vulnerability. The contribution of working group II to the fourth assessment report of the intergovernmental panel on climate change.* Cambridge: Cambridge University Press. Author lists for the contribution of Working Group II. Retrieved April 24, 2016, from https://www.ipcc-wg2.gov/AR4/rev_archive/authors.pdf.

NAPA. (2016). *National Adaptation Programmes of Action (NAPAs) in Samoa.* Retrieved April 15, 2016, from http://www.asiapacificadapt.net/adaptation-practices/national-adaptation-program.

UPRS. (2015). *Universal periodic review submission for Samoa prepared ….*Retrieved April 22, 2016, from https://www.culturalsurvival.org/sites/default/files/media/uprsamoafinal.pdf.

WB. (2013). *Shift ranking of November 12: Countries that suffer most extreme weather events.* Retrieved April 29, 2016, from http://www.dw.com/en/shift-ranking-of-november-12-countries-that-suffer-most-fro.

Chapter 25
Coastal Climate Adaptation at the Local Level: A Policy Analysis of the Gold Coast

Aysin Dedekorkut-Howes and Jordan Vickers

Introduction

The Gold Coast is a booming coastal city in South East Queensland, Australia renowned for its iconic ocean beaches, lifestyle, and tourism opportunities that take place in the coastal zone. It is home to over half a million people in an area stretching along approximately 57 kilometres of coastline, with the vast majority of development within 6 km of the ocean (ABS 2015; Dedekorkut-Howes and Bosman 2015). The health and stability of coastal ecosystems on the Gold Coast (beaches, dunes, and coastal wetlands) is integral to the protection of the city's developed coastal areas that fuel the lifestyle and tourism that is integral to the city (Zeppel 2012). However, with the impacts of climate change impending, these areas are increasingly vulnerable without adaptive plans, strategies or policies in place. The vulnerability of the Gold Coast's beaches to coastal processes exacerbated by climate change (increased erosion, flooding, frequency and intensity of extreme weather events and ensuing storm surges, tides, and swells) is widely known (Castelle et al. 2008; Sano et al. 2011; Zeppel 2012). As boasted by the City of Gold Coast (2013: 9) in its *Ocean Beaches Strategy 2013–2023*, the city 'has been at the forefront of coastal management since the 1960s. Nevertheless, the ability of the city's coastal management practices to address climate change adaptation is questionable. This paper attempts to evaluate the current status of coastal climate adaptation on the Gold Coast through investigating the following questions: What is the Gold Coast currently doing for coastal climate adaptation? How do the city's plans, policies and strategies compare with 'best practice' for coastal climate adaptation?

A. Dedekorkut-Howes (✉) · J. Vickers
Griffith School of Environment and Cities Research Institute,
Griffith University, Queensland, Australia
e-mail: a.dedekorkut@griffith.edu.au

W. Leal Filho (ed.), *Climate Change Adaptation in Pacific Countries*,
Climate Change Management, DOI 10.1007/978-3-319-50094-2_25

In order to answer these questions first a literature review on coastal climate adaptation best practices was conducted. As coastal climate adaptation is the product of both coastal management and climate change adaptation actions, next both of these are reviewed in turn for the Gold Coast. Then, three key policies and plans are evaluated against an evaluation framework developed for this purpose. The paper concludes with a discussion on what needs to be done to improve local coastal adaptation.

Coastal Climate Adaptation Best Practices

The literature review on coastal climate adaptation best practices is conducted following the systematic quantitative approach developed by Pickering and Byrne (2014). This approach aims to achieve reproducible results through systematic research, quantify the findings, identify the gaps in literature through concise data organizing, and highlight the relationships between geographic locations, subjects, and variables discussed by researchers. The literature review was used to answer two key research questions of 'What is 'best practice' for adapting to climate change impacts in the coastal zone?' and 'What factors affect the development of climate change policy for the coastal zone?' Journal articles were identified through a systematic search on Google Scholar where the first 15 articles from four search strings with relevance to coastal management, climate change adaptation, and planning were selected and entered into an excel database. Of the articles identified, only those published in journals on the 2010 Excellence in Research in Australia journal list developed by the Australian Research Council were included in the quantitative review, resulting in a total of 33 articles.

The literature review identified numerous coastal climate adaptation best practices (see Vickers 2015 for details). Table 25.1 illustrates the number of times each strategy was mentioned in the reviewed literature. 'Total references' indicates the total number of times the practice was discussed, defined, or demonstrated. Because some strategies were identified as a best practice for multiple impacts, each reference contributed to the 'Total references'. Strategies that were mentioned more than 30 times in the 33 articles in the review were used as best practices to evaluate the ability of the plans and policies on the Gold Coast to address best practice coastal climate adaptation and are indicated by shading in the table.

Table 25.1 Coastal climate adaptation best practices

	Best Practices	Total references
Adaptation Practices	Soft infrastructure • Beach nourishment • Dune protection and revegetation • Wetland protection and revegetation	**91**
	Reefs	16
	Hard infrastructure • Seawall • Levees • Groynes	13
Adaptation Policy	**Local government action**	
	Land use modification	**37**
	• Land use manipulation	19
	• Down zoning	12
	• Re-building restriction	6
	Mapping and public information[a]	**7**
	Setbacks[a]	**37**
	State Government Action	
	Collaboration	**33**
	• Horizontal political collaboration	10
	• Vertical political collaboration	14
	• Interdisciplinary collaboration	9
	Land buy-back	7
	Building codes	22
	Institutional Body	7
	Mandate Lower Order Plans	6
	Transfer of Development Rights	2

[a] These practices were included in the analysis separately as 'setbacks' was mentioned much more often than other factors and while 'mapping and public information' is strongly affiliated with land use modification, it is inherently separate from it.

Coastal Management and Climate Adaptation Frameworks on the Gold Coast

In Queensland, the state government has the statutory authority over coastal management, however, local governments are empowered by state legislation to exercise particular planning and management controls (Mosadeghi et al. 2009). Therefore, state coastal management documents generally provide overarching policy guidance, with local government being the governing body with most power to influence coastal management procedures. The Queensland Government's *Coastal Management Plan* (CMP) (DEHP 2013) is the current state coastal management document advising coastal management for local government areas in Queensland. The CMP provides broad scale policy guidance with six strategic focus areas: Coastal landforms and physical coastal processes, Nature conservation, Indigenous cultural heritage, Public access and enjoyment of the coast, Management planning, and Knowledge sharing and community engagement.

The *Gold Coast Planning Scheme 2003* was the tool used by the City of Gold Coast for guiding and assessing development until recently.[1] All assessable development within the Gold Coast local area must comply with any applicable code as set out in the Planning Scheme, or provide reasonable justification for non-compliance. Among others, the Planning Scheme regulates aspects of land use and development such as: setback requirements from property boundaries, beaches and canals, building height, building density, and car parking requirements. Although not created for the purpose of addressing coastal climate adaptation, the ability of the Planning Scheme to guide development gives this document considerable means for contributing to coastal climate adaptation.

Coastal management on the Gold Coast is guided by the City's *Ocean Beaches Strategy 2013–2023* and was assisted through Policies 7 and 15 of the *Gold Coast Planning Scheme 2003*. The *Ocean Beaches Strategy* is the City of Gold Coast's overarching coastal management tool and seeks to maintain the health, cleanliness, longevity, and economic benefit that the ocean beaches provide. It provides a strategic vision for the management of the city's beaches and a framework for further operational plans such *as Surf Management Plan*, *Commercial Activity Plan*, and *Gold Coast Shoreline Management Plan*. As a strategic document, the *Ocean Beaches Strategy* does not enforce or bind the City of Gold Coast to implement any actions included, but serves as a guidance tool to indicate how Council proposes to manage the ocean beaches (CGC 2013).

In terms of climate adaptation, with partial funding from the former Australian Government Department of Climate Change, the City of Gold Coast did have a *Climate Change Strategy 2009–2014* until 2014. With the retreat of all levels of governments from climate adaptation since 2012 elections (Howes and Dedekorkut-Howes 2016) this lapsed document has not been replaced, nor has anything similar been created. The *Climate Change Strategy* comprehensively addressed the impacts of climate change on the Gold Coast, provided detailed actions for mitigating and adapting to these impacts, and outlined a risk response matrix that clearly identified the likelihood, consequence, level of risk, priority, context, and response action for each impact (Gold Coast City Council 2009). Among other impacts, those relevant to coastal climate adaptation were sea level rise and coastal flooding and the increase in frequency and intensity of extreme weather events, with medium and high level priority for addressing these impacts respectively. While there is no direct climate change plan or strategy for the Gold Coast currently, the *Ocean Beaches Strategy* does make reference to 'climate variability', however this is drastically insufficient to be considered action on climate change (Howes and Dedekorkut-Howes 2016). Similarly, although climate change is mentioned in the *Gold Coast Planning Scheme 2003* in at least one Constraint Code (Flood Affected Areas), there is no specific focus on it in this planning scheme.

[1]This analysis was conducted before the new city plan came into effect on 2 February 2016.

Evaluation of Gold Coast Coastal Climate Adaptation Policies

Gold Coast's rich history of coastal management has been examined various times before (Macdonald and Patterson 1984; Castelle et al. 2008; Strauss et al. 2009). However, although inherently related, coastal management does not always parallel coastal climate adaptation (Vickers 2015). With no specific climate change plan in place by the local council, it is questionable whether the current coastal management measures adequately address coastal climate adaptation. To this end, this section of the paper evaluates the three key policies in place relevant to coastal management and climate adaptation on the Gold Coast to determine how well they achieve best practice coastal climate adaptation as identified through the literature review.

The evaluation framework used in this paper is an adaptation of that established by Baker et al. (2012) where numeric values are allocated to indicate the extent to which the policy action achieves a best practice on a five-point scale. The plan or policy actions were recorded to either: have no evidence of the practice (0), acknowledge the practice but lack further definition and not provide detail (1), mention the practice and include a moderate level of detail, however remain entirely descriptive and lack application and analysis (2), mention the practice and include a limited level of specific application methods (3), or provide detailed methods for the implementation of the practice (4). The ability for each plan or policy to achieve best practice is then determined by dividing the actual score (cumulative total) of the policy (sum of average scores for each of the 9 best practices) by the highest possible score a policy can get (i.e. 36, 9 best practices each with highest possible score of 4). This allows a comprehensive and quantifiable means for determining how appropriately each plan or strategy is achieving best practice coastal climate adaptation. This percentage figure then represents the adaptive capacity of the plan. One key limitation of this method is that the numerical value allocated to each action is purely reliant on a personal interpretation of how well it achieves a practice.

Queensland Coastal Management Plan

The CMP is prepared under the *Coastal Protection and Management Act 1995* and has two overarching goals: (1) provide for the protection, conservation, rehabilitation, and management of the coastal zone, including its resources and biological diversity, and (2) encourage the enhancement of knowledge of coastal resources and the effect of human activities on the coastal zone (DEHP 2013). The plan includes six strategic objectives regarding how coastal land is to be managed. Four of these objectives whose actions were relevant to coastal management or coastal climate planning were included in the analysis (see Table 25.2). Suggested

Table 25.2 Evaluation of Queensland Coastal Management Plan

	Coastal Climate Adaptation Best Practices									
	Soft infrastructure measures			Land use measures			Collaboration			
Strategic Objectives Suggested Management Actions	Beach Nourishment	Dune protection/ revegetation	Wetland protection/ revegetation	Setback requirements	Mapping and public information	Other land use measures	Horizontal political	Vertical political	Inter-disciplinary	Cumulative Total
Coastal Landforms and Physical Coastal Processes										
Exclusion fencing to protect dunes	1	4	-	-	-	2	-	-	-	
Fenced and hardened access points for pedestrians	1	1	-	1	-	1	-	-	-	
Periodic fertilising to increase stock or plant vigour	-	4	4	-	-	-	-	-	-	
Vegetating banks of rivers, creeks, estuaries and wetlands	1	1	2	-	-	-	-	-	-	
Nature Conservation										
Regulating beach driving	1	3	1	2	-	2	-	-	-	
Rehabilitating damaged vegetation	-	4	4	-	-	-	-	-	-	
Management Planning										
Establish a local management plan	1	1	1	1	1	1	1	1	1	
Consult with the broader community for local plan	-	-	-	-	2	-	-	-	2	
Knowledge and Community Engagement										
Develop key stakeholder lists for consultation	-	-	-	-	4	-	-	-	4	
Use a variety of methods to engage with the community	-	-	-	-	4	-	-	-	4	
Average Score	1	2.57	2.40	1.33	2.75	1.50	1	1	2.75	**16.48**
Highest Possible Score	4	4	4	4	4	4	4	4	4	**36**
Adaptive Capacity (%)										**46**

management actions for these strategic objectives indicate how to achieve the outcomes in a direct policy action. These actions are analysed here to determine the degree to which they achieve best practice coastal climate adaptation as identified in the literature review.

While most actions in the CMP incorporate more than one best management practice most practices are only acknowledged (56%), with no further definition or detail of implementation provided. This is not unusual in a higher order policy document whose purpose is to provide strategic guidance for lower level policy. 25% of the actions provide detailed methods for the implementation of the practice, the remaining actions either only mention the practice and include a moderate level of detail, but remain entirely descriptive and lack application and analysis (13%), or mention the practice and include a limited level of specific application methods (6%). The CMP could undoubtedly improve by including more detail on how to implement these best practices.

The CMP only refers to 'climate variability' as a pressure compounding the vulnerability of Queensland's low-lying coastal areas, noting that degraded environments accelerate the rate of change and instability. The plan distinguishes factors such as changed rainfall patterns and increases in sea levels and storm intensity as resultant of 'climate variability'. Not only does climate change carry greater impacts on coastal areas than those identified by this plan (storm surge, coastal flooding, coastal erosion, wet and dry land degradation), but each carries additional compounding impacts that require individual attention. It is essential in state level coastal frameworks that the impacts of climate change are identified and addressed accordingly with best practice coastal adaptation strategies to ensure local coastal frameworks reflect appropriate adaptive practices concurrently and consistently. Overall, the CMP is underperforming in terms of best practice coastal climate adaptation, attaining only 46% adaptive capacity.

Ocean Beaches Strategy

The *Ocean Beaches Strategy 2013–2023* is comprised of four key Strategic Outcomes: everyone can enjoy a beach experience, our beaches are healthy and clean, our infrastructure is protected from coastal hazards, and there is joint stewardship of the ocean beaches (CGC 2013). These Strategic Outcomes provide an overarching framework that make up the intent for the Strategy. The Strategy outlines 13 key actions to achieve these objectives, as well as 'outcome measures' to evaluate the effectiveness of the actions, and 'deliverables' to summarise and present the results of implementation. These key actions have been analysed according to the evaluation framework to determine their ability to achieve best practice coastal climate adaptation.

Only 35% of the 17 best practices included in the 13 actions provide detailed methods of implementation, most notably for beach nourishment and dune protection/revegetation. Another 35% are only descriptive and lack any further mention of application and analysis. The remaining 30% include a limited level of specific application methods. None of the actions address the protection/revegetation of wetlands, setback requirements, or any modification of land use. Of all practices, mapping and public information and interdisciplinary collaboration are the most mentioned.

Overall, the *Ocean Beaches Strategy* achieved an adaptive capacity of 47% (see Table 25.3). Evidently, by achieving only half of its potential adaptive capacity, the strategy is underperforming in terms of coastal climate adaptation with regard to the best practices identified in this research. While the *Ocean Beaches Strategy* clearly targets a diverse range of practices that could be considered good coastal management (beach nourishment, dune protection), and potentially appropriate measures for coastal climate adaptation, the document only refers to 'climate variability' as a driver for increasing erosion. Similarly, where other impacts of climate change such as sea level rise and frequent and intense storms are mentioned in the document, they too are only noted as influencing erosion. While these impacts of climate change contribute to erosion, they pose additional considerable threats to coastal zones and should be considered independently.

Gold Coast Planning Scheme

In the Planning Scheme, hazards relative to coastal climate impacts are addressed by constraint codes affiliated with overlay maps that are implemented through

Table 25.3 Evaluation of Ocean Beaches Strategy

	Coastal Climate Adaptation Best Practices									
	Soft infrastructure measures			Land use measures			Collaboration			
Key Actions	Beach Nourishment	Dune protection/ revegetation	Wetland protection/ revegetation	Setback requirements	Mapping and public information	Other land use measures	Horizontal political	Vertical political	Inter-disciplinary	Cumulative Total
Monitor and improve beach health	4	4	-	-	4	-	-	-	-	
Undertake planning related to coastal protection	-	-	-	-	3	-	-	2	4	
Implement Shoreline Management Plan	4	4	-	-	-	-	-	-	-	
Develop collaborations to support ocean beach man.	-	-	-	-	-	-	2	2	2	
Undertake and promote research about the ocean beaches	-	-	-	-	2	-	-	-	2	
Actively engage local stakeholders in ocean beach man.	3	3	-	-	3	-	-	-	3	
Report on Ocean Beaches Strategy Outcomes	-	-	-	-	2	-	-	-	-	
Average Score	3.67	3.67	0	0	2.80	0	2	2	2.75	**16.89**
Highest Possible Score	4	4	4	4	4	4	4	4	4	**36**
Adaptive Capacity (%)										**47**

development assessment. Performance Criteria include the actions that seek to achieve the intent for each code. A total of 38 Performance Criteria extracted from four relevant Constraint Codes were analysed here. *Canals and Waterways* code seeks to ensure that development adjacent to canals and waterways positively contributes to the maintenance and improvement of water quality while protecting the banks of estuaries, lakes, canals, rivers, streams and other waterbodies from erosion. *Flood Affected Areas* code aims to ensure that development does not cause or have the potential to cause damage to land or premises, and provides standards for development so as to reduce any potential adverse impacts of flood on the environment. *Natural Wetland Areas and Natural Waterways* code seeks to ensure the long term protection, enhancement and management of natural waterways and wetlands for their ecological, fishery, shoreline and bank stabilisation, hydro-geological, open space, recreational, environmental, scientific, and cultural value. *Ocean Front Land* code seeks to protect ocean front properties and beach environment with a foreshore seawall, protect and replenish sand resources, preserve visual amenity of the foreshore, protect and enhance the coastal environment, and ensure adequate access for foreshore seawall maintenance.

Of the best practices included in the planning scheme only 39% provide detailed methods for the implementation. 33% only acknowledge a practice, lacking further definition, detail, or mention of implementation, and 18% mention the practice and include a moderate level of detail, but remain entirely descriptive and lack application and analysis. Evidently, although a significant percentage of Performance Criteria provide methods for implementation, the majority are lacking detail.

Soft practices, setback requirements, and land use modification are the most mentioned best practices in the Constraint Codes (see Table 25.4). A significant number of the Performance Criteria that addressed these best practices provide detailed methods for the implementation. This was expected of the Constraint Codes, as their purpose is to guide development in areas susceptible to coastal

Table 25.4 Evaluation of Gold Coast Planning Scheme

Constraint Codes Performance Criteria	Coastal Climate Adaptation Best Practices									Cumulative Total
	Soft infrastructure measures			Land use measures			Collaboration			
	Beach Nourishment	Dune protection/ revegetation	Wetland protection/ revegetation	Setback requirements	Mapping and public information	Other land use measures	Horizontal political	Vertical political	Inter-disciplinary	
Canals and Waterways										
Provide for setbacks from the waterway	-	-	-	4	4	-	-	-	-	
Boat ramps etc be designed for waterside location	-	-	-	-	-	4	-	-	-	
Stormwater outlets designed for waterside location	-	-	-	-	-	4	-	-	-	
Ensure the water quality and quantity is maintained	-	-	1	-	-	2	-	-	-	
Areas below high water level kept for public purposes	-	-	-	-	4	-	-	-	-	
Not adversely affect the waterbody or bank	-	1	4	-	-	1	-	-	-	
Flood Affected Areas										
Not detrimentally affect flood storage capacity	-	-	1	1	-	-	-	-	-	
Allowance for hydraulic gradient above main floodway	-	-	-	2	-	-	-	-	-	
Height reflecting acceptable flood risk for their purpose	-	-	-	1	-	-	-	-	-	
Not obstruct the free passage of stormwater	-	-	-	1	-	-	-	-	-	
Not damage, not increase the level of risk to life	-	-	-	1	-	-	-	-	-	
Consider hydrologic and hydraulic impacts	-	-	-	1	-	2	-	-	-	
Avoid causing exposure to undue flood hazard	-	-	-	2	-	4	-	-	-	
Sufficient access or egress available to enable evacuation	-	-	-	2	-	1	-	-	-	
Not cause sedimentation, erosion or impact on drainage	-	2	1	-	-	-	-	-	-	
Not impede a natural watercourse, flood channel etc.	-	-	2	-	-	-	-	-	-	
Not be inundated during a designated flood event	-	-	-	4	-	1	-	-	-	
Natural Wetland Areas and Natural Waterways										
Ecological features, functions be identified and assessed	-	-	2	-	-	-	-	-	-	
Ecologically significant areas protected and maintained	-	-	3	-	-	1	-	-	-	
Ecologically significant areas not be negatively impacted	-	-	1	1	-	1	-	-	-	
Development set back from ecologically significant areas	-	-	2	4	-	1	-	-	-	
Degraded wetland and waterway areas be rehabilitated	-	-	4	-	-	-	-	-	-	
Natural hydrological regimes of wetlands be maintained	-	-	3	-	-	-	-	-	-	
Ecologically significant areas buffered from effluent	-	-	3	-	-	-	-	-	-	
Buffers to ensure no negative impact on significant areas	-	-	4	4	-	-	-	-	-	
Ecological corridors to link ecologically significant areas	-	-	4	3	-	-	-	-	-	
Facilitate protection of ecologically significant areas	-	-	3	1	-	-	-	-	-	
Minimise vehicular and pedestrian crossings	-	-	2	1	-	4	-	-	-	
Ocean Front Land										
Set back from active dunal areas/foreshore seawall line	1	2	-	4	-	-	-	-	-	
Certified foreshore seawall provided in designated areas	-	-	-	4	4	-	-	-	-	
A rear dune fence for dune restoration and protection	-	4	-	1	-	2	-	-	-	
Excavated sand cleaned and placed on an ocean beach	4	4	-	4	4	-	-	-	-	
Beach protection/restoration measures used in excavation	4	4	-	-	2	-	-	-	-	
Maintain or enhance local natural coastal environment	4	4	4	-	-	1	-	-	-	
Public access/beach protection in land subdivisions	-	1	1	-	-	3	-	-	-	
Average Score	3.25	2.75	2.50	2.30	3.60	2.13	0	0	0	**16.53**
Highest Possible Score	4	4	4	4	4	4	4	4	4	**36**
Adaptive Capacity (%)										**46**

impacts that these codes seek to address. Conversely, no collaborative practice was mentioned at all. The Ocean Front Land Constraint Code indicated the greatest adaptive capacity with the majority of Performance Criteria providing detailed methods for the implementation of soft infrastructure practices, while also addressing the requirement for setbacks and land use modifications. The majority of the Performance Criteria in the Natural Wetland Areas and Natural Waterways

Constraint Code addressed the protection/revegetation of wetlands, setbacks and land use modification, identifying some methods for implementation; however no other practices were addressed by the Code.

Overall the Planning Scheme scored an adaptive capacity of 46% and is clearly underachieving in terms of best practice coastal climate adaptation. Although this analysis of the Planning Scheme is limited in that it has only evaluated specific Constraint Codes, these are the specific codes used for guiding and assessing development in vulnerable coastal areas. There are measures of coastal management that could be considered coastal climate adaptation throughout the Planning Scheme such as soft infrastructural protective measures, setback requirements, land use modification and the provision of mapping and public information. Additionally, there is at least some mention of climate change in the document. One Performance Criterion that specifically addressed climate change, PC8 of the Flood Affected Areas Code, states that development must consider hydrologic and hydraulic impacts of development in flood affected areas with regard to future climate change. Nonetheless, as the city's planning and development tool, overall the Planning Scheme's Constraint Codes are underperforming in terms of best practice coastal climate adaptation.

Discussion

All the plans evaluated here scored similar adaptive capacities staying just shy of half of their potential in terms of coastal climate adaptation but their strengths and weaknesses were in different areas. The state *Coastal Management Plan* provided detail on implementing practices for mapping and public information, interdisciplinary collaboration, and dune and wetland protection and revegetation. Regardless, five out of the nine best practices received no further than acknowledgement from the actions in the CMP. As the CMP is a state government plan, local governments are entrusted with implementing coastal management procedures. However, the CMP includes no mention of any form of intergovernmental collaboration, vertical or horizontal. As the literature review highlighted, the higher order plans such as state government plans greatly influence the production of lower order plans such as local government plans. Additionally, both higher order plans that mandate lower order plans and horizontal and vertical collaboration can contribute to enabling policy development (Kaswan 2013; Flood and Schechtman 2014; Gurran et al. 2013; Beirbaum et al. 2013; Tobey et al. 2010; Martinez et al. 2011). In the case of Queensland previous research has shown that only local governments which were located in a region with a statutory regional plan that has specific emphasis on climate change adaptation have prepared climate adaptation strategies (Dedekorkut-Howes and Sloan 2012). Furthermore, whereas previously the state of Queensland had a specific policy to address climate change and the preceding Queensland state coastal management policies have mandated coastal local councils to produce a coastal hazard adaptation strategy and required sea level rise to be taken into account in coastal development (Howes and Dedekorkut-Howes 2012), the state no longer has a climate policy and the CMP, the former conservative Liberal National state government's framework advising the development of coastal management plans for local government areas, does not mandate coastal hazard strategies (Dedekorkut-Howes and Howes 2014). In fact, Moreton Bay Council, which included a 0.8-metre sea level rise in its draft planning scheme effectively barring any development below that water mark, was directed by the former Liberal National Government's deputy premier Jeff Seeney 'to remove any assumption about a theoretical projected sea level rise' to protect the rights of existing property owners (Solomons and Willacy 2014). Aligning with the intent of the Liberal National Party in power at the time to remove 'red tape' for development, the current CMP merely provides guidance over the strategic focus areas for Councils that seek to create coastal plans. Therefore, as a state government plan that does not mandate lower order plans; have no mention of vertical collaboration or overseeing/instigating any horizontal collaboration; nor provide any information further than acknowledging multiple best practices, the CMP is fundamentally underperforming in terms of best practice coastal climate adaptation.

The *Ocean Beaches Strategy* does contain particular best practices for coastal climate adaptation, however is lacking specifics relative to the best practices identified in the literature review. The strategy provides detailed methods for implementing four out of nine best practices: beach nourishment, dune protection and revegetation, inter-disciplinary collaboration, and mapping and public information. Soft infrastructure practices such as beach nourishment and dune protection are considered essential to coastal climate adaptation as a means of protection for coastal settlements (Harman et al. 2015; Coleman 2012; Martinez et al. 2011; Klein et al. 2000). However, the strategy fails to address wetland protection and revegetation, setback requirements, and land use modification and only mentions vertical and horizontal political collaboration with limited detail. Revegetating and protecting wetlands is integral to coastal climate adaptation as a protective measure to impacts like storm surge and coastal flooding (Hanak and Moreno 2012; Spalding et al. 2014; Harman et al. 2015) and setback and land use modification are essential as a means of accommodating the impacts of coastal flooding, sea level rise, and erosion in vulnerable areas while providing the mutual benefit of reducing environmental degradation (Hwang 1991; Grannis et al. 2012). While the *Ocean Beaches Strategy* does not have the authority to influence land use or setbacks, the strategy does discuss the implementation of the *Draft City Plan 2015* suggesting that the document will reflect the desired outcomes of the *Ocean Beaches Strategy*. The best practices missing from the *Ocean Beaches Strategy* are integral to coastal climate adaptation and their absence is detrimental to the adaptive capacity of the strategy. As it is intrinsically a coastal management strategy, it should pay greater attention to climate change.

The *Gold Coast Planning Scheme 2003* is effective in detailing the implementation of six out of nine best practices concerning soft infrastructure (beach nourishment and dune and wetland protection and revegetation), setback requirements, land use modification, and mapping and public information. The literature review highlighted that implementing soft infrastructure defences, down-zoning (including setback requirements and land use modification), and providing mapping and public information are beneficial local government initiatives for coastal climate adaptation. However, the Planning Scheme does not address any form of collaboration. Vertical, horizontal, and interdisciplinary collaboration are essential for coastal climate adaptation, as they create congruent implementation throughout neighbouring coastal areas (Martinez et al. 2011), provide opportunities to learn and exchange expertise gathered in other coastal areas (Flood and Schechtman 2014), and break down the communication and interpretation barriers between researchers and policy makers (Langridge et al. 2014). The Planning Scheme is not a document seeking to achieve coastal climate adaptation, but to guide and assess development, thus more applicable to land use. Therefore, although underachieving in terms of best practice coastal climate adaptation, as a local government tool for guiding and assessing development and land use, the Planning Scheme is fairly effective in addressing some coastal climate adaptation practices relative to its purpose on paper.

Conclusions

While Gold Coast local coastal management and planning documents do address particular coastal climate adaptation best practices, neither document is specifically developed for achieving coastal climate adaptation. Thus both are lacking the fundamental purposive attention to best practice adaptation strategies. Additionally, despite the state government's CMP addressing select best practices, this policy does not acknowledge climate change, provides no guidance on land use modification, setback requirements, or beach nourishment, and most detrimentally as a higher order plan carries no mandate for lower order plans and makes no mention of any form of collaboration, vertical or horizontal. Consequently, the City of Gold Coast is in dire need of direction for coastal climate adaptation.

As supra local level guidance is beneficial for achieving effective coordination of coastal adaptation policy (Verschuuren and McDonald 2012; Boyer 2012), and higher order plans that mandate lower order plans assist policy development (Kaswan 2013; Flood and Schechtman 2014; Gurran et al. 2013), this coordination would best originate in higher levels of government. While some plans addressed interdisciplinary collaboration, no plan even acknowledged vertical or horizontal political collaboration. This is particularly detrimental to the CMP as a state government document. Without horizontal collaboration, the City of Gold Coast is potentially missing out on vital expertise through information sharing and learning from other coastal areas (Flood and Schechtman 2014). Without vertical collaboration, adaptation measures are generally less likely to be employed (Hurlimann et al. 2014), and local action without guidance could be implemented in redundant and inefficient ways (Boyer 2012). Additionally, the presence of an institutional body to coordinate collaboration greatly contributes to the congruent adoption of effective climate adaptation practices (Martinez et al. 2011; Verschuuren and McDonald 2012; Hurlimann et al. 2014).

Whereas the *Gold Coast Planning Scheme 2003* stated that climate change should be considered in development of flood affected areas, the more recent *Ocean Beaches Strategy* and the *Coastal Management Plan* only noted that climate variability exists and contributes to generic coastal processes. While public pressure can create a barrier to climate policy development (Abel et al. 2011; Gurran et al. 2013; Verschuuren and McDonald 2012), action for climate change in local planning is essential (Boyer 2012; Baker et al. 2012; Serrao-Neumann et al. 2014; Dedekorkut et al. 2010), particularly for coastal resort cities such as the Gold Coast (Cooper and Lemckert 2012).

Although the Gold Coast boasts of decades of coastal management expertise, in the age of coastal climate adaptation it is quickly falling behind. This research has reviewed the current state of coastal management and climate adaptation on the Gold Coast and discussed the ability of the city's plans and policies to address best practice coastal climate adaptation. This evaluation points to an urgent need of a local plan, policy or strategy with the specific purpose of informing coastal climate adaptation; horizontal and vertical political collaboration regarding coastal climate adaptation, potentially through an institutional body or department (with guidance that may originate in higher government); and a higher order climate adaptation policy document to provide guidance for general climate adaptation with specific reference to coastal climate adaptation.

References

Abel, N., Gorddard, R., Harman, B., Leitch, A., Langridge, J., Ryan, A., et al. (2011). Sea level rise, coastal development and planned retreat: analytical framework, governance principles and an australian case study. *Environmental Science & Policy, 14*(3), 279–288.

Australian Bureau of Statistics (ABS). (2015). *Gold Coast (SA4).* Retrieved September 11, 2015, from http://stat.abs.gov.au/itt/r.jsp?RegionSummary&region=309&dataset=ABS_REGIONAL_ASGS&geoconcept=REGION&datasetASGS=ABS_REGIONAL_ASGS&datasetLGA=ABS_REGIONAL_LGA&regionLGA=REGION&regionASGS=REGION.

Baker, I., Peterson, A., Brown, G., & McAlpine, C. (2012). Local government response to the impacts of climate change: an evaluation of local climate adaptation plans. *Landscape and Urban Planning, 107,* 127–136.

Beirbaum, R., Smith, J., Lee, A., Blair, M., Carter, L., Stuart Chaplin III, F., et al. (2013). A comprehensive review of climate adaptation in the United States: More than before, but less than needed. *Mitigation and Adaptation Strategies for Global Change, 18*(3), 361–406.

Boyer, M. A. (2012). Adapting to climate change: mapping connecticut's coastal responses to a global problem. *Sea Grant Law and Policy Journal, 5*(1), 15–40.

Castelle, B., Le Corre, Y., & Tomlinson, R. (2008). Can the Gold Coast beaches withstand extreme events? *Geo-Marine Letters, 1,* 23–30.

City of Gold Coast (CGC). (2013). *Ocean Beaches Strategy 2013–2023.* Gold Coast.

Coleman, L. (2012). Making soft infrastructures a reality in New York City: Incorporating unconventional storm defence systems as sea levels rise. *William and Mary Environmental Law and Policy Review, 36*(2), 529–563.

Cooper, J. A. G., & Lemckert, C. (2012). Extreme sea-level rise and adaptation options for coastal resort cities: A qualitative assessment from the Gold Coast Australia. *Ocean & Coastal Management, 64,* 1–14.

Dedekorkut-Howes, A., & Bosman, C. (2015). The Gold Coast: Australia's playground? *Cities, 42* (Part A(0)), 70–84.

Dedekorkut-Howes, A., & Howes, M. (2014). Climate adaptation policy and planning in South East Queensland. In P. Burton (Ed.), *Responding to climate change: Lessons from a hotspot* (pp. 59–67). Collingwood, VIC: CSIRO Publishing.

Dedekorkut-Howes, A., & Sloan, M. (2012, September 24–26). Disaster resilience and local adaptation capacity in Coastal Queensland. *Australian Political Studies Association Conference*, Hobart.

Dedekorkut, A., Mustelin, J., Howes, M., & Byrne, J. (2010). Tempering growth: Planning for the challenges of climate change and growth management. *Australian Planner, 47*(3), 203–215.

Department of Environment and Heritage Protection (DEHP). (2013). *Coastal Management Plan.* Queensland Government.

Flood, S., & Schechtman, J. (2014). The rise of resilience. Evolution of a new concept in coastal planning in Ireland and the US. *Ocean and Coastal Management, 102,* 19–31.

Gold Coast City Council. (2009). *Gold Coast Climate Change Strategy 2009–2014.*

Grannis, J., Wyman, J., Singer, M., Shoaf, J., & Lynch, C. (2012). Coastal management in the face of rising seas: Legal strategies for connecticuit. *Sea Grant Law and Policy Journal, 5*(1), 59–88.

Gurran, N., Norman, B., & Hamin, E. (2013). Climate change adaptation in coastal Australia: An audit of planning practice. *Ocean and Coastal Management, 86,* 100–109.

Hanak, E., & Moreno, G. (2012). California coastal management with a changing climate. *Climatic Change, 111*(1), 45–73.

Harman, B., Heyenga, S., Taylor, B., & Fletcher, C. (2015). Global lessons for adapting coastal communities to protect against storm surge inundation. *Journal of Coastal Research, 31*(4), 790–801.

Howes, M., & Dedekorkut-Howes, A. (2016). The rise and fall of climate adaptation governance on the Gold Coast, Australia. In J. Knieling (Ed.), *Climate adaptation governance theory, concepts and praxis in cities and regions* (pp. 237–250). Hamburg: Wiley Inc.

Howes, M., Dedekorkut-Howes, A. (2012). Climate adaptation and the Australian system of government: The Gold Coast example. In: K. Crowley, & K. J. Walker (Eds.), *Environmental policy failure: The Australian story* (pp. 116–130). Prahran, VIC: Tilde University Press.

Hurlimann, A., Barnett, J., Fincher, R., Osbaldiston, N., Mortreux, C., & Graham, S. (2014). Urban planning and sustainable adaptation to sea-level rise. *Landscape and Urban Planning, 126,* 84–93.

Hwang, D. (1991). Shoreline setback regulations and the takings analysis. *University of Hawaii Law Review, 13*(1), 1–56.

Kaswan, A. (2013). Climate change adaptation and land use: Exploring the federal role. *The John Marshall Law Review, 47*(2), 509–541.

Klein, R., et al. (2000). Coastal adaptation. In B. Metz, O.R. Davidson, J. W. Martens, S. N. M. van Rooijen, & L. van Wie McGrory (Eds.), *Methodological and technological issues in technology transfer* (pp. 349–372). Cambridge, United Kingdom: Cambridge University Press.

Langridge, S., Hartge, E., Clark, R., Arkema, K., Verutes, G., Prahler, E., et al. (2014). Key lessons for incorporating natural infrastructure into regional climate adaptation planning. *Ocean and Coastal Management, 95,* 189–197.

Macdonald, H., & Patterson, D. (1984). Beach response to coastal works Gold Coast, Australia. *Coastal Engineering Proceedings, 1*(19), 1522–1538.

Martinez, G., Bizikova, L., Blobel, D., & Swart, R. (2011). Emerging climate change coastal adaptation strategies and case studies around the world. In G. Schernewski, J. Hofstede &

T. Neumann (Eds.), *Global change and baltic coastal zones* (pp. 249–273). Netherlands: Springer.
Mosadeghi, R., Tomlinson, R., Mirfenderesk, H., & Warnken, J. (2009). Coastal management issues in Queensland and application of the multi-criteria decision making practices. *Journal of Coastal Research, 56,* 1252–1256.
Pickering, C., & Byrne, J. (2014). The benefits of publishing systematic quantitative literature review for Ph.D. candidates and other early-career researchers. *Higher Education Research & Development, 33*(3), 534–548.
Sano, M., Golshani, A., Splinter, K., Strauss, D., Thurston, W., & Tomlinson, R. (2011). A detailed assessment of vulnerability to climate change in the Gold Coast Australia. *Journal of Coastal Research, SI, 64,* 245–249.
Serrao-Neumann, S., Crick, F., Harman, B., Sano, M., Sahin, O., van Staden, R., et al. (2014). Improving cross-sectoral climate change adaptation for coastal settlements: Insights from South East Queensland Australia. *Regional Environmental Change, 14,* 489–500.
Solomons, M., & Willacy, M. (2014, December 9). Jeff Seeney orders Moreton Bay regional council to remove references to climate change-derived sea level rises from regional plan. *ABC News*. Retrieved March 24, 2015, from http://www.abc.net.au/news/2014-12-09/seeney-removes-climate-change-references-from-council-plan/5954914
Spalding, M., Ruffo, S., Lacambra, C., Meliane, I., Hale, L., Shepard, C., et al. (2014). The role of ecosystems in coastal protection: Adapting to climate change and coastal hazards. *Ocean and Coastal Management, 90,* 50–57.
Strauss, D., Tomlinson, R., & Hunt, S. (2009). Profile response and dispersion of beach nourishment: Gold Coast, Australia. *Journal of Coastal Research, 56,* 133–137.
Tobey, J., Rubinoff, P., Robadue, D., Jr., Ricci, G., Volk, R., Furlow, J., et al. (2010). Practicing coastal adaptation to climate change: Lessons from integrated coastal management. *Coastal Management, 38*(3), 317–335.
Verschuuren, J., & McDonald, J. (2012). Towards a legal framework for coastal adaptation: Assessing the first steps in Europe and Australia. *Transnational Environmental Law, 1*(2), 355–379.
Vickers, J. (2015). *Coastal climate adaptation: A policy analysis of the City of Gold Coast*. Thesis Bachelor of Urban and Environmental Planning (Honours). Griffith University.
Zeppel, H. (2012). Local adaptation responses in climate change planning in coastal Queensland. *Australasian Journal of Regional Studies, 18*(3), 341–361.

Chapter 26
Coastal Environments Under a Changing Climate—What If Resilience Building Is Not Enough?

Gregory Wilford Fisk, Philip Edward Haines and Beth Frances Toki

Background

Climate change is expected to have a significant adverse effect on coastal and island natural systems (Walther et al. 2002; Hobday and Lough 2011). The intertidal position of mangroves, salt marshes, beaches and tidal flats, make them susceptible to changes in sea level but also other climate change variables such as increasing intense storms, and changes to rainfall patterns that can affect the physiology, ecology and long term stability of these habitats over time (refer Lovelock and Ellison 2007; Commonwealth of Australia 2009; Australian Government 2015).

Freshwater wetlands on the coast such as perched lakes, dune lakes, palustrine swamps and peat swamps, are particularly vulnerable to saline intrusion and will face more periodic tidal incursions as a result of coastal storms. Prolonged periods of high salinity will cause ecological shifts in these communities away from predominantly freshwater flora and fauna communities, favouring instead those estuarine species that can withstand more variable conditions (Neilsen and Brock 2009).

Subtidal habitats such as nearshore coral and rocky reefs and seagrass will be less affected by sea level rise, but will be more exposed to broader scale changes in ocean temperature, acidity and impacts from more intense land based runoff (affecting salinity, nutrient, light penetration and sedimentation rates).

G.W. Fisk (✉) · P.E. Haines · B.F. Toki
BMT WBM, Level 8, 200 Creek Street,
Brisbane, QLD, Australia
e-mail: greg.fisk@bmtwbm.com.au

P.E. Haines
e-mail: Philip.haines@bmtwbm.com.au

B.F. Toki
e-mail: beth.toki@bmtwbm.com.au

W. Leal Filho (ed.), *Climate Change Adaptation in Pacific Countries*,
Climate Change Management, DOI 10.1007/978-3-319-50094-2_26

Coastal species are also being affected by climate change both locally and at much greater spatial scales. Changes to the distribution of migratory species including usage of habitats for key life cycle stages such as nesting, breeding and roosting are expected for both migratory birds and marine turtles that make use of coastlines as their core habitat.

While there has been considerable effort by scientists to diagnose the effects of future climate change on natural coastal systems, climate change adaptation as a field of study and management practice continues to be heavily focused on the built environment. Many jurisdictions around the world (including all of the States and Territories in Australia) have implemented some form of legislative measure and policy guidance to ensure climate change is considered in future urban development decisions on the coast. Climate change adaptation planning and management for natural environments has been comparatively much slower to evolve.

Well formed climate change strategy documents for natural systems are now only just emerging; usually focused on existing protected areas and reserves. Invariably, these 'first generation' climate change strategies highlight a strong research focus and on building the resilience of these natural systems to either withstand or recover from climate change impacts (Hyder 2008; Australian Government 2015). However, there are challenges and differences of opinion amongst land and water managers to the need to start managing these areas differently because of future climate change.

It is widely recognised that climate change is most likely to exacerbate existing and more conventional threats as well as introduce new challenges for coastal managers in the context of how species, ecosystems, infrastructure and human uses are managed. Both these points question the long term effectiveness of a research and resilience building strategy, particularly facing the budget realities of governments with less public money to spend on park and reserve management and broader natural resource management issues.

The paradox of this situation is that many of the natural coastal systems that are most susceptible to climate change are incredibly valuable public assets. They contribute to both national economic growth and well as local subsistence through tourism, recreational use, and fisheries resources. In many cases, the economic benefits we derive from these systems far outweighs private infrastructure on the coast under similar threat (Buckley 2011), which raises a far more fundamental question: Will Governments and their communities seek to actively protect and manage natural assets that are vulnerable to climate change in the same way as the built environment?

If this is the case, it is asserted that a strategic approach to adaptation planning and management—similar to what is currently being developed for urban built environments—needs to start to be considered for coastal natural areas. Like the urban environment, this new management paradigm needs to be both risk and time based.

Understanding and Assessing Vulnerability

Before the questions posed above can be answered, the extent and severity of climate change on coastal natural systems and habitats must be defined and analysed. Invariably this should take the form of a vulnerability assessment.

IPCC defines vulnerability as 'the degree to which a system is susceptible to, or unable to cope with, adverse effects of climate change. Vulnerability is a function of the character, magnitude, and rate of climate variation to which a system is exposed, its sensitivity, and its adaptive capacity' (IPCC 2001 quoted in Australian Government 2009; Brooks 2003). In summary, vulnerability is a function of exposure, sensitivity and adaptive capacity as shown in Fig. 26.1.

Vulnerability of coastal environments to future climate change is broad given that there are a number of climate factors that potentially influence coastal ecosystems. Nicholls et al. (2008) summarised climate factors and there potential influence on coastal environments. Table 26.1 presents an adaptation of Nicholls et al. (2008).

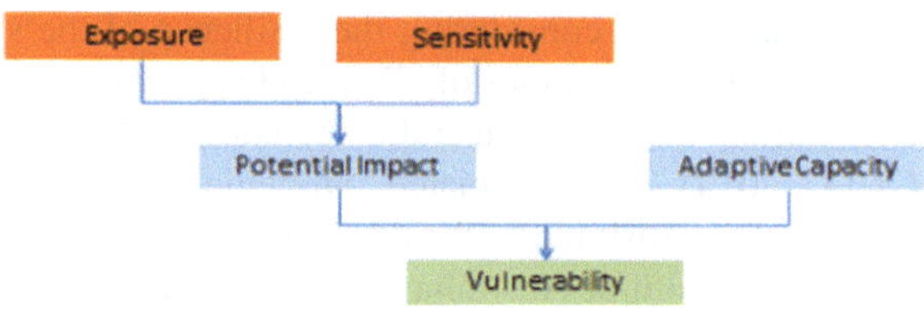

Fig. 26.1 Components of vulnerability (adapted from Australian Government 2009)

Table 26.1 Climate change parameters and potential vulnerability of coastal ecosystems (adapted from Nicholls et al. 2008)

Climate change parameter	Environmental coastal ecosystem response to driver
Sea level rise	Shoreline recession; saline inundation of low-lying lands; saltwater intrusion into freshwater wetlands and groundwater systems; rising groundwater levels; vegetation change in response to hydrology and salinity changes
Increased storminess (intensity) and associated storm tide inundation	Increased extreme ocean water levels; increased storm erosion or shorelines; increased frequency of inundation of low-lying lands
Changes to rainfall patterns	Changes to flood risk in low-lying lands; altered water quality and salinity; changes to alluvial sediment supply; changes to nutrient supply from catchment
Increased temperature	Increase in evaporation, decrease in groundwater levels; decrease in surface water ponding; vegetation change

The assessment of vulnerability can range from a rapid assessment of a large area or network (say across the protected area or World Heritage estate of a national government) to specific habitat types (for instance a climate change assessment of future impacts to coral reefs); to a particular site such as an small island or beach. One thing to consider here is that while assessment of vulnerability can be done at various spatial scales, adaptation management to address vulnerability will be **most effective** at a local scale where:

- the impacts on values and systems can be more perceptible;
- the trajectory of impact is clearer, can be locally monitored, and can be readily communicated to local stakeholders; and
- the development, implementation and evaluation of the effectiveness of management actions can be more readily observed and measured.

Ultimately the vulnerability assessment should assist the coastal manager to understand how various climate change issues and trends may affect the resource values including key parameters such as the extent and condition of habitat components (e.g. systems and species) and the natural and anthropogenic ecosystem services provided by the habitat. In this context, the current condition or exposure of the receiving environment to impact can be critical to understand as a first step, particularly when considering the resilience of the system to change.

Some recent examples of climate change vulnerability studies focused on coastal and marine assets in Australia include the Great Barrier Reef Outlook Reports (2011, 2014) and associated Strategic Assessments; Implications of Climate Change for Australia's World Heritage Properties: A Preliminary Assessment (Commonwealth of Australia 2009); Climate Change Risks to Australia's Coast: A First Pass National Assessment (Australian Government 2009).

In the Pacific, reports prepared under the SPREP PACC (Pacific Adaptation to Climate Change Programme) such as 'Ecosystem-based adaptation and climate change vulnerability for Choiseul Province, Solomon Islands: synthesis report' (Iacovino et al. 2013) and other local studies such as the 'Comparative Analysis of ecosystem-based adaptation and engineering options for Lami Town, Fiji' (UNEP 2012) are relevant. Also, the recently released 'Integrated Vulnerability Assessment Framework for Atoll Islands—A collaborative Approach' (SPC 2016) is informative in providing an overarching approach to undertaking vulnerability assessments for coastal environments.

Using these studies and guidelines as representation of best practice documents, we have formulated an essential seven-step methodology for assessing vulnerability of coastal and marine natural assets, as presented in Table 26.2.

Table 26.2 Methodology for assessing vulnerability of coastal and marine environments

Step 1: Establish the environmental, social and economic values/assets of the coastal system that could be affected by climate change
Step 2: Consider the different aspects of climate change (e.g. sea level rise, changes to rainfall patterns, storminess, etc.) that may affect the values of the asset both in isolation and interactively
Step 3: Determine and map what habitats and species (upon which the values depend) are at particular risk from future climate change aspects
Step 4: Determine the current extent and condition of these habitats and species and how they may change/shift as a result of climate change (e.g. will they succeed landward in response to sea level rise?)
Step 5: Predict responses to climate change using models and other numerical tools to better define spatially and temporally how the habitats and species will respond
Step 6: Examine at broader landscape and bioregional scales, the implications of losing vulnerable species and habitats—are there refugia habitats with similar values that are not going to be affected by climate change impacts
Step 7: Identify key knowledge gaps and recommending monitoring to fill these gaps

Treatment of Risks Through Resilience

Vulnerability assessment plays a critical role in raising our awareness of the issues and provides detailed information and data from which to inform management. However, it still can be very difficult to know where to start in terms of an appropriate management response to climate change vulnerability and risk—particularly where results and findings are highly complex, inconclusive in terms of timing and severity, requiring further research or in some cases, contradictory.

On that basis, building the resilience of coastal natural systems to climate change should be central to any adaptation strategy. In terms of day to day management, this can be achieved by effectively managing non-climate related pressures such as human disturbance, overfishing, edge effects from development, stormwater run-off, fire, weeds and introduced species. At a broad planning level, reserve and protected area planners should be examining the availability of climate refugia (habitats that persist as climate changes), establishing landscape corridors that allow plants and animals to move to more suitable locations and trying to ensure healthy populations with sufficient genetic diversity to adapt (from NPS 2011).

In a Pacific context, climate change resilience and adaptation is increasingly being considered and delivered through an Ecosystem-based Adaptation (EbA) (Munang et al. 2013). EbA is the use of biodiversity and ecosystem services as part of an overall adaptation strategy to the adverse effects of climate change. By taking into account the ecosystem services on which people depend for their livelihoods as well as social and economic security, EbA integrates sustainable use of biodiversity and ecosystem services in a comprehensive adaptation strategy.

The concept of EbA has been common within the literature since about 2009, but as a management paradigm has had much slower uptake in Australia. Notwithstanding, new initiatives in this field are starting to emerge such as the 'Building with Nature Australia' initiative, driven by CSIRO and Ecoshape.

But what if resilience is not enough to protect ecosystem sustainability?
Given the extent and severity of future climate change predictions and the potential failure to achieve global emission reductions, there are questions as to whether an approach of resilience building on its own can be successful in the long term. Even highly resilient habitats and species may not be able to respond effectively to the rate and extent of change predicted for issues such as sea level rise (which is expected to continue to rise for several centuries before stabilizing at a new level).

In timescales of mere decades, as ecosystems change in response to climatic factors such as sea level rise and saltwater intrusion, there is considerable uncertainty to what the response will be. Will Governments and their communities seek to protect natural assets against these processes the same way as the built environment with engineered seawalls, levees, dykes and tidal gates? Can natural defenses using EbA principles be as effective as built solutions to manage the risk? Or will we actively try to reduce the duration of impact by facilitating a planned retreat from climate change? Alternatively, do we simply do nothing and let 'nature' take its course?

Faced with these alternatives, it is clear that more difficult decisions about management of coastal natural assets lie ahead. For the current managers of these areas, the key question is:

> **what practically can/should be done now to assist this future decision before climate change impacts start to become more frequent and severe?**

We argue here that approaches that are being developed for climate change adaptation in built environments need to be considered for managing natural areas on the coast in a way that complements the resilience building approach and paradigm.

Central to this approach are two concepts:

1. In addition to broader scale resilience building activities, what are the practical adaptation tools/actions that can be applied by coastal managers to help coastal natural systems adapt to climate change?
2. Can we develop trigger-based approaches to management such that adaptation actions are not implemented too early or too late to accommodate future climate change in these systems?

These two approaches are discussed in more detail in the following sections.

Natural Environment Climate Change Adaptation Toolkit

Using an example from Eastern Australia, Queensland's "Climate Change Strategy, ClimateSmart Adaptation 2007-2012" (Queensland Government 2007) recognised the need to build the capacity of key sectors to understand, avoid, reduce and manage the impacts of climate change. In the Queensland Strategy, Action 45 (under the heading of Natural Environment and Landscapes), sought to

> Develop climate change information and advice in a form that can be effectively included in conservation and natural resource management programs such as:
>
> - Protected area planning, design and management
> - Corridor and landscape planning
> - Species management and planning
> - Private land conservation initiatives
> - Regional natural resource management planning
> - Pastoral leasehold land management
> - Water quality improvement planning

This action reflected the need to ensure information and research being collected on climate change was effectively translated into the respective management frameworks listed.

It also pointed to the need for better understanding of the adaptation tools that are available for use by coastal natural resource managers.

Building on this approach, BMT WBM (2011) developed a conceptual suite of adaptation tools in consultation with Parks Australia, Parks North staff and other workshop participants as part of the Kakadu National Coastal Vulnerability Assessment study. This work demonstrated there was an urgent need to develop a more comprehensive and applied climate change adaptation toolkit for coastal natural systems built around the 'Protect:Accommodate:Retreat' framework already adopted for urban coastal management studies (Rollason and Haines 2011).

With reference to the climate change parameters discussed previously in Tables 26.1 and 26.3 summarises some of the initial tools for consideration under this framework for coastal natural systems.

In general the following descriptions characterise the three categories

- Protect—this category involves carrying out works or otherwise manipulating the natural environment to achieve a desired outcome. These actions provide longer term protection against climate change impacts but will still have a defined life and require human interaction/effort to be maintained
- Accommodate—this category can be a combination of works or activities that seek to minimise climate change impacts over time. Most involve human involvement or management effort. These actions can provide short and medium term solutions to climate change but may not be effective for avoiding climate change impacts in the longer term
- Retreat—this category involves either avoiding impact or otherwise abandoning areas. It may involve works or activities on areas not vulnerable to climate

Table 26.3 Adaptation actions for coastal natural systems

Climate change parameter		Protect actions	Accommodate actions	Retreat actions
Sea level rise		Engineering works such as seawalls, levy banks, breakwaters, tidal gates, locks and weirs to control ingress of saltwater into coastal habitats and wetlands from sea level rise This category could also include the use of natural defenses such as mangroves, dune enhancement and offshore reef construction	Retrofit/upgrade of infrastructure to accommodate sea level rise Changes to visitor management regime in areas under threat from saltwater intrusion to reduce pressures Modification/rehabilitation of landscapes to accommodate succession of tidal habitats landward Beach nourishment and dune enhancement to maintain beaches over time	Translocate species and habitats that are subject to permanent inundation or loss of habitat or otherwise creation of supplementary habitat elsewhere Progressively rehabilitating areas in view of long term shifts in saltwater inundation patterns from sea level rise Translocate or abandon infrastructure in areas prone to future hazard to reduce cumulative impacts on natural systems Translocate existing human uses and activities to other coastal areas under no/less threat Do nothing—monitor how ecosystems are adapting to new paradigm
Increased storminess (intensity) and associated storm tide inundation		Engineering works such as seawalls, levy banks, and breakwaters to control or reduce the impact on natural assets from storm surges and flooding This category could also include the use of natural defenses such as mangroves, dune enhancement and offshore reef construction	Retrofit/upgrade of infrastructure to accommodate more intense storms and surges Rehabilitate and strengthen existing habitats such as mangroves and dunes that are the first defence against coastal storms Increased budget and workforce for clean up and post-event reconstruction/rehabilitation works Ensuring an adequate buffer is maintained between new development and the hazard zone that allows natural systems to respond	Re-establishing/rehabilitating only those habitats outside of the most likely storm surge zone (sacrifice habitats that cannot be sustained in the long term) Translocate or abandon any infrastructure in the hazard zone to reduce cumulative impacts on natural systems Translocating existing human uses and activities within coastal areas to areas under no/less threat Do nothing—monitor how ecosystems are adapting to new paradigm

(continued)

Table 26.3 (continued)

Climate change parameter		Protect actions	Accommodate actions	Retreat actions
			Change fisheries management regimes, noting the need for recovery of systems following major storm events	
Changes to rainfall patterns—increased rainfall and catchment flooding		Engineering works to control flows (for example dredging of artificial entrances to lagoons and construction of weirs and barrages to control downstream flows)	Change to management practices (weeds, fire, feral animals control) to take into account greater freshwater flows into downstream systems Manage duration, timing and volume of freshwater flows into the coastal systems to simulate more natural (historic) patterns Change fisheries management regimes, noting the influence of more freshwater flows, nutrients and run off on habitat and species abundance and conditions	Abandon uses of areas that are subject to frequent flooding or inundation to reduce cumulative impacts on natural systems Translocate existing human uses and activities within coastal areas to areas not under threat Do nothing—monitor how ecosystems are adapting to new paradigm
Changes to rainfall patterns—decreased rainfall and catchment flooding, increasing salinity and evaporation		Engineering works to control flows (for example weirs and water control structures) to downstream coastal environments	Change to management practices (weeds, fire, feral animals control) to take into account decreased freshwater availability in coastal systems More active management of the duration, timing and volume of freshwater flows into downstream coastal systems to simulate more regular (historic) patterns Change fisheries management regimes, noting the influence of less freshwater flows on habitat and species abundance and conditions	Abandon use of areas due to lack of freshwater flows or water supply (for example, visitor use areas used for recreational boating, swimming and similar) to reduce cumulative impacts on natural systems Do nothing—monitor how ecosystems are adapting to new paradigm

(continued)

Table 26.3 (continued)

Climate change parameter		Protect actions	Accommodate actions	Retreat actions
Increased temperature—decreased water availability		Create new storages and change approaches to capture of stormwater and groundwater Increase rate of groundwater (bore) extraction (including for stock and wildlife)	Ensure the overall water balance takes into account water needed for environmental systems Change water consumption rates through education/regulation Changes to management practices (weeds, fire, feral animals control) to take into account reduced water availability Retrofit existing infrastructure to increase water efficiency	Abandon use of areas where there is insufficient or unsustainable water sources Do nothing—monitor how ecosystems are adapting to new paradigm
Increased temperature—increasing hot days		Construct additional shading/cooling infrastructure (including for wildlife)	Changes to visitor management regime/restrict access regimes to reduce potential temperature impacts Changes to access regimes to reflect temporal and spatial changes to how fauna are using coastal habitats Retrofit existing infrastructure to energy efficient design to reduce emissions and create shading and other cooling benefits	Abandonment of public use/closure of the recreation areas during hottest days Do nothing—monitor how ecosystems are adapting to new paradigm

Icons in Table 27.3 are courtesy of the Integration and Application Network, University of Maryland Center for Environmental Science (ian.umces.edu/symbols/)

change (to facilitate landward migration of habitats) and translocation of key natural and cultural assets.

While greater research and implementation examples are needed to develop and refine these tools in Table 26.3, they provide a basis for consideration of when such tools should be applied in planning and decision making which is the second part of the strategic approach discussed below.

Triggers for Future Action and Limits of Acceptable Change

A key challenge to both attracting funding for planning and implementing measures to building resilience and more active climate change adaptation responses is that the impacts in many cases are not yet apparent or otherwise masked by natural variability in natural systems. The application of the toolkit presented in Table 26.3 can be assisted by a simple decision support system taking into account the risks of climate change impacts over time.

This approach draws from observations made in Australian climate risk literature about the need for triggers for future action with respect to climate change (refer, in particular, Atwater et al. 2008) but also draws on similar trigger approaches that can be found within an increasing number of natural resource management planning and assessment frameworks (Fisk and Kay 2010).

Specifically the tool uses a process of setting of 'limits of acceptable change' for wetland values as outlined by the National Framework and Guidance for the Preparation of Ecological Character Descriptions (ECD) for Australia's Ramsar Wetland (DEWHA 2008). Under the Framework, indicators of change are defined for wetland components and processes such as bird numbers and habitat extent based on an understanding of the natural variability of those populations and systems over time. These indicators are framed as 'limits of acceptable change' that guide the future management of the wetland by defining the quantitative limit of impact that can be accepted before the ecological values of the wetland are diminished or lost. In the future, if there is an exceedance of one or more of the limits of acceptable change set out in the ECD (e.g. there is a change outside of natural variability or a change beyond the limit set for the indicator) then this serves as a trigger that an ecological character change has occurred within the wetland that must be halted or reversed.

This approach of defining undesirable impacts (as opposed to desirable conditions) has some attraction for assessing and addressing future climate change which is also based on the notion of trying to avoid or otherwise minimise future impacts (see Fig. 26.2).

Information from vulnerability studies about current condition and the potential effects of climate change on coastal natural assets can be used to: (a) define a profile for each climate change risk over time (e.g. along a continuum) based on the

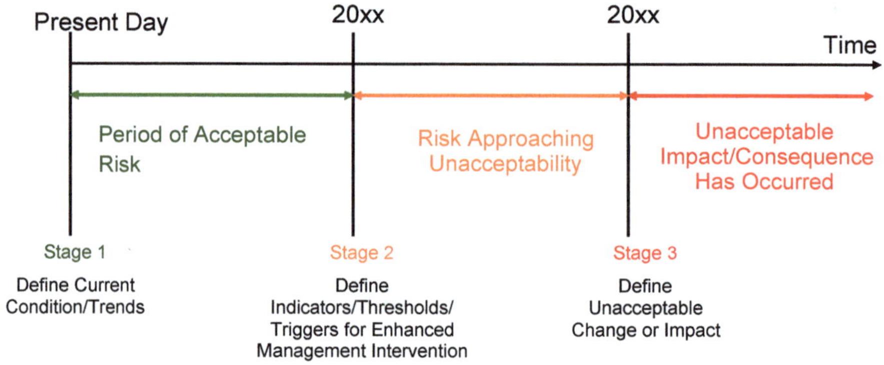

Fig. 26.2 Climate change risk continuum (adapted from Fisk and Kay 2010)

projection or prediction of future climate change impacts; and (b) consider how various adaptation actions developed to treat the risk should be applied over time.

For each climate change risk issue or parameter that is being considered as part of the climate change vulnerability or risk assessment, there is an assumption that three points (or stages) along the continuum can be defined (see Fig. 26.2):

- Stage (1) The baseline (present day/current condition) of the climate change parameter or risk being examined;
- Stage (2) The identification of one or more 'trigger' points along this time continuum that serve as a flag to management agencies that the impacts associated with the risk are occurring and that an undesirable impact is being approached; and
- Stage (3) The undesirable impact or end state of the climate change parameter or risk being examined (e.g. what are the impacts that are trying to be avoided).

Linking to the discussion about the ECD framework above, a key innovation of the tool is the approach of seeking the coastal management agency to define—at the present time—what it believes to be the unacceptable future impact from climate change.

Once this undesirable end state is established and agreed, the planning or management agency can consider, in a more practical sense, what are the most appropriate adaptation actions that can be undertaken to either avoid the impact from occurring or otherwise to minimise or mitigate the future impact and when such interventions need to be considered and implemented.

Use of the tool can also help decision-makers to align perceived risk with the selection of the most appropriate adaptation measures and actions. Importantly, such an approach can guide agencies to differentiate between 'no-regrets' and other resilience building actions that can be undertaken in the short term (period of acceptable risk shown in green in Fig. 26.2) versus the more difficult decisions and options that deliver adaptive action in the medium to long term (shown in orange and red zones in Fig. 26.2).

In a similar fashion, the adaptation toolkit discussed in the previous section of this paper can also be aligned with the continuum. While it may be appropriate to implement ‘accommodation’ actions in the green zone of acceptable risk; ‘protection’ or ‘planned retreat’ actions may be needed to be implemented in the orange zone of the continuum to avoid the undesirable impact shown in the red zone.

Overall, the approach assumes that while a range of resilience building actions may be appropriate to develop and implement while the risk is at an acceptable level, there will need to be consideration of more direct actions to address the risk as the likelihood and severity increases.

Some other advantages to setting some form of a trigger level for adaptation actions are as follows:

- Assuming the trigger is set effectively it should give management agencies adequate time to act prior to realising the undesirable impact. This could include for example, development of a response plan, time to run an adequate public consultation programme, and sufficient time to more quantitatively assess the various adaptation options such the feasibility of constructing works (or other expensive and/or controversial adaptation response);
- It prevents the management agency from acting too early. Premature action in response to climate change can have significant cost and political implications such as exposing the management agency to appeals against development decisions and compensation claims. This is termed ‘over-adaptation’ to climate change (Willows and Connell 2003). Acting too early also does not take advantage of any research and technological advancements that have occurred during the period of acceptable risk (e.g. the period leading up to the trigger point);
- The approach can help to ‘lock’ decision makers into an action plan over a longer period of time and facilitate evaluation of plans and strategies. This is especially important given the invariable turnover of staff from management agencies, election and re-election of political leaders and similar issues; and
- The approach demonstrates to the broader community a level of preparedness and a willingness to act as climate change impacts are realised and as trigger points are approached in the future.

Limitations and Constraints

Research supporting this paper has focused on climate change adaptation practice in the Australasian area. While principles of climate change are generic, relevant tools and approaches are likely to vary across the globe based on varying climatic conditions and natural environments. Geopolitical drivers for climate change adaptation also vary across the globe, and while the recent COP21 agreement in Paris has increased awareness, political focus and investment remains on mitigation rather than adaptation.

Conclusions

We argue that building resilience in coastal natural ecosystems will be insufficient to manage potential future impacts associated with climate change, especially in the longer term. Moving beyond resilience building to more active climate change adaptation for coastal natural ecosystems will require a combination of sound science, clear drivers for change and political will and funding to implement new measures.

To facilitate this move from resilience to adaptation, this paper has presented the foundations for coastal manager to make more difficult future decisions through effective planning and data collection, including:

1. A robust methodology for assessing vulnerability of coastal natural ecosystems to climate change;
2. A toolkit of adaptation options addressing the potential impacts of various climate change parameters, focusing on either Protection, Accommodation, or Retreat; and
3. A time continuum framework for establishing triggers that prompt specific actions and measures to ensure undesirable outcomes are avoided but also that authorities are not leaping to over-adaptation and spending unnecessary resources before the actual impacts of climate change become more certain.

The authors' experiences have been that the use of the foundations above are practical and can help authorities conceptualise their response to climate change and plan for future action in the longer term timeframes. This is important given the uncertainty around the scale and timing of climate change impacts and the need to develop an adaptive approach to management, which entails evaluation and refinement as better information becomes available over time.

References

Atwater, C., Carley J., and Witte, E. (2008). Establishing triggers for adaptive response to climate change. In *Proceedings of National Coast to Coast Conference*, Darwin, Australia.

Australian Government. (2015). *National Climate Resilience and Adaptation Strategy*. Canberra: Department of the Environment.

BMT WBM. (2011). *National coastal vulnerability assessment: south alligator river catchment case study*. Report prepared for the Australian Government Department of Climate Change, BMT WBM Pty Ltd, Brisbane.

Brooks, N. (2003). *Vulnerability, risk and adaptation: A conceptual framework*. Working Paper 38, Tyndall Centre for Climate Change Research, Norwich, UK.

Buckley, R. (2011). Tourism and environment. *Annual Review of Environment and Resources, 36,* 397–416.

Commonwealth of Australia. (2009). *Parks Australia: Climate Change Strategic Overview 2009-2014*. Canberra: Director of National Parks.

Department of the Environment, Water, Heritage and the Arts (DEWHA). (2008). *National Framework and Guidance for the Preparation of Ecological Character Descriptions for*

Australia's Ramsar Wetlands. Water, Heritage and the Arts, Canberra: Department of the Environment.

Fisk, G. W., & Kay R. (2010). Dealing with uncertainty in climate change adaptation planning and developing triggers for future action. In *Proceedings of Practical Responses to Climate Change National Conference 2010*. Melbourne, Australia.

Hobday, A. J., & Lough, J. M. (2011). Projected climate change in Australian marine and freshwater environments. *Marine & Freshwater Research, 62,* 1000–1014.

Hyder Consulting. (2008). *The impacts and management implications of Climate Change for the Australian Government's Protected Areas*. Report prepared for the Department of Environment, Water, Heritage and the Arts and the Department of Climate Change by Hyder Consulting, Sydney.

Iacovino, C., Donohoe, P., Carruthers, T., & Chape, S. (2013). *Ecosystem-based adaptation and climate change vulnerability in Choiseul Province, Solomon Islands. Synthesis report*. Samoa: Secretariat of the Pacific regional Environmental Programme (SPREP), Apia.

Lovelock, C., & Ellison, J. (2007). Chapter 9: Vulnerability of mangroves and tidal wetlands of the GBR to Climate Change. In *Climate Change and the Great Barrier Reef: A Vulnerability Assessment*, Great Barrier Reef Marine Park Authority (GBRMPA).

Munang, R., Thiaw, I., Alverson, K., Mumba, M., Liu, J., & Rivington, M. (2013). Climate change and Ecosystem-based Adaptation: A new pragmatic approach to buffering climate change impacts. *Current Opinion on Environmental Sustainability, 5,* 1–5.

Neilsen, D. L., & Brock, M. A. (2009). Modified water regime and salinity as a consequence of climate change: Prospects for wetlands of Southern Australia. *Climatic Change, 95,* 523–533.

Nicholls, R. J., Wong, P. P., Burkett, V., Woodroffe, C. D., & Hay, J. (2008). Climate change and coastal vulnerability assessment: Scenarios for integrated assessment. *Sustainable Science, 3,* 89–102.

Queensland Government. (2007). *ClimateSmart Adaptation 2007-12 An action plan for managing the impacts of climate change*. Brisbane: Department of Natural Resources and Water.

Rollason, V., & Haines, P. (2011). Outcomes from the Application of ISO 31000:2009 Risk Management Principles to Coastal Zone Management. In *Proceedings of 20th NSW Coastal Conference*, Tweed Heads, Australia.

SPC. (2016). "Integrated Vulnerability Assessment Framework for Atoll Islands: A collaborative approach" Prepared jointly by the Pacific Community (SPC), Secretariat of the Pacific Regional Environment Programme (SPREP) and Deutsche Gesellschaft für Internationale Zusammenarbeit (GIZ) GmbH.

UNEP and SPREP. (2012). *Comparative Analysis of ecosystem-based adaptation and engineering options for Lami Town, Fiji—Synthesis Report*. Samoa: Secretariat of the Pacific regional Environmental Programme (SPREP), Apia.

US National Parks Service (NPS). (2011). [online] Information sourced from website www.nps.gov/climatechange/adaptationplanning.

Walther, G.-R., Post, E., Convey, P., Menzel, A., Parmesan, C., Beebee, T. J. C., et al. (2002). Ecological responses to recent climate change. *Nature, 416,* 389–395.

Willows, R. I., & Connell, R. K. (Eds.). (2003). Climate adaptation: Risk, uncertainty and decision-making. UKCIP Technical Report. UKCIP, Oxford.

Chapter 27
Climate Change Adaptation in the Pacific: Setting-Up Priorities in the Health Sector

Walter Leal Filho

Introduction

Pacific island countries are among the most vulnerable in the world, both to natural disasters and to those influenced by anthropogenic action (Leal Filho 2015a), in a way which raises concerns about the ability to secure the well being of their populations (Tisdell 2008). The latest report (AR5) by the Intergovernmental Panel on Climate (IPCC 2014) has outlined the many risks seen in the region, which include loss of adaptive capacity and damages to ecosystem services critical to lives and livelihoods in small islands. In addition, sea-level rise is mentioned as posing one of the most widely recognized climate change threats to low-lying coastal areas on islands and atolls. Furthermore, given the dependence of island communities on coral reef ecosystems for a range of services including coastal protection, subsistence fisheries and tourism, there is high confidence that coral reef ecosystem degradation will negatively impact island communities and livelihoods. Given the inherent physical characteristics of small islands, AR5 reconfirms the high level of vulnerability of small islands to multiple stressors, both climate and non-climate. These elements illustrate the fact that, in addition to the necessary measures in the field of environmental mitigation, adaptation approaches are urgently needed.

Since the 1950s, natural disasters have directly or indirectly affected the lives, livelihood and property of more than 3.4 million people in the Pacific, and has led to more than 1,700 reported deaths in the region (IFAD 2007). In the 1990s alone, reported natural disasters cost the Pacific islands region US$ 2.8 billion (2004 value). More recent extreme events such as Cyclone Pam and Winston have also created much damage and losses are well in excess of the hundreds of millions of dollars. The Asian Development Bank (2012) produced a document titled "Pacific

W. Leal Filho (✉)
School of the Science and the Environment, Manchester Metropolitan University, Chester Street, Manchester, UK
e-mail: walter.leal@haw-hamburg.de

W. Leal Filho (ed.), *Climate Change Adaptation in Pacific Countries*,
Climate Change Management, DOI 10.1007/978-3-319-50094-2_27

Regional Strategic Program for Climate Resilience" where many impact areas are listed.

But the impacts of climate change are not only related to those caused by extreme events such as storms and cyclones, which particularly affect coastal communities, many of which are already stressed by unsustainable development practices. Vulnerability in the region is quite wide, reflecting trends seen in a developing country context (UNFCCC 2007).

Agriculture —both subsistence and commercial— is also likely to be negatively influenced, and the productivity of a variety of cash crops, including sugar cane and banana which are often exported and are important in gaining foreign exchange, is expected to decrease. Furthermore, some of the projected impacts of climate change such as droughts and reductions in the frequency and intensity of rainfall are expected to lead to drops in the levels of loss of soil fertility which, in turn, may be responsible for drops in the production of cash crops.

No longer a seldom trend in the Pacific, long dry spells such as the ones seen in Samoa in 2016 lead to significant reductions in the amount of water that can be used for agriculture or tourism, reductions in river flow, and to slower rates of recharge of groundwater, which can result in prolonged drought impacts.

Scientific studies have stated that, in the absence of adaptation, a high island such as Fiji, could experience damage of US$ 23 million to US$ 52 million per year by 2050. The overall change in agricultural welfare is expected of a range between −8 and +4 billion USD per year facing with 2–4 °C temperature increase (World Bank?).

Generally speaking, the vulnerability of countries in the Pacific region to climate change may be better understood if four main factors are considered:

- **Geographical**: the Pacific region is, due to its geographical location, periodically visited by tropical storms and cyclones. The most affected sites are low-lying coastal areas, which are often the most populated parts of the islands, with villages, towns, agriculture, infrastructure and tourist development competing for space (UNDP 2015).
- **Economic**: the majority of countries in the regional have fragile economic systems, and do not often have the cash needed or easy access to the financing need to take some adaptative measures.
- **Governance**: despite the pressing need for climate change adaptation measures, they are not as highly emphasized in government policy as they should be. As a result, support to adaptation measures translated by government funding, is rather limited. Here, it is often less a question of lack of awareness on the impacts of climate change. Rather, it is the fact that other issues competing for political attention and funding tend to get the upper hand and be more highly prioritized.
- **Inappropriate infra-structure**: traditional buildings and forms of land occupation have not given due consideration to climate change. As a result, many public and private buildings, as well as private housing are vulnerable to the

impacts of extreme events and rainfall on the one hand, and to salt intrusion from storms on the other

One example is damages to property and to public infrastructure in coastal areas a result of extreme events. These are among the most significant consequences (Leal Filho 2015b) and have an immediate impact on properties, often leasing to substantial human casualties. Article 4 of the United Nations Framework Convention on Climate Change (UNFCCC) requires the assessment and reporting of vulnerability and adaptation, but such a requirement is sometimes not fulfilled in the way it should.

A report issued by the Asian Development Bank has estimated that coastal flooding may potentially affect between 60,000 and 90,000 people by 2050 (ADB 20XX). The impacts of extreme weather events, combined with coastal erosion, and progressive sea-level rise is associated with high human and economic costs. Kiribati, the Marshall Islands, Tokelau and Tuvalu for instance, are particularly vulnerable as their fresh water reserves are limited to a shallow subsurface lens which is susceptible to depletion in drought and susceptible to contamination from salt water. Extreme events such as storm surges and cyclones may cause salt water intrusion, bringing in severe problems to the water supplies of the affected communities.

An item often overlooked when estimating the impacts of extreme events, is in relation to the immediate drop in income from tourism: cruise operators tend to take off some destinations of their list as soon as an extreme event on some islands has taken place, leading to even greater economic losses to the affected islands. Since small island nations are economically fragile (Briguglio 1995), the financial burden of climate change poses an additional problem to them.

This state of affairs illustrates the fact that coordinated efforts of the different sectors are necessary, so as to build resilience among the countries in the region.

Climate Change and the Health Sector: An Ignored Threat

Despite the fact that substantial elements of the literature focus on the physical impacts of climate change, a comparatively low degree of emphasis is seen in respect of the health sector. Yet, there are strong links between climate change and health conditions (Leal Filho et al. 2016).

Many Pacific island countries face the pressures posed by having to deal with the burden caused of communicable diseases, non-communicable diseases and the health impacts of climate change. The effects of climate change on health, are already seen in the Pacific, have emerged as a major for the local populations and to the health systems of Pacific islands. This is particularly so bearing in mind their fragile economies and the pressures already being posed in their health systems by conventional expenditures. It is thus important to understand how climate change affects health in the Pacific region, and discuss the measures needed to optimize their relations.

In an attempt to handle this issue, the World Health Organization (WHO) Regional Office for the Western Pacific issued a report titled "*Human health and climate change in Pacific island countries*", that scientifically analyzes the impact of climate change on the health and adaptation strategies and evidence-based policy options for Pacific island countries.

Thirteen Pacific island countries are included in this report, these being Cook Islands, the Federated States of Micronesia, Fiji, Kiribati, the Marshall Islands, Nauru, Niue, Palau, Samoa, Solomon Islands, Tonga, Tuvalu and Vanuatu. They all participated in the formal Vulnerability Assessment and Adaptation Planning project led by WHO between 2010 and 2013, on environmental health in Pacific island countries.

According to the WHO (2016) three different methods were used to assess health vulnerabilities at the country level. In countries such as the Marshall Islands, the Federated States of Micronesia and Palau, statistical analysis were applied while consultative and interview-based approach was used in Cook Islands, Kiribati, Niue, Samoa, Tonga and Tuvalu. The modified health impact assessment approach was used for Nauru, Solomon Islands and Vanuatu.

The WHO study has identified the fact that the majority of the climate change and health priorities identified in Pacific island countries are largely to some key issues, namely:

- Increased incidences incidence of waterborne and vector-borne diseases
- health impacts of heat waves, especially among children and the elderly
- health impacts of natural disasters on both physical and mental health
- sub-optimal malnutrition due to reduced availability of food crops, whose prices increase when their productivity is low

The reasons why many island nations in the Pacific regions are also vulnerable to the health impacts of climate change include the following factors:

(a) limited information on the expected or likely health problems associated with climate change
(b) lack of training from the side of medical staff to associate clinical conditions with climate and environmental phenomena
(c) lack of historical data on the occurrence of health problems as a result of climate change
(d) lack of a wide awareness on the mental problems resulting from damages or losses to property due to extreme events

From a sub-regional perspective, common priorities identified in the national action plans of Pacific island countries and areas include:

- Perform the vulnerability assessment and adaptation planning processes periodically and updating the NCCHAPs, to reflect improved knowledge and

information regarding climate-sensitive health risks and appropriate management strategies;
- Align climate change and health with activities and ongoing improvements in disaster risk management, food security and water, sanitation and hygiene (WASH);
- Mainstream consideration of climate change and health into existing public health systems, while ensuring that such mainstreaming does not lead to dilution, diversion or diminution of the urgency of the targeted activities;
- Improve understanding of the relationship and diverse causal pathways between climate variability and climate-sensitive diseases (CSDs);
- Explore the previously neglected areas of climate change and health research in the Pacific, including climate change impacts on heat-related illnesses, mental health and non-communicable diseases (NCDs), as well as on communicable diseases; and
- Develop climate-based early warning systems (WHO 2016).

These elements call for cross-sectoral collaboration in relation to the adaptation of the health sector to climate change: national governments, international agencies, nongovernmental organizations, communities and the academic community need to join their efforts to address the effects of climate change on health.

Some Priorities and Action Needed

If history has shown something, then it is the fact that current trends cannot continue, and current problems and gaps in adaptation strategies cannot be left unattended. In order to decrease their vulnerability, countries in the region need to get their priorities more systematically established.

In particular, the need to acknowledge the linkages between vulnerability, resilience, and the adaptive capacity of countries (Gallopín 2006) is imperative if effective measures may be implemented. A better understanding of uncertainty is in this context quite important, for it offers a sound basis upon which a reduction of vulnerability can be achieved.

Adger et al. (2008) on a paper which explores the social limits to adaptation to climate change, have raised the issue that some conditions are often not fully considered when decisions to implement adaptation measures are taken. This does not need to be so. The predominant approach to vulnerability has largely been to measure and assess it, and this information has not always guided climate change adaptation efforts. This trends needs to be changed. In this context, a diagnose of the inherent barriers (Moser and Ekstrom 2010) may provide useful insights and guide the correct sets of action.

Furthermore, it is important that climate change adaptation policies are not only more consistently prepared and have the long term view in mind (and not the short term view and the limitations imposed by immediate costs), but they also need to be properly embraced and their implementation needs to be closely monitored. Table 27.1 outlines some of the pressing priorities in climate change adaptation in the Pacific region, and offers an analysis of the various degrees of complexity in their implementation.

The implementation of the elements outlined in Table 27.1 depends on one pre-condition: the willingness to invest. The range of projects which may be implemented to meet the above priorities require various sets of financial resources, but these vary from expensive ones (e.g. coping with damages to infra-structure) to less cost intensive measures such as the greater engagement of women on climate change adaptation affairs.

It can be seen that further progresses in promoting climate change in the region is not necessarily about inventing new technologies. Quite often, effective results can be achieved by simply applying new ideas, and using available materials and resources without necessarily the need for large expenditures. Ecological practices based on traditional knowledge, may also enable a high degree of resilience, and more use should be made of them.

A further need is seen in the field of education. There is a perceived need to train a new generation of scientists across the Pacific region, who will not only excel in obtaining, processing and climate data, but will also be able to analyse the relationships and dependencies across a complex system. And by doing so, propose concrete climate change adaptation measures in support of global mitigation efforts. A more systematic use of currently available climate data may allow countries to move away from making decisions on what they think, and based more on what they know. Indeed, data and evidence driven policy-making is essential if the countries in the region are able to meet the challenges climate change poses to

Table 27.1 Some priorities to climate change adaptation in the Pacific

Priority	Relevance	Degree of complexity in implementation
Better use of climate data	Make sense of the available data with a greater sensitivity to social concerns	Low since most of the data is already available
Handling of erractic rainfall via rainwater harvesting systems	Increase food security and resilience to droughts	Low since rainwater harvesting systems and simple and cheap to implement
Securing water supplies from salt intrusion from sea level rise	Upkeeping agriculture and reduction of dependence on foreign food assistance	Medium since it may require investments in water systems
Coping with damages from extreme events to infra-structure	Making public buildings and public services less vulnerable to storms and cyclones	High since if may require substantials investments and improvements in existing infra-structure
More engagement from women	Stronger engagement of women in climate change adaptation efforts	Low since women's participation in climate change adaptation efforts is high

them. By doing so, they can also realise new innovation, optimize their investment and —often— incur in cost savings.

Apart from these priorities, a closer integration of the works being performed by the many regional and international organizations working in this field in the Pacific region is highly desirable. There are various examples showing this is already happening. The reduction of the vulnerability of coasts through appropriate adaptation measures is, for example one of the focus areas of the Pacific Adaptation to Climate Change (PACC) programme. Supported by the United Nations Development Programme (UNDP) and the Secretariat of the Pacific Regional Environment Programme (SPREP), the PACC project is the first major climate change adaptation initiative in the Pacific region, with demonstration projects in 14 Pacific island countries piloting on-the-ground adaptation solutions (PACC XX). The project has laid the groundwork for more resilient Pacific communities that can cope with climate variability today, and climate change tomorrow (UNDP 2015).

Possibilities of joint works also exist between countries in the Pacific region and Europe. To date, this happens on an ad hoc basis. Therefore, one further tool which may help Pacific countries in their climate change mitigation and adaptation efforts is the set-up of the "European-Pacific Climate Change Platform" (E-P-CCP). This instrument may help to achieve and catalyse a greater and more systematic cooperation between people and institutions concerned with climate change in Europa and the Pacific.

The Platform meets a perceived need for a greater Integration of climate change adaptation and mitigation efforts between the European Union, which is responsible for substantial amounts of CO_2 emissions, a major driver of climate change, and Pacific countries, which are struggling to cope with its impacts. Apart from seeking solutions in the context of infrastructure, the Platform can facilitate information on different types of sustainability solutions in areas such as energy (especially power generation and energy efficiency) and lighting, up to transport and waste treatment and Management. The Platform will provide information and catalyse actions based on best practices, both on European-Pacific and Pacific-Pacific perspectives. E-P-CCCP is committed to partnerships and will promote projects, policy documents, events and other information which may be of relevance and interest to its network members.

Apart from a special focus on adaptation, the Platform may also promote information to support efforts to mitigate green house gas emissions in both regions. The Platform will be a unique setting, offering an opportunity to showcase know-how, experiences and solutions. It is also a tool to foster collaboration by means of joint projects and joint ventures with the private sector.

The Platform may also serve the purpose of leveraging knowledge about currently available technological solutions in Europe, which may be suitable to Pacific countries, helping them to leap-frog in areas such as CO_2 emissions from landfills, use of renewable energy or planning in coastal zone areas, so as to increase resilience. It may also lead initiatives targeting employment and sustainable development schemes involving people in the most disadvantaged and disaster-prone places.

The Platform will, when established, bring together representatives from the private sector, government organisations, educational institutions, international agencies, women groups and other non-State actors concerned with climate change.

By means of the Platform, the European and Pacific relations can be developed into something bigger, better integrated and more effective, also helping Pacific countries to meet the Sustainable Development Goals. It can also create more connected experiences which may help to improve the quality of life of people in the Pacific.

Conclusions

The climate change challenges faced by the island countries in the Pacific region are manifold. Attempts to face them and increase the resilience of coastal communities and infrastructure of Pacific communities to the impacts of climate change, need to take into account the particularities of individual countries. They also need to be performed in a more systematic way, i.e. not only based on climate data and sea level projections, but also following socio-economic assessments, community consultations, and by taking into account the needs of the various stakeholders, so as to yield the expected benefits. The interactions between social and ecological systems should not be ignored, and need to be used as a basis for well considered decisions. This may, for instance, help to reduce the need for the rapid rural-urban migration which is taking place in some countries.

A special emphasis should be paid to the health sector, an area where the level of attention paid so far leaves much room for improvements. Scientific uncertainty should not be used as an argument not to take action in this central field. This is especially so since some uncertainties are not likely to be reduced in the short term (Barnett 2001). In this context, the identification and replication of successful adaptation interventions as suggested by the European-Pacific Climate Change Platform may provide a good basis for jump-starting efforts in countries where health elements have received little attention to date, helping them to cope with the climate variations of today, and the long-term, future impacts of climate change.

References

Adger, N., Dessai, S., Goulden, M., Hulme, M., Lorenzoni, I., Nelson, D. R., et al. (2008). Are there social limits to adaptation to climate change? *Climatic Change, 93*(3–4), 335–354.

Asian Development Bank. (2012). *Pacific regional strategic program for climate resilience*. Manila: ADB.

Barnett, J. (2001). Adapting to climate change in Pacific Island countries: The problem of uncertainty. *World Development, 29*(6), 977–993.

Briguglio, L. (1995). Small island states and their economic vulnerability. *World Development, 23,* 1615–1632.

Gallopín, G. C. (2006). Linkages between vulnerability, resilience, and adaptive capacity. *Global Environmental Change, 16,* 293–303.

IFAD. (2007). *Climate change impacts in the Asia/Pacific region*. Rome: IFAD.

IPCC. (2014). Climate Change 2014: Impacts, Adaptation and Vulnerability. Working Group II Contribution to the Intergovernmental Panel on Climate Change Fourth Assessment Report. Geneva: IPCC.

Leal Filho, W. (Ed.). (2015a). *Climate change in the Asia-Pacific region*. Berlin: Springer.

Leal Filho, W. (Ed.). (2015b). *Handbook of climate change adaptation*. Berlin: Springer. ISBN 9783642386695.

Leal Filho, W., Azeiteiro, U. M., & Azul, A. (Eds.). (2016). *Climate change and health*. Berlin: Springer.

Moser, S., & Ekstrom, J. (2010). A framework to diagnose barriers to climate change adaptation. *Proceedings of the National Academy of Sciences of the United States of America, 107*(51), 1–6.

Tisdell, C. (2008). Global warming and the future of Pacific Island countries. *International Journal of Social Economics, 35*(12), 889–903.

United Nations Framework Convention on Climate Change (UNFCC). (2007). *Climate change: Impacts*. UNFCCC, Bonn: Vulnerabilities and Adaptation in Developing Countries.

UNDP. (2015). *Climate change in the Pacific: Reducing vulnerability of island coasts*. New York: UNDP.

World Health Organisation. (2016). *Human health and climate change in Pacific island countries.*

Erratum to: Dynamic Adaptive Management Pathways for Drinking Water Security in Kiribati

Pierre Mukheibir, Louise Boronyak-Vasco and Pelenise Alofa

Erratum to:
Chapter 17 in: W. Leal Filho (ed.), *Climate Change Adaptation in Pacific Countries*, Climate Change Management, DOI 10.1007/978-3-319-50094-2_17

In the original version of the book, in Chapter 17, the misspelt author name "Lousie Boronyak-Vasco" should be corrected to read as "Louise Boronyak-Vasco". The erratum chapter and the book have been updated with the change.

The updated original online version for this chapter can be found at
http://dx.doi.org/10.1007/978-3-319-50094-2_17

W. Leal Filho (ed.), *Climate Change Adaptation in Pacific Countries*,
Climate Change Management, DOI 10.1007/978-3-319-50094-2_28